Novel Biotechnologies for Biocontrol Agent Enhancement and Management

NATO Security through Science Series

This Series presents the results of scientific meetings supported under the NATO Programme for Security through Science (STS)

Meetings supported by the NATO STS Programme are in security-related priority areas of Defence Against Terrorism or Countering Other Threats to Security. The types of meeting supported are generally "Advanced Study Institute" and "Advanced Research Workshops". The NATO STS Series collects together the results of these meetings. The meetings are co-organized by scientist from NATO countries and scientists from NATO's "Partner" or "Mediterranean Dialogue" countries. The observations and recommendations made at the meetings, as well as the contents of the volumes in the Series, reflect those of participants and contributors only; they should not necessarily be regarded as reflecting NATO views or policy.

Advanced Study Institutes (ASI) are high-level tutorial courses to convey the latest developments in a subject to an advanced-level audience

Advanced Research Workshops (ARW) are expert meetings where an intense but informal exchange of views at the frontiers of a subject aims at identifying directions for future actions

Following a transformation of the programme in 2004 the Series has been re-named and re-organised. Recent volumes on topics not related to security, which result from meetings supported under the programme earlier, may be found in the NATO Science Series.

The Series is published by IOS Press, Amsterdam, and Springer, Dordrecht, in conjunction with the NATO Public Diplomacy Division.

Sub-Series

A. Chemistry and Biology	Springer
B. Physics and Biophysics	Springer
C. Environmental Security	Springer
D. Information and Communication Security	IOS Press
E. Human and Societal Dynamics	IOS Press

http://www.nato.int/science
http://www.springer.com
http://www.iospress.nl

Series A: Chemistry and Biology

Novel Biotechnologies for Biocontrol Agent Enhancement and Management

edited by

Maurizio Vurro
Consiglio Nazionale delle Ricerche, Bari, Italy

and

Jonathan Gressel
Weizmann Institute of Science, Rehovot, Israel

Springer

Published in cooperation with NATO Public Diplomacy Division

Proceedings of the NATO Advanced Study Institute on
Novel Biotechnologies for Biocontrol Agent Enhancement and Management
held in Gualdo Tadino, Italy
8–19 September 2006

A C.I.P. Catalogue record for this book is available from the Library of Congress.

ISBN-10 1-4020-5798-9 (PB)
ISBN-13 978-1-4020-5798-4(PB)
ISBN-10 1-4020-5797-0(HB)
ISBN-13 978-1-4020-5797-7(HB)
ISBN-10 1-4020-5799-7 (e-book)
ISBN-13 978-1-4020-5799-1(e-book)

Published by Springer,
P.O. Box 17, 3300 AA Dordrecht, The Netherlands.

www.springer.com

Printed on acid-free paper

CONTENTS

PREFACE

The intent of the NATO Advanced Study Institute (ASI) entitled "Novel Biotechnologies for Biocontrol Agent Enhancement and Management" was to permit the meeting of the major exponents in the scientific community working with enhancing different biological control agents (fungi, bacteria, virus, nematodes, and insects) on different targets (pathogens, insects, weeds, and rodents). This multidisciplinary group, having backgrounds in the different aspects of biotechnologies (transgenic enhancement, molecular biology, formulation, genetics, risk assessment, new technology, biochemistry, and physiology), presented highly advanced lectures during the 10-day-ASI, in order to allow students to improve their capability to enhance and manage biological control agents. This approach will allow ASI attendees to bring new ideas, new approaches, or new methodologies coming from different fields of application to their own field of expertise.

A further aim of the NATO ASI was to create a network of young and experienced scientists, with few geographical barriers among countries, who will develop new opportunities to collaborate in this field of science that requires a "global" collaborative approach.

Forty students from twenty countries took part to the NATO ASI.

In addition to the 45 lectures from the 15 lecturers, there were 25 short presentations and 8 posters on cogent research from students in this course, held between September 8- 2006 and September 19, 2006. This book represents a partial distillation of all this material together with the daily workshops on various topics, and long discussions over the excellent meals and breaks, in the very conducive environment of the Borgo Hotel Le Terre del Verde at Gualdo Tadino near Perugia, in Italy.

The editors especially appreciated the efforts of the anonymous peer reviewers who expeditiously reviewed the chapters of this book.

This workshop could not have been possible without the financial assistance of NATO and Valent BioSciences, as well as the lecturers who contributed their time, and in most instances their travel expenses, to assist in allowing the maximum support of students. To these all we have many thanks, along with the knowledge and collaborations engendered by this workshop.

Maurizio Vurro and Jonathan Gressel
Codirectors

October 2006

1. BIOTECHNOLOGY IN CROP PROTECTION: TOWARDS SUSTAINABLE INSECT CONTROL

Martin G. Edwards and Angharad M. R. Gatehouse*
Institute for Research on Environment and Sustainability, Division of Biology, Devonshire Building, Newcastle University, Newcastle upon Tyne, NE1 7RU, UK

Abstract. With a projected increase in world population to 10 billion over the next four decades, an immediate priority for agriculture is to achieve maximum production of food and other products in a manner that is environmentally sustainable and cost effective. Whilst insecticides are very effective in combating the immediate problem of insect attack on crops, nonspecific insecticides are harmful to beneficial organisms including predators and parasitoids of the target pest species. The concept of utilizing a transgenic approach to host plant resistance was realized in the mid 1990s with the commercial introduction of transgenic maize, potato and cotton plants expressing genes encoding the entomocidal δ-endotoxin from *Bacillus thuringiensis* (Bt). Other strategies based on the use of plant-derived genes (enzyme inhibitors, lectins) and those from animal sources, including insects (biotin-binding proteins, neurohormones, enzyme inhibitors), are currently being developed. The use of fusion proteins to increase the spectrum and durability of resistance is also actively being pursued. Biotechnology in crop protection is not restricted to production of transgenic crops, and has been extended to include the modification of baculoviruses for increased efficacy as biopesticides, and arthropod natural enemies (predators and parasitoids) to enhance their capacity to control insect pests, this chapter will only consider the benefits and risks of its role in the context of insect-resistant transgenic crops.

Keywords: sustainability, insect-resistant transgenic crops, insect resistance genes, insecticidal proteins, fusion proteins, pests, natural enemies

1.1. Introduction

1.1.1. NEED FOR SUSTAINABLE AGRICULTURE

The dawn of agriculture occurred some 10,000 years ago with the domestication of cereals, soon to be followed by other crops (Table I). This step

*To whom correspondence should be addressed, e-mail: a.m.r.gatehouse@ncl.ac.uk

M. Vurro and J. Gressel (eds.), Novel Biotechnologies for Biocontrol Agent Enhancement and Management, 1–23.

TABLE I. First attested dates of independent transition to agriculture and the main domesticates (after Olsson[1])

Region	Date	Plants
Near East	8500 BC	Wheat barley
Central Mexico	8000 BC	Maize
South China	7500 BC	Rice
North China	6800 BC	Soybean
South Central Andes	5800 BC	Potato, manioc
Eastern United States	3200 BC	Sunflower
Sub-Saharan Africa	2500 BC	Sorghum

was seen as a necessary condition for the development of civilizations. The evolution of agriculture has been divided into four discrete periods, namely the Prehistoric, Roman, Feudal and Scientific Era, with each being associated with specific advancements or developments (Table II). The Prehistoric, or Neolithic Era (10,000 BCE), was thus recognized as the era of crop domestication originating in the regions of low to middle latitude. The Roman Era (1000 BCE–500 CE) saw the introduction of metal tools, the use of animals for farm work and the development of the manipulation of watercourses for irrigation, while the Feudal Era (height, 1100 CE) saw the beginning of international trade based on exportation of crops. Interestingly, the era known as the Scientific Era started as early as the 16th century and although

TABLE II. The four major eras of agriculture (www.adbio.com/science/agrihistory)

Period	Date	Facts
Prehistoric (Neolithic)	10000 BCE	Domestication of crops 6000 BC: people dependent on domesticated crops
Roman	1000 BCE–500 CE	Metal tools Horses & oxen for farm work Irrigation
Feudal	1100 CE (height)	Export of crops (international trade) 8th Century: rotation
Scientific	1500 CE	15th–19th century: slave labor 16th c: First efforts in plant crop protection 17th–18th Century: Pest control World War II: Mechanization of farming; pesticides 1950s: Mutation breeding, using radioactive isotopes 1970s: Green revolution 1990s: GM crops

there is documentary evidence for the use of pest control from ancient times, its adoption is primarily attributed to this era. Whilst mineral-based pesticides (arsenates and copper salts) had been used previously major advances in the development of synthetic insecticides did not occur until the end of the Second World War and was accompanied by the intensification of farming. Although in recent years there has been a move towards the development and use of more benign pesticides, the next major breakthrough in this area was seen with the development and commercialization of insect-resistant transgenic crops during the 1990s. In addition to advances made in crop protection, this era has also seen the development and use of mutation breeding, and in the 1970s a major landmark was achieved with the Green Revolution.

1.1.2. THE CHALLENGES AHEAD

The human population is ever increasing, with conservative estimates predicting that the population will rise to approximately 10 billion by 2050. Thus the major challenges facing the world are to feed and provide shelter for a world population that is increasing at an exponential rate (Table III; Figure 1).

Furthermore, it is essential to protect human health, and ensure social and economic conditions that are conducive to the fulfillment of the human potential. Agriculture must play a major role in achieving these goals both by providing ever-increasing food yields (Figure 2) and an ever-increasing supply of natural products required by industry. The recent advent of bioethanol further confirms the constraints on agriculture.[2] Thus the challenge in the forthcoming decades is to achieve maximum production of food and other products without further irreversible depletion or destruction of the natural

TABLE III. World population growth

World population in billions	Year	Time needed to reach this level
One	1804	All of human history
Two	1927	123 years
Three	1960	33 years
Four	1974	14 years
Five	1987	13 years
Six	1999	12 years
Seven*	2012	13 years
Eight*	2026	14 years
Nine*	2043	17 years

Source: United Nations Populations Division, World Population Prospects.
*Projected population growth; medium variant.

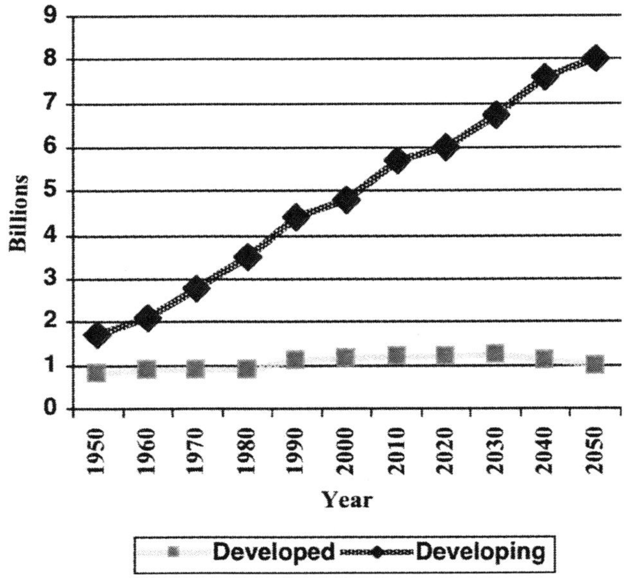

Figure 1. Predicted population growth 1950–2050. The data suggests that there has been a fourfold population increase during the last century. While the population is predicted to remain stable in developed regions of the world, based on current trends. Significant increases are predicted to occur in the least developed nations

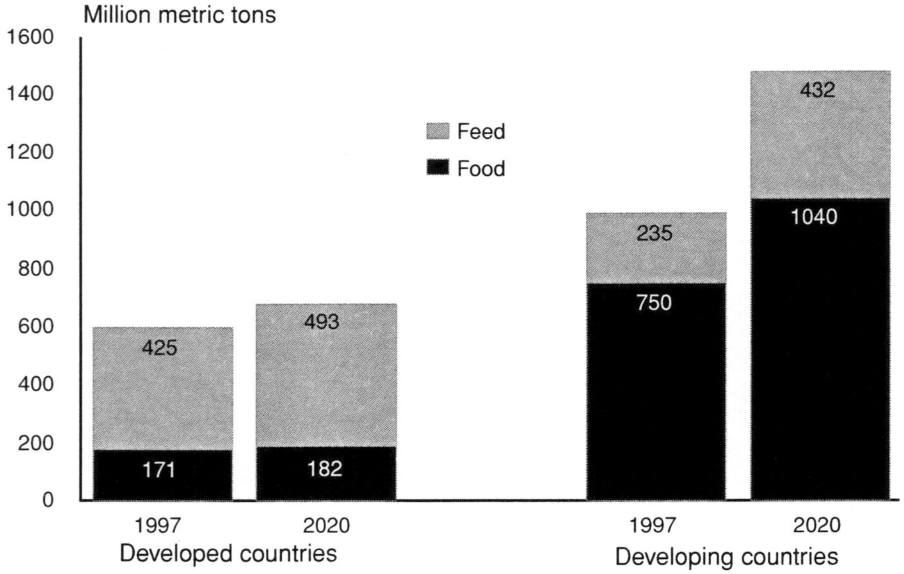

Figure 2. Demand for cereals for human food and animal feed, baseline scenario, 1997–2020 (personal communication A. Cockburn)

environment, against a backdrop of climate change which not only is predicted to result in the loss of agricultural land as a consequence of rising sea levels, but is also likely to have a major impact on the dynamics of pest populations. Agriculture must become an integral part of a sustainable global society.

Agricultural sustainability integrates three major goals, those of environmental health, economic profitability and social and economic equity. It thus rests on the principle that we must meet the needs of the present without compromising the ability of future generations to meet their own needs. Current figures suggest that to feed a world population of 10 billion in 2050 without allowing for additional imports of food, Africa will have to increase its food production by 300%, Latin America by 80% and Asia by 70%. Even North America, which is not usually associated with food shortages, would have to increase its food production by 30% to feed its own projected population of 348 million. Given the current scenario of some 800 million people going hungry on a daily basis and an estimated 30,000 (half of them children) dying every day due to hunger and malnutrition, it is clear that society has many major challenges to address. One step towards achieving sustainability is to identify current major constraints on crop productivity. Simply putting more land into agricultural use, thereby increasing the "agricultural footprint," is not a viable option in the long term.

Currently stress constitutes a major factor in limiting productivity; it can be classified as being either biotic (pests, pathogens and weeds) or abiotic (physical constraints, e.g., temperature, water availability, salinity) where the former can be as can be as high as 40% globally. Insecticides are effective in dealing with the immediate problem of insect attack on crops. They have been responsible for dramatic yield increases in crops that are subject to serious pest problems, but in the longer term severe drawbacks have become apparent. For example, non-specific insecticides are harmful to non-target organisms, many of which play key roles in suppressing the build up of insect populations.[3] Other problems associated with high pesticide application include accumulation of toxic residues in food products and subsequent consequences for human health.[4] The hypothesis that an over reliance on insecticides is non-sustainable, is further supported by the finding that many insect pests have evolved resistance to such compounds.

1.2. Role of Transgenic Crops in Agriculture

Biotechnology offers many opportunities for agriculture and provides the means to address many of the constraints placed to productivity outlined above. It uses the conceptual framework and technical approaches of molecular biology and plant cell culture systems to develop commercial processes

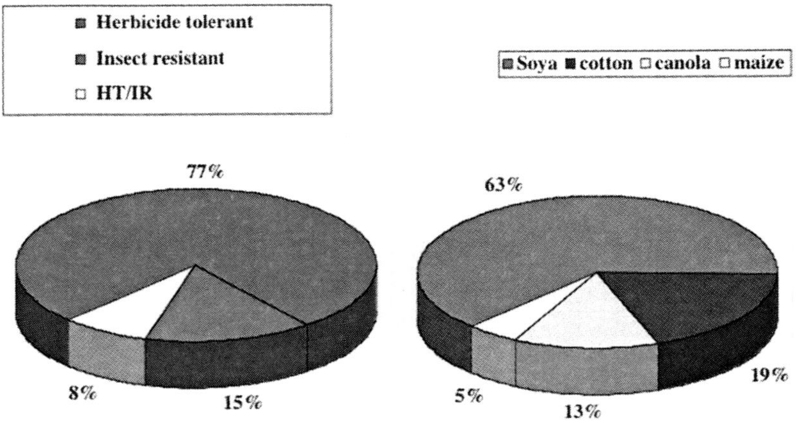

Figure 3. Percentage areas of genetically enhanced crops by trait and by crop. (From ISAAA James 2005)[4]

and products. With the rapid development of biotechnology, agriculture has moved from a resource-based to a science-based industry, with plant breeding being dramatically augmented by the introduction of recombinant DNA technology based on knowledge of gene structure and function. The concept of utilizing a transgenic approach to host plant resistance was realized in the mid 1990s with the commercial introduction of transgenic maize, potato and cotton plants expressing genes encoding the insecticidal δ-endotoxin from *Bacillus thuringiensis*. Similarly, the role of herbicides in agriculture entered a new era with the introduction of glyphosate-resistant soybeans in 1995. Currently the commercial area planted to transgenic crops is in excess of 90 million hectares (22 million acres) with approximately 77% expressing herbicide tolerance, 15% expressing insect resistance genes and approximately 8% expressing both traits (Figure 3; current production by country is illustrated in Table IV). Despite the increasing disquiet over the growing of such crops in Europe and Africa (at least by the media and certain NGOs) in recent years, the latest figures available demonstrate that the market is increasing, with an 11% increase between 2004 and 2005.[5]

Applications of transgene technology in agriculture have clearly defined benefits, not least in providing greater sustainability in terms of improved levels of crop protection resulting in higher yields and reduced pesticide application. However, a major challenge facing this new industry is in the identification of suitable genes for transfer that will confer the desired agronomic traits. In terms of insect resistance, several different classes of bacterial-, plant- and animal derived proteins have been shown to be insecticidal towards a range of economically important insect pests from different orders, with the midgut being the prime target.[6] Of these the Bt toxins are the most

TABLE IV. Major countries growing of biotech crops in 2005

Country	Million hectares	Crop
USA	49.8	Soybean, maize, cotton, oilseed rape, Squash, papaya
Argentina	17.2	Soybean, maize, cotton
Brazil	9.0	Soybean
Canada	6.1	Canola, maize, soybean
China	3.3	Cotton
Paraguay	1.8	Soybean
India	1.3	Cotton
South Africa	0.4	Maize, soybean, cotton
Uruguay	0.3	Soybean, maize
Australia	0.2	Cotton
Mexico	0.1	Cotton, soybean
Romania	0.1	Soybean
Philippines	0.1	Maize
Spain	0.1	Maize
Colombia	<0.05	Cotton
Iran	<0.05	Rice
Honduras	<0.05	Maize
Portugal	<0.05	Maize
Germany	<0.05	Maize
France	<0.05	Maize
Czech Republic	<0.05	Maize

Source: James, 2005.[4]

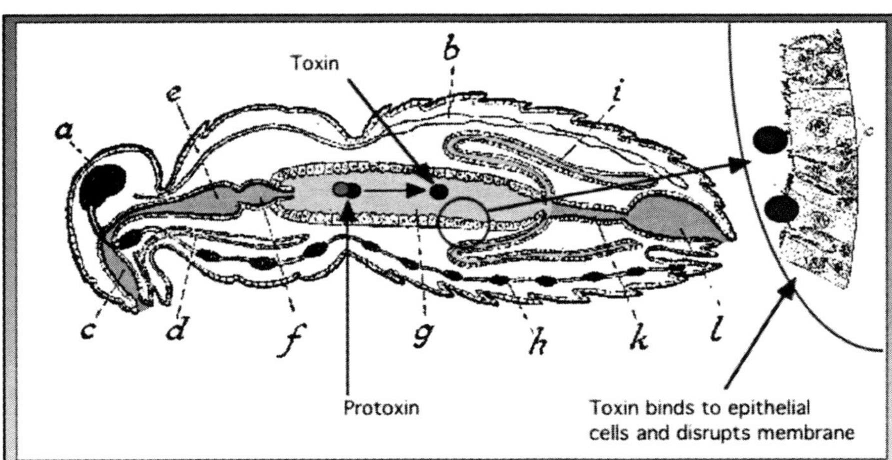

Figure 4. Diagramatic representation of the insect gut showing binding of active Bt toxin to receptors on the midgut epithelial cells (Unpublished figure kindly supplied by J.A. Gatehouse, Durham University, UK)

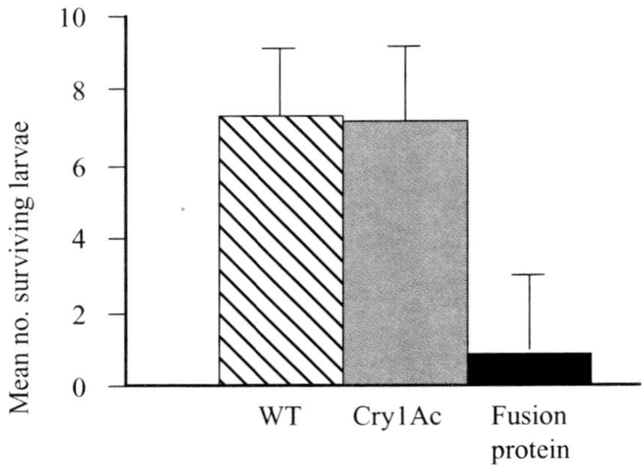

Figure 5. Effect of Bt-Ricin B-Chain Fusion Protein on *Spodoptera littoralis* (after Mehlo[70])

commercially relevant, since, to date, Bt-expressing crops are the only insect-resistant transgenic crops to have been commercialized.

1.2.1. PLANTS EXPRESSING BACILLUS THURINGIENSIS (BT) TOXINS

Bacillus thuringiensis (Bt) is a soil dwelling bacterium of major agronomic and scientific interest (refer to Chapter 2). While the subspecies of this bacterium colonies and kill a large variety of host insects, each strain tends to be highly specific. Toxins for insects in the orders Lepidoptera (butterflies and moths), Diptera (flies and mosquitoes), Coleoptera (beetles and weevils), and Hymenoptera (wasps and bees) have been identified,[7] but interestingly, none with activity towards Homoptera (sap suckers) have, as yet been identified, although a few with activity against nematodes have been isolated.[8] Further there is little evidence of effective Bt toxins against many of the major storage insect pests.

These Bt toxins (also referred to as d-endotoxins; Cry proteins) exert their pathological effects by forming lytic pores in the cell membrane of the insect gut. On ingestion, they are solubilized and proteolytically cleaved in the midgut to remove the C-terminal region, thus generating an "activated" 65–70 kDa toxin. The active toxin molecule binds via domains I and II to a specific high-affinity receptor in the insect midgut epithelial cells. Following binding domain I inserts into the membrane where it forms pores with other toxin molecules; this results in cell death by colloid osmotic lysis, followed by death of the insect.[7] A number of putative receptors have been identified and include aminopeptidase N proteins,[9–12] cadherin-like proteins[13–15] and glycolipids,[16] although Griffiths[17] suggest that a common carbohydrate motif may explain why a single toxin can bind to at least two receptors that are completely unrelated in sequence.

Transgenic plants expressing Bt toxins were first reported in 1987[18] and following this initial study, numerous crop species have been transformed with genes encoding a range of different Cry proteins targeted towards different pests species. Since bacterial *cry* genes (genes encoding Bt toxins) are rich in A/T content compared to plant genes, both the full-length and truncated versions of these *cry* genes have had to undergo considerable modification of codon usage and removal of polyadenylation sites before successful expression in plants.[19] These studies have been extensively reviewed and the reader is referred elsewhere.[20,21] Crops expressing Bt toxins were first commercialized in the mid 1990s, with the introduction of Bt potato and cotton. Currently more than 16 million hectares (equivalent to 18%) are planted to Bt crops, with a further 10 million hectares (equivalent to 10%) planted to crops expressing both Bt and genes conferring herbicide tolerance (Figure 3).[5] To date there are no reports of resistance in pest populations having evolved in the field to transgenic Bt expressing plants.[22] However resistance has evolved to the lower Bt levels found in Bt bacterial sprays used in organic agriculture.[23] The cultivation of Bt expressing crops have brought some substantial gains to the farming community both in terms of increased yields and lower production costs. For example, the costs for producing Bt cotton in China compared to isogenic non-Bt cotton varieties were approximately fivefold less, representing significant savings. This saving was primarily due to reduced pesticide application. Similarly benefits in India include a 70% reduction in insecticide applications in Bt cotton fields, resulting in a saving of up to US$ 30/ha in pesticide costs, with an increase of approx 85% in yield of harvested cotton.[24] Furthermore, expression of Bt has also resulted in improved crop quality as a consequence of decreased levels of *Fusarium* infestation and fumonisin mycotoxin production; this benefit is particularly important in food crops such as maize.

1.2.2. TRANSGENIC PLANTS EXPRESSING INHIBITORS OF INSECT DIGESTIVE ENZYMES

The concept of employing genes encoding Bt toxins to produce insect-resistant transgenic plants arises from the successful use of Bt-based biopesticides A number of other strategies for protecting crops from insect pests actually exploit endogenous resistance mechanisms.[25,26] Genes encoding such defensive proteins were obvious candidates for enhancing crop resistance to insect pests.

Interfering with digestion, and thus affecting the nutritional status of the insect, is a strategy widely employed by plants for defense, and has been extensively investigated as a means of producing insect-resistant crops.[27]

Numerous studies since the 1970s have confirmed the insecticidal properties of a broad range of protease inhibitors from both plant and animal sources.[27,28] Proof of concept for exploiting such molecules for crop protection was first demonstrated with expression of a serine protease inhibitor from cowpea (CpTI), which was shown to significantly reduce insect growth and survival.[29] These studies were subsequently extended to include a greater range of target pests,[30–32] and a broader range of inhibitors and plant species, including economically important crop species.[33,34]

Since many economically important coleopteran pests predominantly utilize cysteine proteases for protein digestion, inhibitors for this class of enzyme (cystatins) have been investigated as a means for controlling pests from this order. Oryzacystatin, a cysteine protease inhibitor isolated from rice seeds, is effective towards both coleopteran insects and nematodes when expressed in transgenic plants.[35–37] Similarly the cysteine/aspartic protease inhibitor equistatin, from sea anemone, is also toxic to several economically important coleopteran pests, including the Colorado potato beetle.[38] More recent studies have included the stacking of different families of inhibitors to increase the spectrum of activity.[39]

A major limitation, however, to this strategy for control of insect pests arises from the ability of some lepidopteran and coleopteran species to respond and adapt to ingestion of protease inhibitors by either over-expressing native gut proteases, or producing novel proteases that are insensitive to inhibition.[40,41] Thus detailed knowledge about the enzyme–inhibitor interactions, both at the molecular and biochemical levels, together with detailed knowledge on the response of insects to exposure to such proteins is essential to effectively exploit this strategy. The concept of inhibiting protein digestion as a means of controlling insect pests has been extended to inhibition of carbohydrate digestion. For example, inhibitors of α-amylase have been expressed in transgenic plants and shown to confer resistance to bruchid beetles.[42–45]

1.2.3. TRANSGENIC PLANTS EXPRESSING LECTINS

Lectins, found throughout the plant and animal kingdoms, form a large and diverse group of proteins identified by a common property of specific binding to carbohydrate residues, either as free sugars, or more commonly, as part of oligo- or polysaccharides. Many physiological roles that have been attributed to plant lectins including defense against pests and pathogens.[46,47]

Although some lectins are toxic to mammals, and are thus not suitable candidates for transfer to crops for enhanced levels of protection, this is by no means universal. Many lectins are not toxic to mammals, yet are effective against insects from several different orders,[48] including homopteran pests such as hoppers and aphids.[49–52] This finding has generated significant interest, not least since no Bt*s* effective against this pest order have been identified to date. One such lectin is the snowdrop lectin (*Galanthus nivalis* agglutinin; GNA). Both constitutive and phloem specific (Rss1 promoter) expression of GNA in rice is an effective means of significantly reducing survival of rice brown plant hopper (*Nilaparvata lugens*), and green leafhopper (*Nephotettix virescens*) both serious economic pests of rice.[49,53,54] GNA has been expressed in combination with other genes encoding insecticidal proteins, including the *cry* genes.[55] When a linear transgene construct lacking vector backbone sequences was used to generate transgenic rice plants, the subsequent levels of transgene expression were two- to fourfold higher than plants transformed with whole plasmids.[54] Although lectins such as GNA, and ConA are not as effective against aphids as they are against hoppers, they nonetheless have significant effects on aphid fecundity when expressed in potato[6,50,56] and wheat.[57]

The precise mode of action of lectins in insects is not fully understood although binding to gut epithelial cells appears to be a pre-requisite for toxicity. In the case of rice brown planthopper, GNA not only binds to the luminal surface of the midgut epithelial cells, but also accumulates in the fat bodies, ovarioles and throughout the haemolymph, suggesting that the lectin is able to cross the midgut epithelial barrier and pass into the insect's circulatory system, resulting in a systemic toxic effect.[58] One of the receptors for GNA in brown planthopper gut is a subunit of ferritin, indicating that GNA may be interfering with metal homeostasis within the insect.[59]

As with protease inhibitors, the levels of protection conferred by expression of lectins in transgenic plants are generally not high enough to be considered commercially viable. However, the absence of genes with proven high insecticidal activity against homopteran pests may well mean that transgenic crops with partial resistance may still find acceptance in agriculture, especially if expressed with other genes that confer partial resistance, or if introduced into partially resistant genetic backgrounds.

1.2.4. TRANSGENIC PLANTS EXPRESSING NOVEL INSECTICIDES

Generating insecticidal transgenic crops harboring genes from non-conventional sources is an extremely active area (refer to Chapter 18), with amongst others, foreign genes from plants (e.g., enzymes inhibitors and novel lectins,[54,60–62] and animal sources including insects (e.g., biotin-binding proteins,[63] neurohormones,[64] venoms and enzyme inhibitors[65] being a major focus.

The development of second-generation transgenic plants with greater durable resistance might result from the expression of multiple insecticidal genes such as the Vip (vegetative insecticidal proteins) produced by *Bacillus thuringiensis* during its vegetative growth. The benefit of such an approached is a broader insect target range than conventional Bt proteins and the proposed expectation to control current Bt resistant pests due to the low levels of homology between the domains of the two proteins classes.[66,67]

With Bt toxins as the classical reference, toxins from other insect pathogens provide a potential repository of novel insecticidal compounds. *Photorhabdus* spp. are bacterial symbionts of entomopathegenic nematodes which are lethal to a wide range of insects.[68] *Photorhabdus* toxin expression in *Arabidopsis* caused significant insect mortality.[69]

1.2.5. TRANSGENIC PLANTS EXPRESSING FUSION PROTEINS

The concept of "gene stacking" has recently been extended to the development and use of fusion proteins. Such proteins not only provide a means of increasing durability, but also provide a "vehicle" for more effective targeting of insecticidal molecules, including peptides. It thus offers an alternative/complementary strategy to address potential limitations in conventional transgenic insect pest control. For example, recognition of binding sites in the insect gut is an important factor determining the toxicity of Bt. Enhancing toxin binding capabilities should thus extend host range and delay resistance. Bt is believed to bind primarily to aminopeptidase N or cadherin membrane proteins, whilst the generation of a fusion protein with the non-toxic B chain of ricin (RB) was shown to extend the binding of Bt to include specific glycoproteins. Transgenic plants expressing the Bt fused RB demonstrated that the addition of the RB binding domain provided a wider repertoire of receptor sites within target species and significantly enhanced the levels of toxicity of Bt. For example, survival of the armyworm *Spodoptera littoralis*, a species of insect not sensitive to Bt, was reduced by approx. 90% when feeding on transgenic maize expressing the fusion, compared to plants expressing either Bt Cry1Ac alone, or the RB binding domain.[70] Expression of the fusion protein resulted in the insect becoming sensitive to Bt.

Not only do fusion proteins have potential for use in transgenic crops, but also to improve the efficacy of biopesticide-based sprays. Neuropeptides potentially offer a high degree of biological activity, and thus provide an attractive alternative pest management strategy. There are major drawbacks to their use, particularly as topical sprays. They are unlikely to be rapidly absorbed through the insect cuticle to their site of action, and are prone to proteolysis and rapid degradation in the environment. Should they survive the application process and are then taken up by the insect, they are then unlikely to survive the conditions of the insect gut or be delivered to the correct targets within the insect. The discovery that snowdrop lectin (GNA) remains stable and active within the insect gut after ingestion, and that it is able to cross the gut epithelium, provides an opportunity for its use as a "carrier molecule" to deliver other peptides to the circulatory system of target insect species. This strategy effectively delivered the insect neuropeptide hormone, allatostatin, to the haemolymph of the tomato moth *Lacanobia oleracea*.[64] Subsequent expression of the fusion protein in potato further provided proof of concept for the efficacy of fusion proteins, as a means of delivery. The results demonstrated significant reduction in mean larval weight when compared to the controls. GNA can be used to deliver insecticidal peptides isolated from the venom of the spider *Segestria florentina* (SFI1) to the haemolymph of *L. oleracea*.[71] Neither the GNA nor the SFI1 moieties alone were acutely toxic the SFI1/GNA fusion, was insecticidal to first stage larvae, causing 100% mortality after 6 days. This spider venom neurotoxin is believed to irreversibly block the pre-synaptic neuromuscular junctures. Such venom toxins show high degrees of specificity and thus lend themselves to environmentally benign pest management strategies.

1.3. Exploitation of Endogenous Defense Mechanisms

Plants routinely face sustained periods of stress, sufficient to limit their growth and reproductive capacity. One mode of achieving increased crop productivity is through a greater understanding of the complex adaptive responses that plants have evolved to cope with the various forms of stress that they encounter. Stress is also a major force leading to genetic change, as mutation frequencies typically increase during stress. Those individuals within a population with superior stress tolerance characteristics produce more progeny in subsequent generations. It has long been understood that plants exhibit multi-mechanistic resistance towards herbivores, but the molecular mechanisms underpinning these complicated responses are just being elucidated.[72] The plant's herbivore-induced transcriptome is being studied using microarrays and differential display technologies. Such investigations have provided novel insights into plant–insect interactions, with the jasmonic acid cascade playing a central role in transcript accumulation in plants exposed to herbivory.[73,74]

Phytophagous insects have an additional effect on the plant response, above and beyond that caused by mechanical tissue damage.[75] Analysis of timing, dynamics, and regulation of the expression of 150 genes in leaves of *Arabidopsis* showed that many genes strongly induced by mechanical damage were induced less, or not at all, when the plant was attacked by the lepidopteran pest *Pieris rapae*. Whereas chewing insects cause extensive damage to plant tissues when feeding, many insects of the order Homoptera feed from the contents of vascular tissues by inserting a stylet between overlying cells, thus limiting cell damage and minimizing induction of a wound response. They thus elicit a more pathogenic like response. Recent work by Zhang et al.[76] with rice brown plant hopper (BPH) suggests that the plant response is very complex as it appears to induce responses that would participate in a jasmonic acid independent pathway, and crosstalk with those genes related to abiotic stress, pathogen invasion and phytohormone signaling pathways.

Differential response of plants to pest attack to identify insect resistance genes, with a view to their over-expression in crops, is an area currently receiving much attention. Analysis of transcripts provides evidence for the simultaneous activation of salicylic acid, ethylene, cytokinin and jasmonic acid-regulated pathways during herbivore attack. Similar co-activation of numerous signaling cascades in response to various stresses occurs in *Arabidopsis*[77] and supports the idea of a network of interacting signal cascades. Microarray analysis has also indicated direct plant defensive responses through a dramatic increases in protease inhibitor transcripts, and increases in transcripts encoding putrescine *N*-methyl transferase (which catalyze, amongst many, the first committed step in the biosynthesis of nicotine),[78] as well as metabolic commitment to terpenoid based indirect defenses. Deciphering the signals regulating herbivore-responsive gene expression will afford many opportunities to manipulate the response. Knowledge of these interactions can be exploited in the rational design of transgenic plants with increased disease/insect resistance.[79] However, engineering natural pathways for plant improvement is limited by a lack of understanding of the underlying biochemistry and by the need for co-ordinate regulation of multiple gene activities.[80]

The concept of identifying genes involved in crop protection at the transcriptome level is still very much in its infancy. More recent, are studies are investigating gene expression at the proteome level. Proteomics provides detailed information on the analysis of expression, localization, function, and interactions of the proteins expressed in a given organism. It also provides information on post-translational modification providing slightly different information to that provided by transcriptomics. This should assist in forming a greater understanding of the bases of plant–insect interactions, with a view to subsequent use in crop protection.

1.4. Environmental Impact

Transgenic crops provide clear benefits to both the grower and consumer not least is the significant reduction in chemical pesticides, thus making them more environmentally sustainable. However, it must be recognized that alongside these benefits, many concerns are being expressed, particularly regarding their wide-scale growing (now >90 million ha, globally). These concerns include effects on human health and the environmental at large, particularly through the gene flow of transgenes to wild relatives and the potential of increasing invasiveness of weeds. Since the above mentioned topics are outside the scope of the present chapter, which focuses specifically on the role of biotechnology in crop protection, including their impact on beneficial insects, the reader is referred to the following reviews and articles for the broader issues associated with this technology.[81–90]

1.4.1. IMPACT OF INSECT-RESISTANT TRANSGENIC CROPS ON NATURAL ENEMIES

Assessing the environmental consequences of transgenic crop species is an important precursor to their becoming adopted in agriculture. The expression of transgenes that confer enhanced levels of resistance to insect pests is of particular significance as it is aimed at manipulating the biology of organisms in a different trophic level to that of the plant. Recent research has identified potential risks to beneficial non-target arthropods via; bitrophic interactions, involving the plant and a herbivorous insect, and tritrophic interactions, those involving the plant, the pest insect and its natural enemy, particularly in relation to arthropod biodiversity.[91–95]

There are two major routes for insecticidal transgene products to impact on exposed natural enemies (predators and parasitoids) at higher trophic levels through the tritrophic interaction: (1) through direct exposure to the product as it accumulates in the pest; and (2), through indirect effects on the growth and development of the pest that influence subsequent growth and developmental processes in the parasitoid or predator. The distinction between these two mechanisms is of considerable environmental significance, but discriminating between them is not straightforward.[96,97] Several lepidopteran species tested contained Bt toxin after consumption of transgenic tissue and therefore provided a potential route of secondary Bt exposure.[98] The level of toxin was uniformly low, but it was still biologically active. Concerns over the use of plant-derived insecticidal proteins, such as protease inhibitors and lectins, are perhaps greater as they usually do not cause rapid and complete mortality of the target insect pest. There is no doubt that they do have a significant effect on insect survival, but their major contribution in crop protection is to reduce the build up of pest populations on plants. Thus they will be readily available for subsequent parasitism and predation, although many predators, such as carabids, do scavenge dead prey. Despite this opportunity for exposure, most studies to date have demonstrated that although the predator/parasitoid is exposed to the transgene product, which in many cases can be detected in the natural enemy, it has little effect. For example, exposure of the parasitoid *Eulophus pennicornis* to GNA via parasitism of *Lacanobia oleracea* larvae reared on GNA expressing plants failed to cause any deleterious effects, and in some instances parasitoid performance was actually improved.[99] Furthermore, GNA had no deleterious effects on the parasitoid *Meterous gyrator.*[100] Conversely cowpea trypsin inhibitor, was deleterious at the third trophic level, but these effects were considered to be indirect, as a result of poor performance of the pest larvae on CpTI expressing potato plants. There is also little evidence that predators are much affected. The exposure of the predatory stinkbug *Podisus maculiventris* to pest larvae (*L. oleracea*) reared on either GNA expressing or CpTI expressing potato plants had no significant effects on nymphal survival or weight.[101] Those insects reared on GNA did show a significant lengthening of preadult development. GNA had no deleterious effects on two spot ladybird *Adalia bipunctata* when fed GNA-dosed aphids from artificial diets or aphids colonizing GNA expressing potato plants.[102,103] Interestingly, the cysteine protease inhibitor OC-1 has no effect on *Harmonia axyridis* predating diamond back moth larvae (DBM, *Plutella xylostella*) reared on OC-1 expressing oilseed rape plants, despite these predators relying predominantly on cysteine proteases for proteolytic digestion. In the early stages of development the predators performed better on the DBM fed with the transgenic oilseed rape than the controls. The predators were able to modulate enzyme activity in response to dietary protease inhibitors.[104] Carabid beetles could circumvent the inhibitory effects the serine protease inhibitor MTI-2 expressed in oilseed rape and delivered through the prey by modulation of their digestive proteases profile[105]—in this study expression of MTI-2 was selected to target serine proteases, since carabids rely predominantly on this class of protease for protein digestion.

1.5. Conclusions

Time has demonstrated that biotechnology can provide very clear benefits to agriculture, not least with the increasing contribution it can make towards sustainability (Table V). Indeed, globally there has been a steadily increasing market for genetically modified (enhanced) crops, particularly for production

TABLE V. Benefits of transgenic insect resistant crops

Benefits of transgenic insect-resistant crops
Promotes greater sustainability of natural resources by reducing use of energy and chemicals (more targetuse of pesticides and reduction in use of fossil fuels)
Reduction in land/water contamination through reduced pesticide usage
Preserving natural habitats for biodiversity (more efficient use of land)
Reduced impact on non-target organisms, including beneficial insects
Enhancing safety of food crops by reducing mycotoxin contamination
Increased yield

of cotton and animal feeds, with the acreage now in excess of 90 million hectares. The fears voiced over the environmental impact of the technology, particularly in terms of deleterious consequences for biodiversity, including effects on natural enemies such as predators and parasitoids have not been realized. There is no room for complacency, and it is essential that all novel technologies are thoroughly investigated before their release. It is important that these investigations are carried out in comparison with their conventional counterparts to ensure a meaningful evaluation. Thus in agricultural terms, biotechnology must be evaluated in comparison with current conventional practices, e.g., chemical and biopesticide application for pest control. It is not suggested that biotechnology is necessarily used as a stand-alone technology, but rather that it is used as a component of integrated pest management.

References

1. O. Olsson, The rise of Neolithic agriculture, *Working Paper in Economics No 57* (2001).
2. J. Hill, E. Nelson, D. Tilman, S. Polasky, and D. Tiffany, Environmental, economic, and energetic costs and benefits of biodiesel and ethanol biofuels, *Proc. Natl. Acad. Sci. U S A* 103(30), 11206–11210 (2006).
3. S. A. Hassan, F. Bigler, H. Bogenschutz, E. Boller, J. Brun, J. N. M. Calis, J. Coremanspelseneer, C. Duso, A. Grove, U. Heimbach, N. Helyer, H. Hokkanen, G. B. Lewis, F. Mansour, L. Moreth, L. Polgar, L. Samsoepetersen, B. Sauphanor, A. Staubli, G. Sterk, A. Vainio, M. Vandeveire, G. Viggiani, and H. Vogt, Results of the 6th Joint Pesticide Testing Program of the Iobc/Wprs Working Group Pesticides and Beneficial Organisms, *Entomophaga* 39(1), 107–119 (1994).
4. A. G. Renwick, Pesticide residue analysis and its relationship to hazard characterisation (ADI/ARfD) and intake estimations (NEDI/NESTI), *Pest Manag. Sci.* 58, 1073–1082 (2002).
5. C. James, ISAAA Brief 34 (2005).
6. J. Gatehouse and A. Gatehouse, in *Biological and Biotechnological Control of Insect Pests*, edited by J. Reichcigl and N. Reichcigl (CRC Press, Boca Raton, FL, 1999), pp. 211–241.
7. R. A. de Maagd, A. Bravo, and N. Crickmore, How *Bacillus thuringiensis* has evolved specific toxins to colonize the insect world, *Trends Genet.* 17(4), 193–199 (2001).
8. A. M. R. Gatehouse, N. Ferry, and R. J. M. Raemaekers, The case of the monarch butterfly: A verdict is returned, *Trends Genet.* 18(5), 249–251 (2002).
9. S. S. Gill, E. A. Cowles and V. Francis, Identification, isolation, and cloning of a *Bacillus-Thuringiensis* Crylac toxin-binding protein from the midgut of the Lepidopteran insect *Heliothis-Virescens*, *J. Biol. Chem.* 270(45), 27277–27282 (1995).
10. P. J. K. Knight, N. Crickmore, and D. J. Ellar, The receptor for *Bacillus-Thuringiensis* Cryla(C) delta-endotoxin in the brush-border membrane of the Lepidopteran *Manduca-Sexta* is aminopeptidase-N, *Mol. Microbiol.* 11(3), 429–436 (1994).
11. K. Luo, S. Sangadala, L. Masson, A. Mazza, R. Brousseau, and M. J. Adang, The *Heliothis virescens* 170 kDa aminopeptidase functions as "receptor A" by mediating specific *Bacillus thuringiensis* Cry1A delta-endotoxin binding and pore formation, *J. Biochem. Mol. Biol.* 27 (8/9), 735–743 (1997).

12. S. Sangadala, F. S. Walters, L. H. English, and M. J. Adang, A mixture of *Manduca-Sexta* aminopeptidase and phosphatase enhances *Bacillus-Thuringiensis* insecticidal Cryla(C) toxin binding and (Rb+-K+)-Rb-86 efflux in vitro, *J. Biol. Chem.* 269(13), 10088–10092 (1994).
13. L. J. Gahan, F. Gould, and D. G. Heckel, Identification of a gene associated with bit resistance in *Heliothis virescens*, *Science* 293(5531), 857–860 (2001).
14. Y. Nagamatsu, S. Toda, T. Koike, Y. Miyoshi, S. Shigematsu, and M. Kogure, Cloning, sequencing, and expression of the *Bombyx mori* receptor for *Bacillus thuringiensis* insecticidal CryIA(a) toxin, *Biosci. Biotechnol. Biochem.* 62(4), 727–734 (1998).
15. R. K. Vadlamudi, E. Weber, I. H. Ji, T. H. Ji, and L. A. Bulla, Cloning and expression of a receptor for an insecticidal toxin of *Bacillus-Thuringiensis*, *J. Biol. Chem.* 270(10), 5490–5494 (1995).
16. P. Denolf, Isolation, cloning and characterisation of *Bacillus thuringiensis* delta-endotoxin receptors in Lepidoptera, Ph.D. thesis (University of Gent, 1996).
17. J. S. Griffitts, J. L. Whitacre, D. E. Stevens, and R. V. Aroian, Bt toxin resistance from loss of a putative carbohydrate-modifying enzyme, *Science* 293(5531), 860–864 (2001).
18. M. Vaeck, A. Reynaerts, H. Hofte, S. Jansens, M. Debeuckeleer, C. Dean, M. Zabeau, M. Vanmontagu, and J. Leemans, Transgenic plants protected from insect attack, *Nature* 328(6125), 33–37 (1987).
19. R. A. de Maagd, D. Bosch, and W. Stiekema, *Bacillus thuringiensis* toxin-mediated insect resistance in plants, *Trends Plant Sci.* 4(1), 9–13 (1999).
20. M. Peferoen, in *Advances in Insect Control: The Role of Transgenic Plants*, edited by N. Carozzi and M. Koziel (Taylor and Francis, London, pp. 21–38 (1997).
21. A. M. Shelton, J. Z. Zhao, and R. T. Roush, Economic, ecological, food safety, and social consequences of the deployment of Bt transgenic plants, *Annu. Rev. Entomol.* 47, 845–881 (2002).
22. N. Ferry, M. Edwards, J. Gatehouse, T. Capell, P. Christou, and A. Gatehouse, Transgenic plants for insect pest control: A forward looking scientific perspective, *Transgenic Res.* 15(1), 13–19 (2006).
23. B. E. Tabashnik, F. R. Groeters, N. Finson, Y. B. Liu, M. W. Johnson, D. G. Heckel, K. Luo, and M. J. Adang, in *Molecular Genetics and Evolution of Pesticide Resistance*, edited by T. Brown (Oxford University Press, USA, 1996), pp. 130–140.
24. P. Christou, T. Capell, A. Kohli, J. A. Gatehouse, and A. M. R. Gatehouse, Recent developments and future prospects in insect pest control in transgenic crops, *Trends Plant Sci.* 11(6), 302–308 (2006).
25. J. A. Gatehouse, Plant resistance towards insect herbivores: A dynamic interaction, *New Phytol.* 156(2), 145–169 (2002).
26. J. Harborne, *Introduction to Ecological Chemistry* (Academic Press, London, 1988).
27. J. Gatehouse, A. Gatehouse, and D. Bown, in *Recombinant Protease Inhibitors in Plants*, edited by D. Michaud (Landes Bioscience, Austin, TX, 2000), pp. 9–26.
28. L. Jouanin, M. Bonade-Bottino, C. Girard, G. Morrot, and M. Giband, Transgenic plants for insect resistance, *Plant Sci.* 131(1), 1–11 (1998).
29. V. A. Hilder, A. M. R. Gatehouse, S. E. Sheerman, R. F. Barker, and D. Boulter, A novel mechanism of insect resistance engineered into tobacco, *Nature* 330(6144), 160–163 (1987).
30. A. M. R. Gatehouse, V. A. Hilder, K. S. Powell, M. Wang, G. M. Davison, L. N. Gatehouse, R. E. Down, H. S. Edmonds, D. Boulter, C. A. Newell, A. Merryweather, W. D. O. Hamilton, and J. A. Gatehouse, Insect-resistant transgenic plants—Choosing the gene to do the job, *Biochem. Soc. Trans.* 22(4), 944–949 (1994).

31. J. Graham, R. J. McNicol, and K. Greig, Towards genetic based insect resistance in strawberry using the Cowpea trypsin inhibitor gene, *Ann. Appl. Biol.* 127(1), 163–173 (1995).
32. D. P. Xu, Q. Z. Xue, D. McElroy, Y. Mawal, V. A. Hilder, and R. Wu, Constitutive expression of a cowpea trypsin inhibitor gene, CpTi, in transgenic rice plants confers resistance to two major rice insect pests, *Mol. Breed.* 2(2), 167–173 (1996).
33. R. M. Broadway, Dietary regulation of serine proteinases that are resistant to serine proteinase inhibitors, *J. Insect Physiol.* 43(9), 855–874 (1997).
34. F. De Leo, M. Bonade-Bottino, L. R. Ceci, R. Gallerani, and L. Jouanin, Effects of a mustard trypsin inhibitor expressed in different plants on three lepidopteran pests, *J. Biochem. Mol. Biol.* 31(6/7), 593–602 (2001).
35. J. C. Leple, M. Bonadebottino, S. Augustin, G. Pilate, V. D. Letan, A. Delplanque, D. Cornu, and L. Jouanin, Toxicity to *Chrysomela-Tremulae* (Coleoptera, Chrysomelidae) of transgenic poplars expressing a cysteine proteinase-inhibitor, *Mol. Breed.* 1(4), 319–328 (1995).
36. C. Pannetier, M. Giband, P. Couzi, V. LeTan, M. Mazier, J. Tourneur, and B. Hau, Introduction of new traits into cotton through genetic engineering: Insect resistance as example, *J. Biochem. Mol. Biol.* 96(1), 163–166 (1997).
37. P. E. Urwin, H. J. Atkinson, D. A. Waller, and M. J. McPherson, Engineered oryzacystatin-I expressed in transgenic hairy roots confers resistance to *Globodera-Pallida*, *Plant J.* 8(1), 121–131 (1995).
38. N. S. Outchkourov, B. Rogelj, B. Strukelj, and M. A. Jongsma, Expression of sea anemone equistatin in potato. Effects of plant proteases on heterologous protein production, *Plant Physiol.* 133(1), 379–390 (2003).
39. A. Abdeen, A. Virgos, E. Olivella, J. Villanueva, X. Aviles, R. Gabarra, and S. Prat, Multiple insect resistance in transgenic tomato plants over-expressing two families of plant proteinase inhibitors, *Plant Mol. Biol.* 57(2), 189–202 (2005).
40. D. P. Bown, H. S. Wilkinson, and J. A. Gatehouse, Differentially regulated inhibitor-sensitive and insensitive protease genes from the phytophagous insect pest, *Helicoverpa armigara*, are members of complex multigene families, *J. Biochem. Mol. Biol.* 27(7), 625–638 (1997).
41. M. A. Jongsma and C. Bolter, The adaptation of insects to plant protease inhibitors, *J. Insect Physiol.* 43(10), 885–895 (1997).
42. S. C. Dias, O. L. Franco, C. P. Magalhaes, O. B. de Oliveira-Neto, R. A. Laumann, E. L. Z. Figueira, F. R. Melo, and M. F. Grossi-de-Sa, Molecular cloning and expression of an alpha-amylase inhibitor from rye with potential for controlling insect pests, *Protein J.* 24(2), 113–123 (2005).
43. A. L. Marsaro, S. M. N. Lazzari, E. L. Z. Figueira, and E. Y. Hirooka, Arnylase inhibitors in corn hybrids as a resistance factor to *Sitophilus zeamais* (Coleoptera: Curculionidae), *Neotrop. Entomol.* 34(3), 443–450 (2005).
44. H. E. Schroeder, S. Gollasch, A. Moore, L. M. Tabe, S. Craig, D. C. Hardie, M. J. Chrispeels, D. Spencer, and T. J. V. Higgins, Bean alpha-amylase inhibitor confers resistance to the pea weevil (*Bruchus pisorum*) in transgenic peas (*Pisum sativum* L) (Vol 107, Pg 1233, 1995), *Plant Physiol.* 109(3), 1129–1129 (1995).
45. R. E. Shade, H. E. Schroeder, J. J. Pueyo, L. M. Tabe, L. L. Murdock, T. J. V. Higgins, and M. J. Chrispeels, Transgenic pea-seeds expressing the alpha-amylase inhibitor of the common bean are resistant to bruchid beetles, *Bio-Technology* 12(8), 793–796 (1994).
46. M. J. Chrispeels and N. V. Raikhel, Lectins, lectin genes, and their role in plant defense, *Plant Cell* 3(1), 1–9 (1991).

47. W. J. Peumans and E. J. M. Vandamme, Lectins as plant defense proteins, *Plant Physiol.* 109(2), 347–352 (1995).
48. A. Gatehouse, K. Powell, W. Peumans, E. V. Damme, and J. Gatehouse, in *Lectins Biomedical Perspectives*, edited by A. Pusztai and S. Bardocz (Taylor and Francis, London, 1995), pp. 35–57.
49. X. Foissac, N. T. Loc, P. Christou, A. M. R. Gatehouse, and J. A. Gatehouse, Resistance to green leafhopper (*Nephotettix virescens*) and brown planthopper (*Nilaparvata lugens*) in transgenic rice expressing snowdrop lectin (*Galanthus nivalis* agglutinin; GNA), *J. Insect Physiol.* 46(4), 573–583 (2000).
50. A. M. R. Gatehouse, G. M. Davison, C. A. Newell, A. Merryweather, W. D. O. Hamilton, E. P. J. Burgess, R. J. C. Gilbert, and J. A. Gatehouse, Transgenic potato plants with enhanced resistance to the tomato moth, *Lacanobia oleracea*: Growth room trials, *Mol. Breed.* 3(1), 49–63 (1997).
51. K. S. Powell, A. M. R. Gatehouse, V. A. Hilder, and J. A. Gatehouse, Antifeedant effects of plant-lectins and an enzyme on the adult stage of the rice brown planthopper, *Nilaparvata-Lugens*, *Entomol. Exp. Appl.* 75(1), 51–59 (1995).
52. N. Sauvion, Y. Rahbe, W. J. Peumans, E. J. M. VanDamme, J. A. Gatehouse, and A. M. R. Gatehouse, Effects of GNA and other mannose binding lectins on development and fecundity of the peach-potato aphid *Myzus persicae*, *Entomol. Exp. Appl.* 79(3), 285–293 (1996).
53. K. V. Rao, K. S. Rathore, T. K. Hodges, X. Fu, E. Stoger, D. Sudhakar, S. Williams, P. Christou, M. Bharathi, D. P. Bown, K. S. Powell, J. Spence, A. M. R. Gatehouse, and J. A. Gatehouse, Expression of snowdrop lectin (GNA) in transgenic rice plants confers resistance to rice brown planthopper, *Plant J.* 15(4), 469–477 (1998).
54. P. Tinjuangjun, N. T. Loc, A. M. R. Gatehouse, J. A. Gatehouse, and P. Christou, Enhanced insect resistance in Thai rice varieties generated by particle bombardment, *Mol. Breed.* 6(4), 391–399 (2000).
55. S. B. Maqbool, S. Riazuddin, N. T. Loc, A. M. R. Gatehouse, J. A. Gatehouse, and P. Christou, Expression of multiple insecticidal genes confers broad resistance against a range of different rice pests, *Mol. Breed.* 7(1), 85–93 (2001).
56. R. E. Down, A. M. R. Gatehouse, W. D. O. Hamilton, and J. A. Gatehouse, Snowdrop lectin inhibits development and decreases fecundity of the glasshouse potato aphid (*Aulacorthum solani*) when administered in vitro and via transgenic plants both in laboratory and glasshouse trials, *J. Insect Physiol.* 42(11/12), 1035–1045 (1996).
57. E. Stoger, S. Williams, P. Christou, R. E. Down, and J. A. Gatehouse, Expression of the insecticidal lectin from snowdrop (*Galanthus nivalis* agglutinin; GNA) in transgenic wheat plants: Effects on predation by the grain aphid *Sitobion avenae*, *Mol. Breed.* 5(1), 65–73 (1999).
58. K. S. Powell, J. Spence, M. Bharathi, J. A. Gatehouse, and A. M. R. Gatehouse, Immunohistochemical and developmental studies to elucidate the mechanism of action of the snowdrop lectin on the rice brown planthopper, *Nilaparvata lugens* (Stal), *J. Insect Physiol.* 44(7/8), 529–539 (1998).
59. J. P. Du, X. Foissac, A. Carss, A. M. R. Gatehouse, and J. A. Gatehouse, Ferritin acts as the most abundant binding protein for snowdrop lectin in the midgut of rice brown planthoppers (*Nilaparvata lugens*), *J. Biochem. Mol. Biol.* 30(4), 297–305 (2000).
60. L. R. Ceci, M. Volpicella, Y. Rahbe, R. Gallerani, J. Beekwilder, and M. A. Jongsma, Selection by phage display of a variant mustard trypsin inhibitor toxic against aphids, *Plant J.* 33(3), 557–566 (2003).
61. N. T. Loc, P. Tinjuangjun, A. M. R. Gatehouse, P. Christou, and J. A. Gatehouse, Linear transgene constructs lacking vector backbone sequences generate transgenic rice plants

which accumulate higher levels of proteins conferring insect resistance, *Mol. Breed.* 9(4), 231–244 (2002).

62. Y. Rahbe, C. Deraison, M. Bonade-Bottino, C. Girard, C. Nardon, and L. Jouanin, Effects of the cysteine protease inhibitor oryzacystatin (OC-I) on different aphids and reduced performance of *Myzus persicae* on OC-I expressing transgenic oilseed rape, *Plant Sci.* 164(4), 441–450 (2003).
63. E. P. J. Burgess, L. A. Malone, J. T. Christeller, M. T. Lester, C. Murray, B. A. Philip, M. M. Phung, and E. L. Tregidga, Avidin expressed in transgenic tobacco leaves confers resistance to two noctuid pests, *Helicoverpa armigera* and *Spodoptera litura*, *Transgenic Res.* 11(2), 185–198 (2002).
64. E. Fitches, N. Audsley, J. A. Gatehouse, and J. P. Edwards, Fusion proteins containing neuropeptides as novel insect control agents: Snowdrop lectin delivers fused allatostatin to insect haemolymph following oral ingestion, *J. Biochem. Mol. Biol.* 32(12), 1653–1661 (2002).
65. J. T. Christeller, E. P. J. Burgess, V. Mett, H. S. Gatehouse, N. P. Markwick, C. Murray, L. A. Malone, M. A. Wright, B. A. Philip, D. Watt, L. N. Gatehouse, G. L. Lovei, A. L. Shannon, M. M. Phung, L. M. Watson, and W. A. Laing, The expression of a mammalian proteinase inhibitor, bovine spleen trypsin inhibitor in tobacco and its effects on *Helicoverpa armigera* larvae, *Transgenic Res.* 11(2), 161–173 (2002).
66. R. A. de Maagd, A. Bravo, C. Berry, N. Crickmore, and H. E. Schnepf, Structure, diversity, and evolution of protein toxins from spore-forming entomopathogenic bacteria, *Annu. Rev. Genet.* 37, 409–433 (2003).
67. C. G. Yu, M. A. Mullins, G. W. Warren, M. G. Koziel, and J. J. Estruch, The *Bacillus thuringiensis* vegetative insecticidal protein Vip3A lyses midgut epithelium cells of susceptible insects, *Appl. Environ. Microbiol.* 63(2), 532–536 (1997).
68. A. Chattopadhyay, N. B. Bhatnagar, and R. Bhatnagar, Bacterial insecticidal toxins, *Crit. Rev. Microbiol.* 30(1), 33–54 (2004).
69. D. Liu, S. Burton, T. Glancy, Z. S. Li, R. Hampton, T. Meade, and D. J. Merlo, Insect resistance conferred by 283 kDa *Photorhabdus luminescens* protein TcdA in *Arabidopsis thaliana*, *Nat. Biotechnol.* 21(10), 1222–1228 (2003).
70. L. Mehlo, D. Gahakwa, P. T. Nghia, N. T. Loc, T. Capell, J. A. Gatehouse, A. M. R. Gatehouse, and P. Christou, An alternative strategy for sustainable pest resistance in genetically enhanced crops, *Proc. Natl. Acad. Sci. U S A* 102(22), 7812–7816 (2005).
71. E. Fitches, M. G. Edwards, C. Mee, E. Grishin, A. M. R. Gatehouse, J. P. Edwards, and J. A. Gatehouse, Fusion proteins containing insect-specific toxins as pest control agents: Snowdrop lectin delivers fused insecticidal spider venom toxin to insect haemolymph following oral ingestion, *J. Insect Physiol.* 50(1), 61–71 (2004).
72. I. T. Baldwin, R. Halitschke, A. Kessler, and U. Schittko, Merging molecular and ecological approaches in plant–insect interactions, *Ecol. Appl.* 4(4), 351–358 (2001).
73. B. A. Bailey, M. D. Strem, H. H. Bae, G. A. de Mayolo, and M. J. Guiltinan, Gene expression in leaves of *Theobroma cacao* in response to mechanical wounding, ethylene, and/or methyl jasmonate, *Plant Sci.* 168(5), 1247–1258 (2005).
74. D. Hermsmeier, U. Schittko, and I. T. Baldwin, Molecular interactions between the specialist herbivore *Manduca sexta* (Lepidoptera, Sphingidae) and its natural host *Nicotiana attenuata*, I: Large-scale changes in the accumulation of growth- and defense-related plant mRNAs, *Plant Physiol.* 125(2), 683–700 (2001).
75. M. J. Stout, A. L. Fidantsef, S. S. Duffey, and R. M. Bostock, Signal interactions in pathogen and insect attack: Systemic plant-mediated interactions between pathogens

and herbivores of the tomato, *Lycopersicon esculentum*, *Plant Pathol.* 54(3/4), 115–130 (1999).

76. F. Zhang, L. Zhu, and G. C. He, Differential gene expression in response to brown planthopper feeding in rice, *J. Plant Physiol.* 161(1), 53–62 (2004).
77. W. Q. Chen, N. J. Provart, J. Glazebrook, F. Katagiri, H. S. Chang, T. Eulgem, F. Mauch, S. Luan, G. Z. Zou, S. A. Whitham, P. R. Budworth, Y. Tao, Z. Y. Xie, X. Chen, S. Lam, J. A. Kreps, J. F. Harper, A. Si-Ammour, B. Mauch-Mani, M. Heinlein, K. Kobayashi, T. Hohn, J. L. Dangl, X. Wang, and T. Zhu, Expression profile matrix of Arabidopsis transcription factor genes suggests their putative functions in response to environmental stresses, *Plant Cell* 14(3), 559–574 (2002).
78. D. D. Schmidt, C. Voelckel, M. Hartl, S. Schmidt, and I. T. Baldwin, Specificity in ecological interactions. Attack from the same lepidopteran herbivore results in species-specific transcriptional responses in two solanaceous host plants, *Plant Physiol.* 138(3), 1763–1773 (2005).
79. E. Rojo, R. Solano, and J. J. Sanchez-Serrano, Interactions between signaling compounds involved in plant defense, *J. Plant Growth Regul.* 22(1), 82–98 (2003).
80. R. A. Dixon, Engineering of plant natural product pathways, *Ecol. Appl.* 8(3), 329–336 (2005).
81. Cropgen (2002); available at http://www.cropgen.org/.
82. GM Science Review (2003); available at http://www.gmsciencedebate.org.uk.
83. E. Abergel and K. Barrett, Putting the cart before the horse: A review of biotechnology policy in Canada, *J. Can. Stud/REC* 37(3), 135–161 (2002).
84. E. A. Clark, Environmental risks of genetic engineering, *J. Biochem. Mol. Biol.* 148(1/2), 47–60 (2006).
85. M. J. Crawley, S. L. Brown, R. S. Hails, D. D. Kohn, and M. Rees, Biotechnology—Transgenic crops in natural habitats, *Nature* 409(6821), 682–683 (2001).
86. P. J. Dale, B. Clarke, and E. M. G. Fontes, Potential for the environmental impact of transgenic crops (*Nat. Biotechnol.* vol 20, p. 567, 2002), *Nat. Biotechnol.* 20(8), 843 (2002) (erratum).
87. J. Davison, Risk mitigation of genetically modified bacteria and plants designed for bioremediation, *J. Ind. Microbiol. Biotechnol.* 32(11/12), 639–650 (2005).
88. D. Lee and E. Natesan, Evaluating genetic containment strategies for transgenic plants, *Trends Biotechnol.* 24(3), 109–114 (2006).
89. D. Michaud, Environmental impact of transgenic crops, I: Transgene migration, *Phytoprotection* 86(2), 93–105 (2005).
90. A. A. Snow, D. A. Andow, P. Gepts, E. M. Hallerman, A. Power, J. M. Tiedje, and L. L. Wolfenbarger, Genetically engineered organisms and the environment: Current status and recommendations, *Ecol. Appl.* 15(2), 377–404 (2005).
91. A. T. Groot and M. Dicke, Insect-resistant transgenic plants in a multi-trophic context, *Plant J.* 31(4), 387–406 (2002).
92. P. A. M. Hogervorst, N. Ferry, A. M. R. Gatehouse, F. L. Wackers, and J. Romeis, Direct effects of snowdrop lectin (GNA) on larvae of three aphid predators and fate of GNA after ingestion, *J. Insect Physiol.* 52(6), 614–624 (2006).
93. L. B. Obrist, A. Dutton, J. Romeis, and F. Bigler, Biological activity of Cry1Ab toxin expressed by Bt maize following ingestion by herbivorous arthropods and exposure of the predator *Chrysoperla carnea*, *Biocontrol* 51(1), 31–48 (2006).
94. T. H. Schuler, R. P. J. Potting, I. Denholm, and G. M. Poppy, Parasitoid behaviour and Bt plants, *Nature* 400(6747), 825–826 (1999).

95. E. Vojtech, M. Meissle, and G. M. Poppy, Effects of Bt maize on the herbivore *Spodoptera littoralis* (Lepidoptera: Noctuidae) and the parasitoid *Cotesta marginiventris* (Hymenoptera: Braconidae), *Transgenic Res.* 14(2), 133–144 (2005).
96. E. B. Dogan, R. E. Berry, G. L. Reed, and P. A. Rossignol, Biological parameters of convergent lady beetle (Coleoptera: Coccinellidae) feeding on aphids (Homoptera: Aphididae) on transgenic potato, *J. Econ. Entomol.* 89(5), 1105–1108 (1996).
97. R. E. Down, L. Ford, S. J. Bedford, L. N. Gatehouse, C. Newell, J. A. Gatehouse, and A. M. R. Gatehouse, Influence of plant development and environment on transgene expression in potato and consequences for insect resistance, *Transgenic Res.* 10(3), 223–236 (2001).
98. G. Head, C. R. Brown, M. E. Groth, and J. J. Duan, Cry1Ab protein levels in phytophagous insects feeding on transgenic corn: Implications for secondary exposure risk assessment, *Entomol. Exp. Appl.* 99(1), 37–45 (2001).
99. H. A. Bell, E. C. Fitches, G. C. Marris, J. Bell, J. P. Edwards, J. A. Gatehouse, and A. M. R. Gatehouse, Transgenic GNA expressing potato plants augment the beneficial biocontrol of *Lacanobia oleracea* (Lepidoptera: Noctuidae) by the parasitoid *Eulophus pennicornis* (Hymenoptera: eulophidae), *Transgenic Res.* 10(1), 35–42 (2001).
100. M. E. Wakefield, H. A. Bell, E. C. Fitches, J. P. Edwards, and A. M. R. Gatehouse, Effects of *Galanthus nivalis* agglutinin (GNA) expressed in tomato leaves on larvae of the tomato moth *Lacanobia oleracea* (Lepidoptera: Noctuidae) and the effect of GNA on the development of the endoparasitoid *Meteorus gyrator* (Hymenoptera: Braconidae), *Bull. Entomol. Res.* 96(1), 43–52 (2006).
101. H. A. Bell, R. E. Down, E. C. Fitches, J. P. Edwards, and A. M. R. Gatehouse, Impact of genetically modified potato expressing plant-derived insect resistance genes on the predatory bug *Podisus maculiventris* (Heteroptera: Pentatomidae), *Biocontrol Sci. Technol.* 13(8), 729–741 (2003).
102. R. E. Down, L. Ford, S. D. Woodhouse, G. M. Davison, M. E. N. Majerus, J. A. Gatehouse, and A. M. R. Gatehouse, Tritrophic interactions between transgenic potato expressing snowdrop lectin (GNA), an aphid pest (peach-potato aphid; *Myzus persicae* (Sulz.) and a beneficial predator (2-spot ladybird; *Adalia bipunctata* L.), *Transgenic Res.* 12(2), 229–241 (2003).
103. R. E. Down, L. Ford, S. D. Woodhouse, R. J. M. Raemaekers, B. Leitch, J. A. Gatehouse, and A. M. R. Gatehouse, Snowdrop lectin (GNA) has no acute toxic effects on a beneficial insect predator, the 2-spot ladybird (*Adalia bipunctata* L.), *J. Insect Physiol.* 46(4), 379–391 (2000).
104. N. Ferry, R. J. M. Raemaekers, M. E. N. Majerus, L. Jouanin, G. Port, J. A. Gatehouse, and A. M. R. Gatehouse, Impact of oilseed rape expressing the insecticidal cysteine protease inhibitor oryzacystatin on the beneficial predator *Harmonia axyridis* (multicoloured Asian ladybeetle), *Mol. Ecol.* 12(2), 493–504 (2003).
105. N. Ferry, L. Jouanin, L. R. Ceci, A. Mulligan, K. Emami, J. A. Gatehouse, and A. M. R. Gatehouse, Impact of oilseed rape expressing the insecticidal serine protease inhibitor, mustard trypsin inhibitor-2 on the beneficial predator *Pterostichus madidus*, *Mol. Ecol.* 14(1), 337–349 (2005).

2. BACTERIA AS BIOLOGICAL CONTROL AGENTS FOR INSECTS: ECONOMICS, ENGINEERING, AND ENVIRONMENTAL SAFETY

Brian A. Federici*
Department of Entomology and Graduate Programs in Molecular Biology, University of California, Riverside, CA 92521, USA

Abstract. Pathogens of insects have been under evaluation as biological control agents for more than a century. With few exceptions, they are not effective as classical biological control agents. Moreover, even as insecticides, only *Bacillus thuringiensis* (Bt) has been a commercial success. Bt's success, in essence, is due to its ease of mass production by fermentation on inexpensive media, which facilitated commercialization. Viruses, fungi, and protozoa are used in only a few niche markets, and thus have largely failed as microbial insecticides, and will continue to fail until more efficacious mass production methods are developed. Despite these failures, research on insect pathogens led to the development of transgenic insect-tolerant Bt crops, arguably the most important advance in pest control technology of the latter half of the 20th century. Numerous laboratory and field studies have shown that these crops are cost-effective and much safer than synthetic chemical insecticides for the environment and non-target organisms. The high specificity of Bt crops provides a new cornerstone for biological control and sustainable agriculture that will enable both to expand during this century.

Keywords: Insect pathogens; biological control; microbial insecticides; insecticidal proteins; Cry proteins; genetically modified organisms; Bt crops

2.1. Introduction

Pathogens of insects have been promoted for their pest control potential for more than a century. Despite this, only a few have been successful in biological control, and are used routinely for large-scale insect control in industrialized countries. At present, less than 1% of the insect control agents used worldwide are based on insect pathogens. Those used most widely are different subspecies of the bacterium, *Bacillus thuringiensis* (Bt), which constitute

*To whom correspondence should be addressed, e-mail: brian.federici@ucr.edu

M. Vurro and J. Gressel (eds.), Novel Biotechnologies for Biocontrol Agent Enhancement and Management, 25–51.

approximately 80% of the microbials used for insect control. Of the remaining 20%, a few baculoviruses have been successful as classical biological control agents, or have achieved moderate success as viral insecticides in several developing countries. Fungi, protozoa, and parasitic nematodes have been much less successful, and essentially have only been used in niche markets in a few countries,[1] or where control programs are subsidized by governments or international agencies. Examples of the latter include the use of fungi for locust control in the Middle East and Africa, and baculoviruses for control of forest pests in the United States and Canada.

The reason why pathogens are not used more widely as biological control agents or microbial insecticides in industrialized countries is due primarily to insufficient methods for cost-effective mass production. This becomes apparent by considering the different ways pathogens can be used for insect control, and by understanding the expectations used to evaluate their performance. Such an assessment identifies the key features required for pathogens to be successful as control agents. Though pathogens are used as the examples here, these principles apply to other biological control agents such as parasitic and predatory insects, as well as to organisms used to control other crop pests such as plant pathogens and nematodes. Following this discussion, insecticidal bacteria are discussed, both conventional and genetically engineered, after which the extension to this technology to transgenic crops, specifically Bt crops, is summarized, especially with respect to environmental safety.

2.2. Economics of Biological Control

The guiding principle in pest control remains primarily one of economics; control strategies and agents used are those that are the most *cost-effective* in the short or long term. This would appear to be obvious, but too often is overlooked in the literature on biological control agents. The effectiveness of a control strategy based on a pathogen, for example, will always be compared with that obtained with other available control strategies, including synthetic chemical insecticides. Importantly, cost-effectiveness can vary with such factors as crop, season, the species complex to be controlled, the cost of production of the control agent, geographical location, governmental regulations affecting registration and use, and the status of economic development in the country where the pathogen is used. The latter two factors are particularly relevant because the costs for development, production, and use of pathogens in developing countries for insect control are much less than in highly industrialized countries. This is due to lower material and labor costs in developing countries, as well as to the smaller size of most farms, lower levels of agricultural mechanization, and cheaper and less cumbersome registration procedures, for example, as exist in China and India. It is for these reasons that baculoviruses, especially nuclear polyhedrosis viruses (NPVs) are used more in developing countries than in the highly industrialized nations, where they account for less than 0.01% of operational pest control agents.[2] As a result, positive assessments of the cost-effectiveness of baculoviruses and other pathogens based on studies carried out in developing countries cannot be directly translated into corresponding levels of cost-effectiveness in developed countries.

2.2.1. STRATEGIES FOR PATHOGEN USE

The most cost-effective strategy for using a pathogen, as with other biological control agents, is as a *classical biological control* agent. In this strategy, introduction of the pathogen results in outbreaks of disease (epizootics) and maintenance of the pathogen in the target population. Within a few years, the pest population is reduced below the economic threshold on a permanent basis. Pathogens may establish and become endemic in a target population within a few years of introduction. Although classical biological control has proven very successful with many predaceous insects, and especially with parasitic wasps, there is only one good example of a classical biological control success with a pathogen—the control the European spruce sawfly, *Gilpinia hercyniae*, in North America by its nuclear polyhedrosis virus (NPV).[3] It is possible that the fungus, *Entomophthora miamiaga*, may also eventually prove to be a classical biological control success for the gypsy moth, *Lymantria dispar*, in the United States, but assessment of this will require at least another decade.[4]

Another strategy is to use a pathogen as an *augmentative control agent*. In this strategy, a pathogen endemic in a population, but at a low level, is applied against a pest population at the beginning of the season or early in the development of the pest population. The pathogen reduces the population below the economic threshold or reduces pest damage substantially sooner than might occur naturally. The effect usually only lasts one or a few seasons, and must be repeated. The granulosis virus (GV) of the grape leaf skeletonizer, *Harrisina brillians*,[5] and the NPV of the Douglas fir tussock moth, *Orgyia pseudotsugata*, are used successfully in this manner. Again, however, this tactic has proven of only limited utility.

The most common strategy for using a pathogen as an insect control agent is as a *microbial insecticide*. In this tactic, pathogens such as various subspecies of *Bacillus thuringiensis*, NPVs, or occasionally fungi such as *Beauveria bassiana* or *Metarhizium anisopliae* are formulated and applied against a target pest on a periodic basis, as needed, much as are chemical insecticides.[1] Depending on the target pest, applications may often be fewer than those required with a chemical insecticide because pathogens are quite specific and typically do not kill predatory and parasitic insects, which typically belong to other insect orders. Natural enemies, therefore, remain in the ecosystem, retarding increases in the pest population after the initial mortality caused by the pathogen. In addition, pathogen reproduction in the target insect adds to the amount in the crop environment, and this can extend control and thus cost-effectiveness. In pests with only one or a few generations per season, a single application can yield effective season-long control where a combination of these factors is in operation.

2.2.2. ECONOMICS OF PERFORMANCE EXPECTATIONS

Performance expectation is an issue of major importance affecting the adoption of pathogens as control agents. In most cases, a successful pathogen is one that can routinely and reliably reduce the pest or vector population to below an economic or disease transmission threshold at a cost that is economical in proportion to the value of the crop or health. Clearly, pathogens that are effective as classical biological control or augmentative agents, would be the most cost-effective because of the limited number of applications required. But very few pathogens can be used successfully in either of these strategies. As a result, the efficacy of most pathogens is evaluated in terms of their utility as microbial insecticides.

Since World War II, the performance of chemical insecticides has set very high expectations against which all alternative pest control strategies must be compared. Traditionally, chemicals have been fast-acting, broad-spectrum control agents with substantial residual activity that are relatively inexpensive, and easy to produce, formulate, and use. Thus, under most circumstances pathogens are evaluated on the basis of how they compare with chemical insecticides, in particular, how quickly they kill the target insect, and at what cost for an acceptable level of crop protection. This leads to a paradox for microbial control agents. The two properties of chemical insecticides originally considered their best attributes, i.e., a broad spectrum of activity and significant residual activity, are now viewed as detrimental because they can result in the destruction of natural enemy populations and insecticide resistance in the target population. Yet, most pathogens have a narrow spectrum of activity and relatively poor residual activity. Though now considered beneficial properties, until recently, with the exception of Bt and a few viruses, these attributes have discouraged interest in pathogens by industry because of the relatively high costs of development and registration in comparison to the likely return on investment.

The costs for development and registration of a naturally occurring pathogen, nevertheless, are much cheaper than those for a chemical insecticide, e.g., approximately half a million Euros for a baculovirus or new Bt compared to at least several millions for most chemicals in the United States. But a company still must see the potential for making its investment pay off within a few years. For viruses, most of which have a very narrow cost-effective target spectrum—often limited to a single species—unless their target is a pest of a major commodity, or a polyphagous pest causing damage to a variety of crops, registration is simply not justified given the current regulatory environment and market size in most industrialized countries. Industrial interest in developing and registering pathogens, whether on a small or large scale, is important because it is industry that will produce most products. This is particularly true in developed countries where the farms tend to be large. In fact, most farmers whether large or small want a reliable supply of control agents, and though willing to change cultural practices, because of numerous other responsibilities, they typically are not willing to manufacture their own insecticides.

2.2.3. PRODUCTION TECHNOLOGY AND COSTS AS CONSTRAINTS

Given that most the effective tactic for using a pathogen as a control agent is as a microbial insecticide, if pathogens are to be widely used, methods must exist for their mass production. Therefore, the status of mass production technology for each of the major pathogen groups is briefly summarized below.

2.2.3.1. *Viruses*

All viruses are obligate intracellular parasites, and as such must be produced in living cells. This limits mass production options to producing viruses in their natural hosts (e.g., caterpillars in the case of baculoviruses), or by using large-scale cell culture (in vitro) technology. Production in caterpillars has been used successfully in many developing countries, but companies in industrialized countries have been reluctant to pursue such technologies owing to problems with quality control and scale-up for making multiple applications to large commodity crops such as corn, cotton, soybeans, and rice. With respect to in vitro culture, problems with cost-effective scale-up have not been resolved for most baculoviruses. The *Autographa californica* multinucleocapsid nuclear polyhedrosis virus (AcMNPV) is the principal viral insecticide candidate because it has a broad host range among caterpillar pests. Against many of these pests, it is not cost-effective in comparison to other registered control agents (Bts and chemicals). Recombinant AcMNPVs have been developed that kill target pest species faster, but serious doubts remain regarding whether even these can compete with existing Bts and chemicals, and new ones coming to market.

There are numerous naturally occurring viruses that could be useful in IPM programs for crops of smaller areas. For example, the MNPV of the beet armyworm, *Spodoptera exigua*, currently produced in caterpillars, is a registered virus that is proving successful for control of the beet armyworm in the glasshouse industry. The NPV of *Heliothis virescens* is registered in several countries for control of this pest in cotton and other crops. The area treated traditionally with this virus has been very limited, and with the exception of Australia, where it is required as a component of resistance management for Bt cotton, its recent use has declined owing to the wide scale adoption of the latter crop. There are many other viruses that could prove useful in niche markets, but in general there is little interest in industry in registering and producing these viruses as because the markets are small, and in most cases the competition from new Bts and chemicals remains strong, providing little incentive for virus development.

2.2.3.2. *Bacteria*

The success of *Bacillus thuringiensis* results from the relative ease of mass-producing products based on this bacterium.[6] Fermentation is typically carried out in 40,000 to 120,000 l fermentors, enabling yearly production for large markets to be accomplished within a few months. Fermentation technology continues to improve. In addition, improved products based on naturally occurring subspecies and genetically engineered strains continue to emerge for control of pests in forestry, field and vegetable crops, as well as for mosquito and blackfly control.

2.2.3.3. *Fungi*

The principal fungi considered for use as microbial insecticides remain strains of imperfect fungi such as *M. anisopliae*, *B. bassiana*, and *Paecilomyces fumosoroseus*. Effective control of target lepidopteran, coleopteran, or homopteran pests with these fungi in the United States typically requires in the range of 10^5–10^6 conidia or colony forming units per cm^2 of leaf surface or cm^3 of soil. Generating these levels of infectious materials requires large amounts of substrate, generally in the range of or higher than 10–15 kg of substrate per hectare treatment. New fermentation systems are coming on line, but few products are currently marketed. Products available are based primarily on *B. bassiana*, and are targeted for use in glasshouses, especially for high value cash crops, such as flowers. As in the case of viruses, serious doubts remain with respect to whether fungi can be cost-effective against major field crop pests.[7] The primary reason for this remains a lack of cost-effective methods for fungal mass production.

Nevertheless, some fungi could prove useful in niche markets of high cash value crops. For example, the glassy-winged sharpshooter, *Homalodisca coagulata* (GWSS), a serious sucking-insect pest that recently invaded California vineyards, is susceptible to several fungal diseases. This pest would be difficult if not impossible to control with viruses, protozoa, or even parasitic nematodes because these usually enter the host by being ingested, and in sufficient quantities to cause disease. With the most expensive wines wholesaling at 30–50 Euros per bottle, and yields of 1–2 bottles per grape plant, it becomes cost-effective to use a fungus like *B. bassiana*, even if the cost of treatment is 3–4 Euros per plant. Opportunities exist for such a use because, even though imidocloprid, a relatively new chemical insecticide, can be used to control the GWSS, wine lovers do not want chemical insecticides in their wines, making use of a fungal insecticide acceptable, even at a high cost.

2.2.3.4. *Protozoa*

The major types of protozoans considered for use in pest control are members of the phylum Microspora, commonly known as microsporidia (now known to be unusual parasitic fungi). These are all obligate intracellular parasites, and thus, like viruses, must be produced in living hosts or cell cultures. They have an additional disadvantage that viruses do not have in that most microsporidia cause chronic diseases. Thus, if hosts are not killed during early instars, they are capable of consuming more plant material than uninfected pests. For these reasons, microsporidia have no apparent potential as microbial insecticides.

To summarize briefly, lack of cost-effective methods of mass production has limited and continues to limit the use of viruses, fungi, and protozoa as pest control agents for most major field, vegetable, and forest crops. Some pathogens may prove successful in the future as classical biological control or augmentative agents, but cases of true success with a widespread economic impact, based on our experience to date, will be rare.

2.3. Conventional and Engineered Bacterial Insecticides

Bacteria are relatively simple unicellular microorganisms that lack internal organelles such as a nucleus and mitochondria, and reproduce by binary fission. With a few exceptions, most of those that cause disease in insects grow readily on various inexpensive substrates, a characteristic greatly facilitating their mass production. A wide variety of bacteria are capable of causing diseases in insects, but, as noted above, those that have received the most study are spore-forming bacilli (family Bacillaceae), especially *Bacillus thuringiensis* (Bt). Many subspecies of Bt are used as bacterial insecticides and as a source of genes for insecticidal proteins used in recombinant bacteria and Bt crops. Other bacteria that have been developed as insecticides are *B. sphaericus*, *Paenibacillus popilliae*, and *Serratia entomophila*. These will be discussed briefly in order of importance, after which key aspects Bt's molecular biology will be reviewed as a prelude to discussion of recombinant bacterial insecticides and transgenic crops based of this species.

2.3.1. *BACILLUS THURINGIENSIS* (BT)

Bt is a complex of bacterial subspecies that occur commonly in such habitats as soil, leaf litter, on the surfaces of leaves, in insect feces, and as a part of the flora in the midguts of many insect species.[8,9] Bts are characterized by the production of a parasporal body during sporulation that contains one or more protein endotoxins in a crystalline form (Figure 1). Many of these are highly

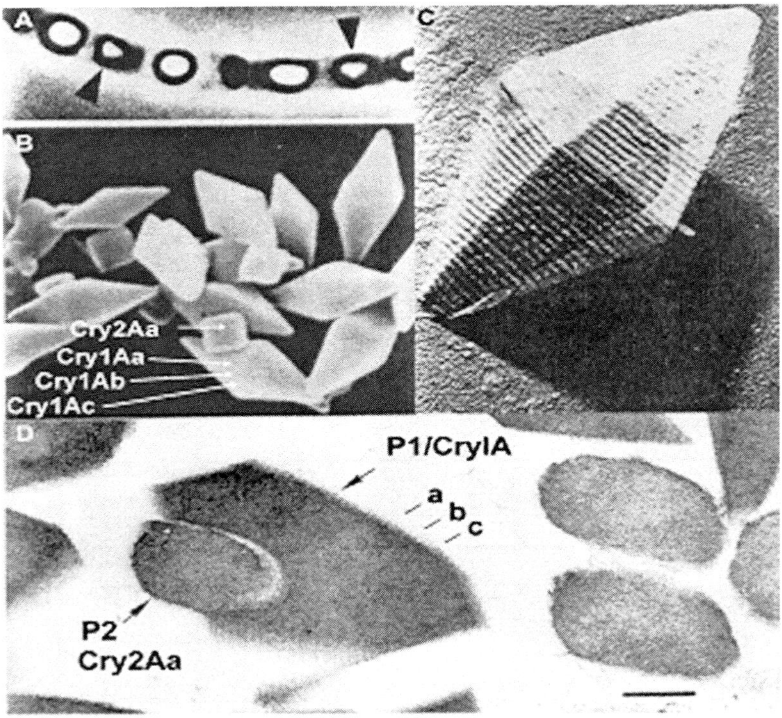

Figure 1. Sporulated cells of *Bacillus thuringiensis* and parasporal protein crystals. **A** Phase contrast micrograph of cells from a sporulated culture of *B. thuringiensis* just prior to lysis. Parasporal protein crystals (arrowheads) lie adjacent to oval spores. **B** Scanning electron micrograph of typical Cry1 and Cry2 crystals purified from a sporulated culture of *B. thuringiensis* subsp. *kurstaki*, isolate HD1. The parasporal body of this isolate consists of a bipyramidal crystal that contains Cry1Aa, Cry1Ab, and Cry1Ac, which co-crystallize, and a separate "cuboidal" crystal composed of Cry2Aa molecules. **C** Carbon replica of a typical bipyramidal Cry1 type protein crystal exhibiting the lattice of Cry1A molecules that compose the crystal. The HD73 isolate of *B. thuringiensis* subsp. *kurstaki* only expresses a single *cry1Ac* gene, and its parasporal body contains only a single crystal, such as this one, which measures approximately 1 mm from point-to-point along the longitudinal axis. **D** Transmission electron micrograph through a parasporal body of the HD1 isolate of B. *thuringiensis* subsp. *kurstaki* illustrating the embedment of the cuboidal Cry2A crystal (P2) in the bipyramidal crystal (P1). Bar in D = 200 nm. Micrograph in **C** by C. L. Hannay

insecticidal to certain insect species. These endotoxins are actually protoxins activated by proteolytic cleavage in the insect midgut after ingestion. The activated toxins destroy midgut epithelial cells, killing sensitive insects within a day or two of ingestion. In insect species only moderately sensitive to the toxins, such as *Spodoptera* species (caterpillars commonly known as armyworms), the spore contributes to pathogenesis by germinating and producing vegetative insecticidal proteins (Vips), proteases and phopholipases. Bt also produces other insecticidal compounds including β-exotoxin and Zwittermicin A. Some of these are synergistic, and thus their combined actions often result in death of recalcitrant lepidopteran species.

The most widely used Bt is the HD1 isolate of *B. thuringiensis* subsp. *kurstaki* (Btk), an isolate that produces four major endotoxin proteins, Cry1Aa, Cry1Ab, Cry1Ac, and Cry2Aa packaged into two crystalline parasporal body. The three Cry1 proteins co-crystallize forming a bipyramidal crystal, whereas Cry2Aa forms a separate cuboidal crystal (Figure 1). This isolate is used as the active ingredient in numerous commercially available bacterial insecticides (Dipel, Foray, Thuricide) used to control many lepidopteran pests in field and vegetable crops, and forests.[9] Another successful Bt is *B. thuringiensis* subsp. *israelensis* (Bti), which is highly toxic to the larvae of many mosquito and blackfly species. This isolate produces a parasporal body that contains four major endotoxins, Cry4Aa, Cry4Ba, Cry11Aa, and Cyt1Aa (Figure 2). The three Cry proteins are related to those of Btk, but have insect spectra limited to mosquitoes, blackflies, and related dipteran species. Cyt proteins are unrelated to Cry proteins.[9,10]

Several commercial products based on Bti are available and are used to control both nuisance and vector mosquitoes and black flies. Formulations of Bti (VectoBac, Teknar) proved particularly important in the World Health Organization's Onchocerciasis Control Program in West Africa, were they were used during the 1980's and 1990's in rotation with the chemical insecticide, temiphos, to virtually eliminate larval populations of the black fly vector, *Simulium damnosum*, of the filarial worm that causes this disease. A third isolate of Bt that has been developed commercially is the DSM2803 isolate of *B. thuringiensis* subsp. *morrisoni* (strain tenebrionis). This isolate produces Cry3Aa, which forms a cuboidal parasporal body toxic to many coleopterous insects, and is available in some countries as an insecticide.

All of the above isolates are essentially used as bacterial insecticides, applied as needed. These are available in a variety of formulations including emulsifiable concentrates, wettable powders, and granules for use against a wide range of pests and vectors in different habitats. On a worldwide basis, millions of hectares are treated annually with products based on Bt. Recent estimates indicate the worldwide market is about 60–80 million Euros.

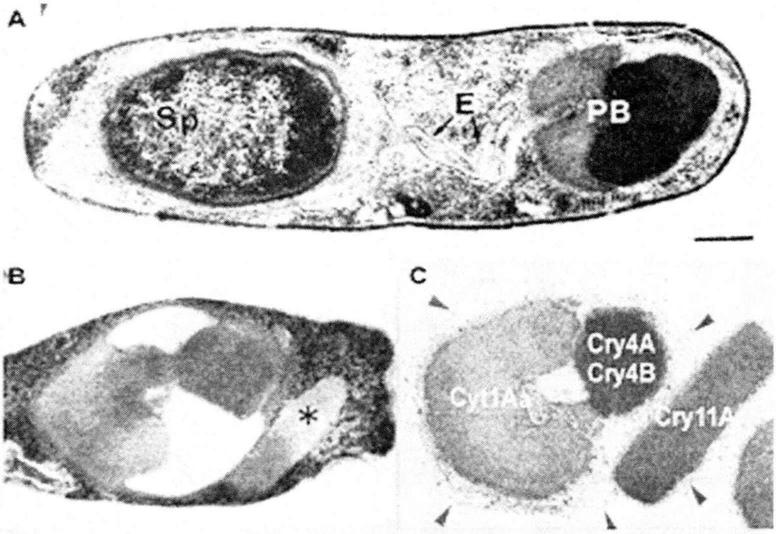

Figure 2. Sporulating cell of *Bacillus thuringiensis* subspecies israelensis and parasporal bodies characteristic of this subspecies as revealed by transmission electron microscopy. **A**: Sporulating cell illustrating the developing spore (Sp) and parasporal body. The parasporal body (PB), composed primarily of Cry4A, Cry4B, Cry11A, and Cyt1A proteins, is assembled outside the exosporium membrane (E). **B**: Portion of sporulating cell just prior to lysis. The Cry11A crystal (*) lies adjacent to the Cyt1A and Cry4A and Cry4B inclusions. **C**: Purified parasproal body showing the components of the parasporal body. In this subspecies, the individual protein inclusions are enveloped in a multilamellar fibrous matrix (arrowheads) of unknown composition, which also surrounds the crystals holding them together. A typical mature parasporal body of this subspecies measures 500–700 nm in diameter. Bar in A = 100 nm

2.3.2. *BACILLUS SPHAERICUS*

Since the mid-1960s it has been known that many isolates of *B. sphaericus* (Bs) are toxic to mosquito species. Over the past three decades, three isolates have been evaluated for mosquito control, 1593 from Indonesia, 2297 from Sri Lanka, and 2362 from Nigeria.[11] The 1593 and 2297 isolates were obtained from soil and water samples at mosquito breeding sites, whereas 1593 was isolated from a dead adult blackfly.

The toxicity of Bs, like Bt, is the result of protein endotoxins that are produced during sporulation and assembled into a parasporal body. Bs is unusual in that the main toxin is a binary toxin, i.e., composed of two protein subunits (Bin A and Bin B). These are proteolytically activated in the mosquito midgut to release peptides of, respectively, 43 and 39 kDa, that associate to form the binary toxin, with the former protein constituting the binding domain, and the latter the toxin domain. The toxins bind to microvilli of the midgut epithelium, causing hypertrophy and lysis of cells, destroying the midgut and killing the mosquito larva.

Recently, a commercial product known as VectoLex (Valent BioSciences, Libertyville, IL) has come to market for control of *Culex* mosquito larvae, and certain species of *Anopheles* mosquitoes. More recently, a strain known as C-41, which is similar to 2362, has been isolated in China and since 2000 has been used to control *Culex* species, as well as certain anophelines sensitive to the Bs Bin toxin.

2.3.3. *PAENIBACILLUS POPILLIAE*

This highly fastidious species is the primary etiologic agent of the so-called milky diseases of scarab larvae. The immature stages of beetles, such as the Japanese beetle, *Popillia japonica*, are important grass and plant pest belonging to the coleopteran family Scarabaeidae. The term "milky disease" is derived from the opaque white color that characterizes diseased larvae and results from the accumulation of sporulating bacteria in larval blood (hemolymph). The disease is initiated when grubs feeding on the roots of grasses or other plants ingest the bacterial spores. The spores germinate in the midgut, and vegetative cells invade the midgut epithelium where they grow and reproduce, changing in form as they progress toward invasion of the body cavity. After passing through the midgut, bacteria colonize the blood over a period of several weeks and then sporulate, reaching populations of 100 million cells per milliliter. The disease is fatal, providing that the larvae ingest a sufficient number of spores early in development. Dead larvae in essence become foci of spores that serve as a source of infection for up to 30 years.

Despite decades of research, suitable media for the growth and mass production of *P. popilliae* in vitro have not been developed. Thus, the spores used in commercial formulations are produced in living, field-collected scarab larvae. Nevertheless, a niche market exists for *P. popilliae* in the U.S. due to the serious damage to turf grass caused by larvae the Japanese beetle. The limited use of this bacterium due to lack of a cost-effective growth medium well illustrates the role that suitable mass production methods play in the commercialization of biological control agents.

2.3.4. *SERRATIA ENTOMOPHILA*

A novel bacterium named *Serratia entomophila* causes amber disease in the grass grub, *Costelystra zealandica*, an important pest of pastures in New Zealand, and has been developed as a biological control agent for this pest.[12] This bacterium adheres to the chitinous lining of the foregut, were it grows extensively, eventually causing the larvae to develop an amber color, resulting in death. This species is easily grown and mass produced in vitro to densities as high as 4×10^{10} cells/ml, leading to its rapid commercialization. It is now used to treat infested pastures with in New Zealand at a rate of 1 liter of product per hectare. Liquid formulations of this living, non-spore-forming bacterium are applied with subsurface application equipment. The rapid development and commercialization of the bacterium, even though the use is rather restricted, shows how microbials can be successful in niche markets were there are few alternatives, and mass production methods, the most critical factor, are available.

2.3.5. GENERAL MOLECULAR BIOLOGY OF BT INSECTICIDAL PROTEINS

Parasporal body proteins account for Bt's activity for most insect pests. These proteins are referred to as δ-endotoxins, the δ referring to their early designation in a series of insecticidal factors, and endotoxin referring to their assembly into inclusions within the cell after synthesis.[9] In the early 1980s, shortly after the development of recombinant DNA techniques, it was discovered that Bt endotoxins were encoded by genes carried on plasmids. This discovery quickly led to a general understanding of endotoxin genetics and molecular biology, including mode of action, through the cloning and sequencing of numerous genes, along with characterization of the toxicity and target spectrum of the protein encoded by each gene. These studies revealed that Bt endotoxins fall into two broad classes, Cry (for crystal) and Cyt (for cytolytic) proteins. Mode of action studies show that each type, after ingestion and proteolytic activation, bind to and cause lysis of midgut epithelial cells, which results in insect death. Cry proteins typically require surface glycoproteins[13] on midgut microvilli for initial binding to exert toxicity, whereas Cyt proteins bind directly to the lipid portion of the microvillar membrane.

2.3.5.1. *Cry and Cyt Protein Structure*

There are two subgroups of Cry proteins based on mass—proteins 130–150 kDa, and 65–70 kDa. Cry1 and Cry4 proteins mentioned above are characteristic of the first subgroup. The N-terminal half of these contains the toxic region. The C-terminal half facilitates crystallization after synthesis. Typical examples of the second subgroup are Cry2A, Cry3A and Cry11A, which lack the C-terminal domain characteristic of Cry1 proteins. Thus, the proteins of 65–70 kDa are essentially naturally truncated versions of the larger proteins, composed primarily of the toxin region.[10,13]

Four Cry protein structures have been solved—Cry1Ac, Cry2A, Cry3A and Cry4B. All consist of three domains. Domain I contains five to seven antiparallel α-helices in which helix 5 is encircled by the other helices. The long hydrophobic and amphipathic helices of Domain I suggest that this domain forms the lytic pores in insect midgut microvilli that cause death. Domain II consists of three antiparallel β-sheets. The loops at the tips of these are thought to bind the toxin to receptors on microvilli. Domain III consists of two antiparallel β-sheets, which form a β-sandwich thought to maintain structural integrity of the molecule, and to assist in receptor binding and membrane penetration.[10,13]

Cyt proteins are highly hydrophobic and have a mass of 24–28 kDa. They insert into lipid portion of the microvillar membrane without the need of a protein receptor. Cyt proteins share no significant amino acid sequence identity with Cry proteins, and have only been reported from mosquitocidal strains. These protein are not very toxic alone, but synergize the toxicity of Cry proteins. This synergy accounts for the high toxicity of Bti. In addition, several recent studies show that Cyt1 proteins, such as Cyt1A, can delay and overcome resistance to *B. thuringiensis* subsp. *israelensis* Cry proteins and the *B. sphaericus* Bin toxin, and can extend the mosquito target spectrum of *B. sphaericus*.

2.3.5.2. *Bt endotoxin Mode of Action*

As noted above, Cry and Cyt proteins are actually protoxins that must be ingested and processed by midgut enzymes to yield active toxins. Most have evolved to dissolve from the environmentally stable endotoxin crystals and be activated under the alkaline conditions, pH 8–10, that are, characteristic of the midgut lumen of caterpillars and mosquito larvae. Once activated, Cry molecules bind to glycoprotein or glycolipid receptors. In general, the former are aminopeptidases, alkaline phosphatases, or cadherins on the midgut epithelial cell microvilli. Toxin molecules oligomerize and insert into the membrane causing cell lysis. Cyt proteins are thought to have a similar mode of action, with the exception of requiring protein receptors for binding. Instead, they bind directly to the microvillar lipid bilayer.

The underlying hypothesis for Cry and Cyt protein mode of action is known as colloid-osmotic lysis. Toxin oligomers are thought to form cation-selective pores that cause an influx of cations, especially K, into the cell. The cell then takes in water, compensating for the cation influx to maintain ionic balance, and subsequently swells and lyses. The actual cause of larval death is not known, but is thought to be nerve paralysis that results from a rise in blood pH due to the inflow of alkaline midgut juices into the hemolymph. While this is the current paradigm, there is some evidence that neither toxin type forms cationic pores. If fact, evidence is quite strong that Cyt proteins act as membrane detergents.

2.3.5.3. *Genetic Regulation of Cry and Cyt Protein Synthesis*

The principal genetic factors controlling the yield of endotoxin synthesis in Bt are promoters, a 5′ mRNA stabilizing sequence and 3′ transcriptional termination sequences. The relative stability of each endotoxin is also a factor that affects yield.

With respect to promoters, Bt endotoxin synthesis is typically under the control of two strong sporulation-dependent promoters, BtI and BtII. BtI is transcribed by sigma-35 complexed with the RNA polymerase, whereas BtII transcription is regulated by sigma-28. While this is the typical state for Bt promoters, Cry3A synthesis is under the control of a weak promoter active during vegetative growth. Moderate levels of Cry3A synthesis occur in the bacterium due to the presence of a mRNA stabilizing sequence of 9 nucleotides referred to STAB-SD present in the 5′ region of the *cry3A* transcript.[14] Endotoxin synthesis can be increased by as much as 10-fold,[15] when this sequence is spliced into expression constructs for many proteins, and placed under the control of Bt sporulation-dependent promoters. The 3′ terminus non-coding terminus of most Bt genes contains a stem-loop structure that acts as a transcription terminator, but these structures also stabilize the transcript, apparently by retarding 3′ exonuclease degradation. This extends transcript half-life, leading to higher endotoxin synthesis than would occur in the absence of these terminators.

Several other factors enhance synthesis of Bt endotoxins during or after translation. For example, a 20-kDa protein encoded as the third ORF (open reading frame) of the *cry11A* operon enhances net synthesis of many endotoxins, apparently acting as a chaperone. A 29-kDa protein encoded by the *cry2Aa* operon facilitates crystallization, and therefore yield of Cry2A. Lastly, different endotoxin proteins vary in their stability, some, such as Cry3A are much more stable than others, for example, than Cry4A. Generally, the more stable a protein, the higher the yield when these are synthesized at high levels using expression vectors.

2.3.6. RECOMBINANT BACTERIAL INSECTICIDES AND BT CROPS

This section deals with how recombinant DNA techniques have been used to construct Bt strains that are significantly more insecticidal than the parental strains from which various endotoxins have been derived and recombined. A brief section on the construction and use of Bt crops follows.

2.3.6.1. *Recombinant Bacterial Insecticides Based on Bt*

The most common strategy for constructing recombinant Bt strains is using a shuttle expression vector that contains replication origins for both *B. thuringiensis* and *E. coli*, a multiple cloning site, and genes for antibiotic resistance, for example to ampicillin and erythromycin for easy selection of transformants. A shuttle vector such as pHT3101 containing a gene of interest is amplified in *E. coli*, isolated, and then introduced into a candidate Bt strain by electroporation.

In many cases, *cry* and *cyt* genes of *B. thuringiensis* inserted into shuttle vectors were expressed under the control of their own promoters, which generally results in a high yield of the encoded protein. In terms of promoter strength, *cyt1A* promoters are among the strongest known among *cry* and *cyt* genes. In addition, as mentioned above, the *cry3A* upstream 5′ mRNA stabilizing sequence (STAB-SD) improves stability of *cry3A* transcripts and concomitantly the yield of certain Cry proteins. To optimize Cry protein yields in Bt, a recombinant expression vector, pSTAB was developed. This vector was constructed by inserting the 660-bp DNA fragment containing *cyt1A* promoters combined with the STAB-SD sequence into the multi-cloning site of pHT3101.

Using the p*cyt1A*/STAB expression vector, which combined these different genetic elements, we were able to significantly increase yields of several Cry proteins. For example, by expressing the *cry3A* gene using this vector, we were able to obtain yields 12-fold greater than those obtained with the wild type strain of *B. thuringiensis* subsp. *morrisoni* (isolate DSM2803) from which this gene was cloned (Figure 3). Cry3A yield obtained per unit medium using *cyt1A* promoters alone, i.e., lackingthe STAB-SD sequence, was only about twofold higher than thatof the wild-type DSM280 strain. This demonstrates that most of the enhancement was due to inclusion of the STAB-SD sequence.[15]

The significant increase in Cry3A yield obtained using *cyt1A* promoters combined with the STAB-SD sequence led us to test this expression vector for enhancing synthesis of other Bt endotoxins. The level of enhancement using this expression system varies depending upon the candidate protein. For example, yields of Cry11B and the Bs Bin binary toxin were increased substantially, as much as eight-fold, whereas yields of proteins such as Cry11A and Cry2A increased only 1.5- to twofold.

As our research is primarily directed toward improving mosquitocidal bacteria, our best examples of the successful use of pSTAB/*cyt1A* come from engineering recombinant Bti strains. We have used this vector to produce several different recombinant strains that vary in complexity, ranging from a strain that produces only a single endotoxin to strains that produce as many as five endotoxins. In the simplest case, we used p*cyt1A*/STAB to synthesize the Bin toxin of Bs 2362. Using this construct, Bin synthesis was eight-fold higher than that obtained with wild type Bs 2362. The toxicity of this strain was much better than wild type Bti and Bs against *Culex* species.

To improve toxicity while at the same time preventing or delaying the evolution of resistance, we constructed several strains in which we increased toxin complexity and added Cyt1A for resistance management. One strain constructed using this strategy was a recombinant that synthesized the Bin toxin, Cyt1A and Cry11B. In this recombinant, the mosquitocidal proteins

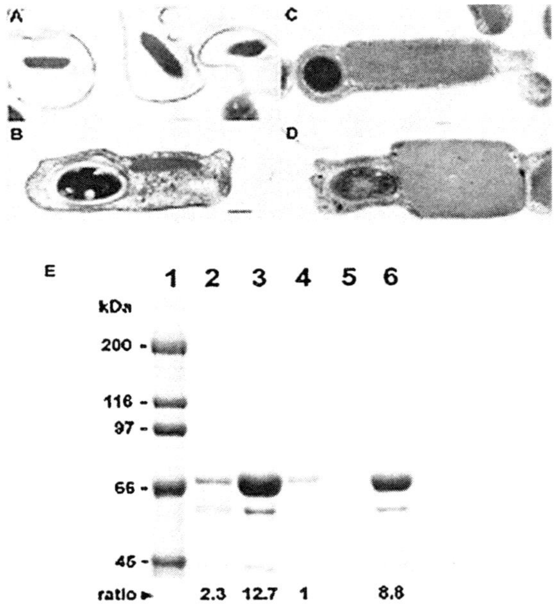

Figure 3. Enhanced synthesis of Cry3A through use of sporulation dependent promoters and the STAB-SD mRNA stabilizing sequence. **A**: Size of wild type Cry3A crystals in sporulated cells of *B. thuringiensis* subsp. *morrisoni* strain *tenebrionis*. **B**: Sporulated Bt cell in which expression of *cry3A* is controlled by the three *cyt1A* sporulation-dependent promoters. **C** and **D**, respectively, longitudinal and cross-sections through Cry3A crystals in sporulated Bt cells in which expression of *cry3A* was under the control of *cyt1A* promoters, and the transcript included the STAB-SD sequence for transcript stabilization. The combination of *cyt1A* promoters and STAB-SD yielded at least 10-fold more protein per cell than the wild type DSM 2803 isolate. Aside from the significant increase in Cry3A yield, these results show that the small size of the crystals in the wild type strain is due primarily to the control of expression by s^A, not an inherent property of Cry3A. All micrographs are the same magnification; bar in B = 300 nm. **E** Analysis of Cry3A production of wild-type, mutant, and engineered strains of Bt by SDS-PAGE. Sedimented crystals, spores, and cellular debris obtained from equal volumes of culture medium at the end of sporulation were loaded into each lane. Lanes: 1, molecular mass markers; 2, Bt 4Q7 transformed with pPFT3A (*cry3A* without the STAB-SD sequence under the control of *cyt1A* a promoters); 3, 4Q7 transformed with pPFT3As (*cry3A* with the STAB-SD sequence under the control of cytA promoters); 4, wild-type Btm (strain *tenebrionis*) DSM 2803; 5, 4Q7 transformed with pHT3101; 6, NB176, a mutant *Bt tenebrionsis* with a higher *cry3A* copy number. The ratios at the bottom of the lanes were determined by densitometry scanning of the gel; they indicate the ratio of Cry3A per unit of GYS (glucose-yeast-salts) medium in comparison to that produced by the DSM 2803 strain. Bar in B = 400 nm

were from three different species; Bin from Bs 2362, Cry11B—a protein more toxic than Cry11A—from Bt subsp. *jegathesan*, and Cyt1A from Bti. This recombinant was constructed using a dual-plasmid expression system with two different plasmids, each with a different antibiotic resistance gene for selection.

The resulting recombinant *B. thuringiensis* produced three distinct crystals and was three to fivefold more toxic to *Culex* species than either Bti IPS-82 or Bs 2362.To construct a recombinant with an even greater range of endotoxins for both increased toxicity and resistance management, we transformed the IPS-82 strain of Bti, which produces the complement of toxins characteristic of this species, with pPHSP-1, the pSTAB plasmid that produces a high level of Bs Bin toxin.[16,17] This recombinant was as more than as ten-fold more toxic than either of the parental strains to larvae of *Cx. quinquefasciatus* and *Cx. tarsalis*. Aside from high efficacy, as noted above, this new bacterium is much less likely to select for resistance in target populations, as it combines Cyt1A from Bti with Bti Cry toxins and Bs Bin. The resistance management properties of this bacterium are currently under evaluation. The markedly improved efficacy and resistance-delaying properties of this new bacterium make it an excellent candidate for development and use in vector control programs, especially to control *Culex* vectors of West Nile and other viruses as well as species of this genus that transmit filarial diseases. Moreover, the larvae of certain species of anopheline mosquitoes that are important malaria vectors, such as *An. gambiae* and *An. arabiensis*, should be highly sensitive to this recombinant, as they are not only sensitive to the toxins of Bti, but Bs Bin as well.

The improvements in activity noted above result from the increase in toxicity per unit weight of fermentation medium. These increases significantly reduce production costs for obtaining the same level of pest or vector control. The extent to which these savings are potentially passed along to consumers, as opposed to being used to increase company profits has not been determined, as the recombinant strains discussed have not yet been commercialized.

2.3.6.2. *Construction and Use of Bt Crops*

The Bt crops constructed to date are primarily based on Cry proteins active against lepidopteran pests, Cry1 proteins in the case of Bt cotton and Bt maize (corn). In addition, Cry3 proteins were used to construct Bt potatoes, and more recently Bt maize to control rootworms, which are major coleopteran pests. The principal crops and target pests are listed in Table I.

Two methods are used to construct Bt crops. In the first, plant tissue is infected with the bacterium, *Agrobacterium tumifaciens* containing a disarmed Ti plasmid containing a Bt *cry* gene. The plasmid integrates the gene into a crop plant chromosome, and transgenic plants are regenerated from the transformed tissue. In the second method, a particle gun is used to basically blast the Bt gene into the plant along with selectable markers, and the constructs are subsequently integrated into the plant chromosome. In either case, the Bt gene is engineered using plant codon usage to optimize the endotoxin protein in plant tissues.

TABLE I. Cry proteins produced by Bt crops registered in the United States

Crop	Cry protein	Target insects
Cotton	Cry1Ac	Tobacco budworm, *Heliothis virescens* Cotton bollworm, *Helicoverpa zea* Pink bollworm, *Pectinphora gossypiella*
Maize	Cry1Ab	European corn borer, *Ostrinia nubilalis* Southwestern corn borer, *Diatraea grandiosella* Corn earworm, *Helicoverpa zea*
Maize	Cry1Ac	European corn borer, *Ostrinia nubilalis* Southwestern corn borer, *Diatraea grandiosella*
Maize	Cry1F	Fall armyworm, *Spodoptera frugiperda*
Maize	Cry3Bb	Western corn root worm, *Diabrotica virgifera*
Potato	Cry3Aa	Colorado potato beetle, *Leptinotarsa decemlineata*

Source: U.S. Environmental Protection Agency Web site.

The efficacy and environmental safety of the two major Bt transgenic crops, Bt cotton and Bt maize, has led to their widespread adoption in the U.S. At present, approximately three million hectares (50%) of the cotton grown in the U.S. is Bt cotton, and about 16 million hectares (40%) is Bt maize. These crops produce Cry proteins such as Cry1Ac (in cotton) and Cry1Ab and others (in corn) to control lepidopteran pests such as, respectively, the cotton budworm, *Heliothis virescens*, and the European corn borer, *Ostrinia nubilalis*. New varieties of Bt maize have now reached the market, including varieties that control corn root worms. Bt cotton and maize varieties in which genes are stacked for resistance management, such as Bt cotton that contains Cry1Ac and Cry2Ab, which came to market within the last 2 years, and Bt maize that produces Cry1Ab (for control of lepidopteran pests) and Cry3 proteins (to control corn root worms) will likely also increase adoption rates by farmers in many regions of the world. Some of these varieties are showing unexpected benefits, that is benefits in addition to controlling the target pests and reducing the use of synthetic chemical insecticides. For example, Bt maize that produces Cry1Ab and Cry3Bb is not only effective at controlling the European corn borer and corn root worms, but under drought conditions, such as those that occurred in the central corn belt (e.g., much of Illinois) in the U.S. in 2005, resulted in increased maize yields in comparison to conventional corn treated with chemical insecticides. The apparent reason for these increased yields was protection of the roots from significant damage by corn root worms, which enabled the transgenic maize to acquire more water and thus better survive the drought conditions.

2.4. Environmental Safety

Knowledge of Cry protein mode of action, as summarized below, provides a basis for understanding the specificity and thus safety of Bt insecticides and Bt crops. Specificity actually exists at several different levels. The following are the key levels:

1. Endotoxin crystals must be ingested to have an effect; there is no "contact" activity, as occurs with chemical insecticides. This is the reason sucking insects and other invertebrates such as spiders and mites are not sensitive to Bt.
2. After ingestion, Bt endotoxin crystals must be activated to be toxic. Activation requires that the crystals dissolve. This requires alkaline conditions, generally a midgut pH in the range of 8 or higher. Most non-target invertebrates have neutral or only slightly acidic or basic midguts. Under the highly acidic conditions in stomachs of many vertebrates, including humans, endotoxin crystals may dissolve, but the solubilized proteins are rapidly degraded to non-toxic peptides by gastric juices, typically within minutes.
3. Once solubilized, activation requires that Cry proteins be cleaved by midgut proteases at both the C-terminus and N-terminus.
4. Once activated, the toxin must bind to glycoprotein "receptors" on midgut microvillar membrane. Recent evidence indicates that the specific arrangement of the sugar residues on these receptors, critical for binding, only occurs in invertebrates.[18] Most chewing insects that ingest toxin crystals, even those with alkaline midguts, including many lepidopterans, do not have the appropriate receptors, and thus are not sensitive to activated Cry proteins. Even insects sensitive to one class of Bt proteins, such as lepidopterans sensitive to Cry1 proteins, are not sensitive to Cry3A active against coleopterans, as they lack receptors for these. Moreover, no binding of Cry protein has been detected in mammalian stomach epithelial cells.
5. After binding to a midgut receptor, the toxin must enter the cell membrane, change conformation in the process, and oligomerize to form pores, leading to toxicity.

With respect to level 5, the specific conformational changes that must take place to exert toxicity are not known. It is known, however, that high affinity irreversible binding can occur in some insects, yet not lead to toxicity. This implies that a specific type of processing, i.e., another level of specificity, is required for toxicity that occurs as or after the toxin inserts into the membrane. In Bt crops, only a portion of the second level, i.e., level 2, of the first five levels of specificity has been circumvented. When synthesized in plants, full length and truncated Cry proteins do not form crystals, and even if quasi-crystalline inclusions do form, the toxin remains in solution within the plant cells. Nevertheless, whether produced in plants as a full length or truncated protoxin, Cry proteins must still be properly activated after ingestion, i.e., cleaved properly at the C and N termini. Additionally, they must meet the criteria for binding and membrane insertion defined above by levels 4 and 5 to be toxic. Furthermore, with the exception of Cry9C, which was engineered to resist rapid proteolytic cleavage, most Bt proteins produced in Bt crops are degraded rapidly—within seconds—under conditions that mimic the mammalian digestive system. Thus, most of the inherent levels of specificity that account for the safety of Cry proteins used in commercial bacterial insecticides apply to these same proteins when used to make Bt crops resistant to insects.

Lastly, an important concept for evaluating safety is to consider the route by which an organism is likely to encounter a toxin. Even though pulmonary (inhalation) and intraperitoneal injection studies are done with microbial Bt insecticides and proteins, their normal route of entry by target and non-target organisms is by ingestion. This is true for Bt proteins produced in Bt crops, as inhalation of plants and plant parts is less likely than bacterial insecticides.

2.4.1. SAFETY OF INSECTICDES BASED ON *BACILLUS THURINGEINSIS*

In addition to their insecticidal efficacy, a major impetus for using Cry proteins in Bt-crops was their long history of safety to non-target organisms, especially to vertebrates, when used in the form of bacterial insecticides. The most important levels of Bt endotoxin specificity described above, i.e., activation, binding, and membrane insertion, apply equally to evaluating the safety of Cry proteins whether used in Bt crops or bacterial insecticides. Therefore, the tests and data that support a very high degree of safety for bacterial insecticides containing Cry proteins are relevant to assessing the safety of Bt-crops. Extensive testing has been and remains required to meet rigorous safety requirements established by governmental agencies. Data from these tests is valid as the major approach to evaluate Bt crop safety, especially considering that many hundreds of safety tests have been conducted over several decades to register numerous bacterial insecticides based on different subspecies of Bt.

In determining what types of tests should be done to evaluate the safety of bacterial insecticides, early tests were based primarily on those used to evaluate chemical insecticides. However, the tests have evolved over the decades and are now designed to evaluate the risks of Bt, specifically the infectivity of the bacteria and toxicological properties of proteins used as active ingredients.

The tests are grouped into three tiers, I–III.[18] Tier I consists of a series of tests aimed primarily at determining whether an isolate of a Bt subspecies, as the unformulated material, poses a risk if used at high levels, typically at least 100 times the amount recommended for field use, to different classes of non-target organisms. The principal tests include acute oral, acute pulmonary (inhalation), and acute intraperitoneal evaluations of the material against different vertebrate species, with durations from a week to more than a month, the length depending on the organism. In the most critical tests, the mammals are fed, injected with, and forced to inhale millions of Bt cells in a vegetative or sporulated form. Against invertebrates, the tests are primarily feeding and contact studies. Representative non-target vertebrates and invertebrates include mice, rats, rabbits, guinea pigs, various bird species, fish, predatory and parasitic insects, beneficial insects such as the honeybee, aquatic and marine invertebrates, and plants. If there is clear infectivity or acute toxicity in any of these tests, then the candidate bacterium would be rejected. If uncertainty exists, then Tier II tests must be conducted. These tests are similar to those of Tier I, but require multiple consecutive exposures, especially to organisms where there was evidence of toxicity or infectivity in the Tier I tests, as well as tests to determine if and when the bacterium was cleared from non-target tissues. If infectivity, toxicity, mutagenicity, or teratogenicity is detected, then Tier III tests must be undertaken. These consist of tests such as two-year feeding studies and additional testing of teratogenicity and mutagenicity. The tests can be tailored to further evaluate the hazard based on the organisms in which hazards were detected in the Tier I and II tests. It must be realized that the tests for Bt crops are much more strict than for many synthetic chemical insecticides on the market, as many of these are known to be toxic to non-target invertebrate, as well as vertebrates such as fish and humans, especially if not used properly.

To date, *none* of the registered Bt insecticides or Bt crops based on Cry proteins has had to undergo Tier II testing.[8,20] In other words, no moderate or significant hazards or risks have been detected with any Bt subspecies used commercially or any Bt crop against any of the non-target organisms studied, including mammals. As a result, all Bt insecticides and Bt crops registered in the U.S. are exempted from a tolerance requirement, i.e., a specific level of insecticide residue allowed on a crop just prior to harvest. Moreover, no washing or other requirements to reduce levels consumed by humans are required. In fact, Bt insecticides can be applied to crops such as lettuce, cabbage, and tomatoes just prior to harvest, and Bt crops have no restrictions for human consumption. It is important to realize that such a statement cannot be made for any chemical insecticide.

Aside from the various safety studies required by the U.S. Environmental Protection Agency, programs have been mounted to monitor the health effects of spraying Bt insecticides directly on human populations. Two such recent studies, for example, were conducted during the late 1990's, one in Victoria, Canada, and another in Auckland, New Zealand.[21,22] In both cases the Bt spray programs were undertaken to eliminate lepidopteran forest pests that had invaded these countries. To eliminate these pests, suburban residential areas inhabited by thousands of people were spayed periodically for several weeks, until the pests were eradicated. During the spray programs, and for months thereafter, the human populations were monitored for the presence of the Bt applied, and for symptoms of disease. Bacteria were easily recovered from nasal samples, for example, and from monitoring particulates in the air. In Auckland, New Zealand, some discomfort followed the sprays, but "most residents saw their health as unaffected by the spray program, and there was no significant increase in visits to general practitioners or alternative health care providers."[21] Similar results were obtained in the populations monitored in Victoria during the Bt spray program—the "human health surveillance program failed to detect any correlation between the aerial application of *B. thuringiensis* subsp. *kurstaki* HD1-like bacteria and short-term health effects in the general adult population."[22] This evidence of little or no significant health effects on human populations subjected to Bt sprays is in sharp contrast to the well-known toxic effects many chemical pesticides have on humans.[23,24]

Despite these and previous studies, there are several putative cases in the literature where it is claimed that certain isolates of *B. thuringiensis* have caused infections in humans. The evidence in support of these claims is very weak, and has been review recently in the context of the high degree of safety that Bt exhibits toward mammals.[25]

2.4.2. SAFETY OF BT CROPS TO NONTARGET ORGANISMS

The safety of Bt crops to non-target organisms including humans was based initially on a combination of experimental studies on vertebrates and invertebrates in the laboratory and field, and on the long record of safety that had accompanied the use of Bt insecticides. The rationale behind these studies was that the Cry proteins produced by Bt crops were likely to be similar to those produced by Bt insecticides. Moreover, whereas Bt insecticides contained a wide variety of Bt components, such as live spores, fermentation products, a multiplicity of Cry proteins and enzymes, Bt crops would only produce one or a few Cry proteins. Thus, the insecticidal component within a Bt crop was in essence simpler than that of a Bt insecticide. This provided a basis for viewing Bt crops as inherently safe for most non-target organisms.[26] Initial laboratory studies and field trials verified the laboratory studies showing Bt crops were safe for mammals and nontarget organisms. Then two papers appeared that caused an uproar over potential nontarget effects. The first was a study that claimed Bt crops could be toxic to the beneficial predatory insect,

Chrysoperla carnea, an insect used as a biological control.[27] The second study, which caused an even broader uproar, and attracted a great deal of media attention, reported that pollen from Bt maize could kill larvae of the monarch butterfly.[28]

Those opposed to Bt crops used these two reports as rallying points to argue against the use of these crops until more environmental impact studies, preferably long-term, were carried out. These studies had a beneficial effect on the field in that they resulted is more in-depth studies on the possible environmental effects of Bt crops on non-target organisms. In the short-term, both studies were shown to be poorly designed and the conclusions unwarranted. In more detailed studies carried out on the predatory insect, *C. carnea*, the same laboratory that published the initial report[27] found that the effects they reported were due to nutritional differences, not to the toxic action of a Cry protein.[29] The potential effects on monarch larvae were investigated in a series of six comprehensive studies published in 2001 in the *Proceedings of the National Academy of Sciences USA*. The key finding of these studies was that if there were any effects of Bt pollen on monarch larvae, they would be "negligible."[30] These studies were followed by numerous others, reviewed recently,[31] all of which provide a large body of evidence from field and laboratory studies that indicate Bt crops have a high degree of safety for the overwhelming majority of non-target organisms. This is especially true in comparison to the effects of synthetic chemical insecticides.

Nevertheless, as Bt crops are a new technology, and the public throughout the world remains concerned about the potential nontarget effects of these crops, numerous long-term field studies are underway. Several of these were appeared in a special issue of *Environmental Entomology* published in October 2005. While some minor effects were observed on certain nontarget populations, the majority of these studies found that Bt crops were remarkably safe for nontarget invertebrates.[32–37] In general, no significant negative effects were observed on nontarget populations when Bt crops were compared with control non-Bt crops that were not treated with chemical insecticides. In crops treated with chemical insecticides, however, the nontarget invertebrate populations suffered very significant declines shortly after treatment.[32–37] An example of the types of effects observed in the long-term field studies is shown in Figure 4.

While it is always possible that some negative nontarget effects will be detected in even longer term studies, many of the field studies cited above have been underway for more than six years. This makes it unlikely that significant nontarget effects will be encountered in the future. While some[38] may wish to continue to criticize the continued use of Bt crops, and require more detailed studies based on local agricultural environments and specific transgenic crops to be grown in these areas, current evidence provides no

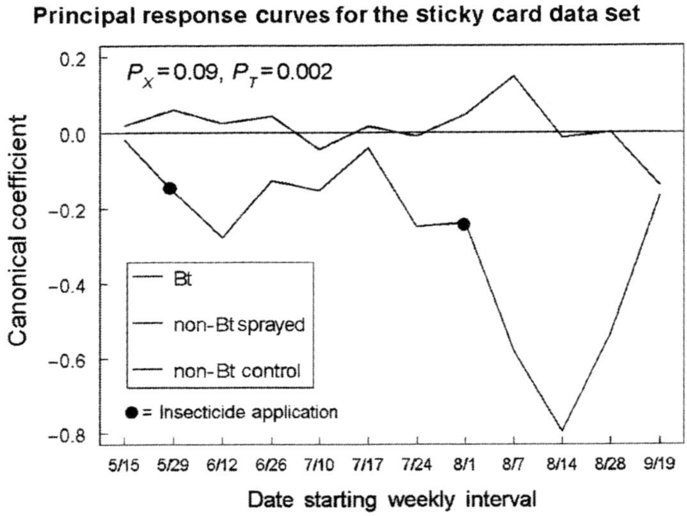

Figure 4. Effects of Bt maize (upper line) on composite nontarget invertebrate populations in comparison to conventional untreated maize (straight horizontal line) and maize treated with a synthetic chemical insecticide. Courtesy G. Dively (University of Maryland). See also Dively[35]

justification for such studies. Bt crops may have some minor negative impacts, but the overwhelming majority of evidence from laboratory and field studies indicates they are a marked environmental and human health improvement over the use of synthetic chemical insecticides.

With respect to direct tests on humans, this is not done. However, it must be realized that in the United States, Bt maize and transgenic soybeans have been used in processed foods eaten by millions of Americans for well over five years. There is no evidence that eating these transgenic crops has had any noticeable negative effects on human health.

2.5. Conclusions

The search over the past century for pathogens effective as insect control agents has demonstrated that the overwhelming majority are not yet cost-effective as classical biological control agents. A few, however, are effective as microbial insecticides, for example, certain subspecies of *Bacillus thuringiensis*, provided that efficient methods are available for their mass production. While these results are considered by some to be very disappointing after more than a century of research, it must be realized that the development of transgenic insect-protected crops, specifically Bt crops, now a multibillion Euro industry that offers great promise for biological control and sustainable agriculture, emerged from the research on insect pathogens. This occurred despite a very low investment in research compared to that spent on the development of synthetic chemical insecticides. Aside from better biocontrol and integrated pest management programs that can be built upon this new Bt crop technology, its development demonstrates the value of basic research, which has been under attack recently in many governmental programs. As the world population grows and resources are diminished, new technologies such as Bt crops offer great promise for the development of improved, environmentally safe pest control programs, especially with the crop needs that have yet to be met. Based on the substantial body of evidence that exists currently, these transgenic crops will be safe for non target organisms including humans.

References

1. B. A. Federici, in *Handbook of Biological Control*, edited by T. S. Bellows and T. W. Fisher (Academic Press, San Diego, 1999), pp 517–548.
2. F. Moscardi, Assessment of the application of baculoviruses for control of Lepidoptera. *Annu. Rev. Entomol.* 44, 257–289 (1999).
3. R. E. Balch and F. T. Bird, A disease of the European spruce sawfly, *Gilpinia hercyniae* [Htg.] and its place in natural control. *Sci. Agricult.* 25, 65–80.
4. R. M. Weseloh, *Entomophaga maimaiga* (Zygomycetes; Entomophthorales) resting spores and biological control of the gypsy moth, *Lymantira dispar* (Lepidoptera; Lymantriidae). *Environ. Entomol.* 28, 1162–1171 (1999).
5. V. M. Stern and B. A. Federici, Granulosis virus: Biological control of the grapeleaf skeletonizer. *Calif. Agricult.* 44, 21–22 (1990).
6. B. A. Federici, Insecticidal bacteria: An overwhelming success for invertebrate pathology. *J. Invertebr. Pathol.* 89, 30–38.
7. M. G. Feng, T. J. Poprawski, and G. G. Khachatourians, Production, formulation and application of the entomophathogenic fungus *Beauveria bassiana* for insect control. *Biocontr. Sci. Technol.* 4, 3–34 (1994).
8. T. R. Glare and M. O'Callaghan. *Bacillus thuringiensis*: Biology, Ecology and Safety (Wiley, Chichester, UK, 2000).
9. B. A. Federici, in *Handbook of Biological Control*, edited by T. S. Bellows and T. W. Fisher (Academic Press, San Diego, CA, 1999), pp. 575–593.
10. E. Schnepf, N. Crickmore, J. Van Rie, D. Lereclus, J. Baum, J. Feitelson, D. R. Zeigler, and D. H. Dean, *Bacillus thuringiensis* and its pesticidal proteins. *Microbiol. Mol. Biol. Rev.* 62, 775–806 (1998).
11. J.-F. Charles, C. Nielsen-LeRoux, and A. Delecluse, *Bacillus sphaericus* toxins: Molecular biology and mode of action. *Ann. Rev. Entomol.* 41, 451–472 (1996).
12. T. A. Jackson, J. F. Pearson, M. O. O'Callaghan, H. K. Mahanty, and M. J. Willocks, in *Use of Pathogens in Scarab Pest Management*, edited by T. A. Jackson and T. R. Glare (Andover, Intercept. Andover, 1992) pp. 191–198.
13. R. de Maagd, A. Bravo, C. Berry, N. Crickmore, and H. E. Schnepf, Structure, diversity, and evolution of protein toxins from spore-forming entomopathogenic bacteria. *Ann. Rev. Genet.* 37, 409–433 (2003).

14. H. Agaisse and D. Lereclus, STAB-SD: A Shine-Dalgarno sequence in the 5′ untranslated region is a determinant of mRNA stability. *Mol. Microbiol.* 20, 633–643 (1996).
15. H.-W. Park, B. Ge, L. S. Bauer, and B. A. Federici, Optimization of Cry3A yields in *Bacillus thuringiensis* by use of sporulation-dependent promoters in combination with the STAB-SD mRNA sequence. *Appl. Environ. Microbiol.* 64, 3932–3938 (1998).
16. B. A. Federici, H.-W. Park, D. K. Bideshi, M. C. Wirth, and J. J. Johnson, Recombinant bacteria for mosquito control. *J. Exp. Biol.* 206, 3877–3885 (2003).
17. H.-W. Park, D. K. Bideshi , M. C. Wirth, J. J. Johnson, W. E. Walton, and B. A. Federici, Recombinant larvicidal bacteria with markedly improved efficacy against *Culex* vectors of West Nile Virus. *Am. J. Trop. Med. Hyg.* 72, 732–738 (2005).
18. J. S. Griffiths, S. M. Haslam, T. Yang, S. F. Garczynski, B. Mulloy, H. Morris, P. S. Cremer, A. Dell, M. J. Adang, and R. V. Aroian, Glycolipids as receptors for *Bacillus thuringiensis* crystal toxin. *Science* 5711, 922–925 (2005).
19. F. S. Betz, S. F. Forsyth, and W. E. Stewart, in *Safety if Microbial Insecticides*, edited by M. Laird, L. A. Lacey, and E. W. Davidson (CRC Press, Boca Raton, FL, 1990), pp. 3–10.
20. B. A. Federici, Effects of Bt on non-target organisms. *J. New Seeds* 5, 11–30 (2003).
21. K. Petrie, M. Thomas, and E. Broadbent, Symptom complaints following aerial spraying with the biological insecticide Foray 48B. *New Zealand Med. J.* 116, 1–7 (2003).
22. V. de Amorim, B. Whittome, B. Shore, and D. B. Levin, Identification of *Bacillus thuringiensis* subsp. *kurstaki* strain HD1-like bacteria from environmental and human samples after aerial spraying of Victoria, British Columbia, Canada, with Foray 48B. *Appl. Environ. Microbiol.* 67, 1035–1043 (2001).
23. W. R. Snodgrass, in *Handbook of Pesticide Toxicology*, edited by R. Krieger (Academic Press, San Diego, CA, 2001), pp. 589–602.
24. G. M. Calvert, W. T. Sanderson, M. Barnett, J. M. Blondell, and L. N. Mehler, in *Handbook of Pesticide Toxicology*, edited by R. Krieger (Academic Press, San Diego, CA, 2001), pp. 603–641.
25. J. P. Siegel, The mammalian safety of *Bacillus thuringiensis*-based insecticides. *J. Invertebr. Pathol.* 77, 13–21 (2001).
26. F. S. Betz, B. G. Hammond, and R. L. Fuchs, Safety and advantages of *Bacillus thuringiensis*-protected plants to control insect pests. *Reg. Tox. Pharmacol.* 32, 156–173 (2000).
27. A. Hilbeck, W. J. Moar, M. Pustai-Carey, A. Filippini, and F. Bigler, Toxicity of *Bacillus thuringeinsis* Cry1A(b) toxin to the predator *Chrysoperla carnea* (Neuroptera: Chrysopidae). *Environ. Entomol.* 27, 1255–1263 (1998).
28. J. J. Losey, L. Raynor, and M. E. Cater, Transgenic pollen harms monarch larvae. *Nature* 399, 214 (1999).
29. J. Romeis, A. Dutton, and F. Bigler, *Bacillus thuringeinsis* toxin (Cry1Ab) has no direct effect on larvae of the green *lacewing Chrysoperla carnea* (Neuroptera: Chrysopidae). *J. Insect Physiol.* 50, 175–183 (2004).
30. M. K. Sears, R. L. Helmich, D. E. Stanley-Horn, K. S. Oberhauser, J. M. Pleasants, H. R. Mattila, S. D. Siegfried, and G. P. Dively, Impact of Bt corn pollen on monarch butterfly populations: A risk assessment. *Proc. Natl. Acad. Sci. USA* 98, 11937–11942 (2001).
31. M. O'Callaghan, T. R. Glare, E. P. J. Burgess, and L. A. Malone. 2005. Effects of plants genetically modified for insect resistance on nontarget organisms. *Annu. Rev. Entomol.* 50, 271–292.
32. S. Naranjo, Long-term assessment of the effects of transgenic Bt cotton on the abundance of the nontarget natural enemy community. *Environ. Entomol.* 34, 1193–1210 (2005).

33. M. E. A. Whitehouse, L. J. Wilson, and G. P. Fitt, A comparison of arthropod communities in transgenic *Bt* and conventional cotton in Australia. *Environ. Entomol.* 34, 1224–1241 (2005).
34. G. Head, W. Moar, M. Eubanks, B. Freeman, J. Ruberson, A. Hagerty, and S. Turnipseed, A multiyear, large-scale comparison of Arthropod populations on commercially managed *Bt* and non-*Bt* cotton fields. *Environ. Entomol.* 34, 1257–1266 (2005).
35. G. P. Dively, Impact of transgenic VIP3A X Cry1Ab lepidopterna-resistant field corn on the nontarget arthropod community, *Environ. Entomol.* 34, 1267–1291 (2005).
36. M. A. Bhatti, J. Duan, G. Head, C. Jiang, M. J. McKee, T. E. Nickson, C. L. Pilcher, and C. D. Pilcher, Field evaluation of the impact of corn rootworm (Coleoptera: Chrysomelidae)-protected *Bt* corn on ground dwelling invertebrates. *Environ. Entomol.* 34, 1325–1335 (2005).
37. M. A. Bhatti, J. Duan, G. Head, C. Jiang, M. J. McKee, T. E. Nickson, C. L. Pilcher, and C. D. Pilcher, Field evaluation of the impact of corn rootworm (Coleoptera: Chrysomelidae)-protected *Bt* corn on foliage-dwelling arthropods, *Environ. Entomol.* 34, 1336–1345 (2005).
38. D. A. Andow and A. Hilbeck, A science-based risk assessment for non-target effects of transgenic crops. *Bioscience* 54, 637–649 (2004).

3. BENEFITS AND RISKS OF USING FUNGAL TOXINS IN BIOLOGICAL CONTROL

Maurizio Vurro*
Institute of Sciences of Food Production, National Research Council, via Amendola 122/O, 70125 Bari, Italy

Abstract. Fungal pathogens are an enormous source of metabolites, mostly still unknown, differing in chemical structure, biological activity, mechanism of action, specificity. Metabolites from agriculturally important fungi have been intensively studied mainly due to the risks posed to human and animal health when these toxins accumulate in agricultural commodities and are eaten. Thus, the use of fungal metabolites produced by pathogens is thought to pose risks instead of benefits. Often very promising fungal biocontrol agents have been discarded in evaluation because they produce powerful and dangerous toxins in vitro. The evaluation of the risk should be ascertained by considering the global environmental impact, i.e., determining the exact production of those metabolites when fungi are formulated, or when they are applied against, and grown on targets; the toxicity to non-target organisms; their fate in the environment; and the risk of water drift. Conversely, toxins could be used to directly or indirectly enhance the efficacy of biocontrol agents, depending on their biological and chemical characteristics, through: their use as sources of natural pesticides; their syntheses; the selection of better biocontrol agents overproducing toxins; their synergistic use with biocontrol agents; their use as biomarkers. Those aspects are described with particular reference to the metabolites produced by weed fungal pathogens and to the recent results obtained by our research group.

Keywords: fungal metabolites, biological control, biopesticides, bioherbicides

3.1. Introduction

Biodiversity is a wonderful instrument used by fungal species to produce an enormous number of natural compounds, differing in chemical structure, biological activity, mechanism of action, specificity and environmental

*To whom correspondence should be addressed, e-mail: maurizio.vurro@ispa.cnr.it

M. Vurro and J. Gressel (eds.), Novel Biotechnologies for Biocontrol Agent Enhancement and Management, 53–74.

impact. The possible use of those compounds as pharmaceuticals has been widely studied, but there have been limited efforts to evaluate and understand their potential use in plant protection. Most of those compounds probably have not yet been discovered let alone chemically and biologically identified. The sources of variability are numerous. For example, species belonging to the same genus often are able to produce a wide variety of metabolites. *Alternaria*, *Claviceps* or *Fusarium* species produce at least 100 different toxic metabolites. Toxins belonging to the same structural group can be produced by different microorganisms belonging to many different genera. This is the case for trichothecenes, a family of mammalian toxic tetracyclic sesquiterpenoid substances (more than 50) produced by different genera, including *Fusarium* (producing at least 25 different trichothecenes), *Myrothecium* (producing roridins and verrucarins), *Stachybotrys* (satratoxins) and *Trichoderma* (trichodermins);[1] or destruxins, a family of cyclic peptide toxins known for their insecticidal and herbicidal properties, produced in many different variants by the entomopathogenic fungus *Metarhizium anisopliae*[2] and by three unrelated plant pathogenic fungi, *Alternaria brassicae, Trichothecium roseum* and *Ophiosphaerella herpotricha*. Ophiobolins, a group of sesterterpenoids that includes at least 23 biogenic analogs, are produced by phytopathogenic species of genera such as *Bipolaris* or *Drechslera*.[3]

Further sources of variation in metabolic biosyntheses are biodiversity intraspecific, origin of strains, environmental conditions, and nutrients.

The public usually looks at bioactive compounds produced by pathogenic fungi with suspicion and tends to consider them a risk. This is because those compounds have been intensively studied mainly in relation to the risks posed to human and animal health when these toxins accumulate in agricultural commodities and are ingested. Besides the aflatoxins, ochratoxins, fumonisins, trichothecenes, zearalenols, or alkaloids of *Claviceps*, which are responsible for severe human and animal poisonings, there are many other metabolites that are not dangerous, but are phytotoxic and could be interesting tools for improving the efficacy of biological control agents.

3.2. Benefits

3.2.1. SOURCES OF NEW NATURAL PESTICIDES

Only a limited number of natural products are used directly as active ingredients in crop protection. They must be sufficiently active against the target species, safe, and biologically selective, standardized for formulation and composition, and produced by easy and rapid processes, such as synthesis, extraction or fermentation.

New bioactive metabolites have often been obtained by screening extracts from different microbes randomly chosen in the environment, as in the case of the herbicide bialaphos,[4] produced by soil *Streptomyces* spp.; or the fungicide strobilurin, produced by *Strobilurus tenacellus*, a wood-rotting fungus.[5] This approach can be useful if applied to a general, and not to a focused, screening for novel bioactive metabolites, but it has a low percentage of success due to their different biological activities and the constraints to evaluate them, and the almost infinite number of organic compounds with low molecular weights that could be produced. A more focused approach could increase the percentage of success in finding useful metabolites. In the case of searching for potential natural herbicides, for instance, the observation of symptoms on diseased plants in the field can be an effective method in choosing the promising organisms. In the case of potential insecticides or antibiotics, the screening among pathogens of insects or antagonists belonging to genera already known as toxin producers can increase the probability of success. Phytotoxins often act as virulence factors and are responsible for symptoms, such as chlorosis or necrosis. Selecting pathogens which cause those kind of symptoms can increase the probability of choosing interesting and novel toxin producers.

3.2.1.1. *Toxins from Fungal Pathogens of Dicot Weeds*

Natural compounds responsible for the bleaching and the loss of photosynthetic pigments in treated plants can be particularly attractive for the development of commercial herbicides. Among them, those acting by inhibiting HPPD (hydroxyphenyl pyruvate dioxygenase) are of particular significance and interest. Recently *Phoma macrostoma* was isolated from diseased thistle plants showing severe chlorotic symptoms. Two new compounds, macrocidin A and B, were isolated from the liquid culture filtrate of this fungus. They are the first representatives of a new family of cyclic tetramic acids.[6] The compounds caused bleaching and chlorosis as did the pathogen when applied to the leaves of the host plant, as well as to many other broadleaf species.

Our group has been studying the production of toxins produced by some interesting phytopathogenetic species of the genus *Ascochyta,* for their potential use as mycoherbicides. This genus includes tens of phytopathogenic fungi, some of which are responsible for severe diseases. They usually cause necrotic lesions to leaves and stems.[7] Many of these pathogens produce phytotoxins involved in symptom appearance.[8–10]

The main phytotoxin ascaulitoxin was isolated from the culture filtrates of *Ascochyta caulina*, a potential mycoherbicide for *Chenopodium album* control. It was characterised as the β-*N*-glucoside of the unusual *bis*-amino acid 2,4,7-triamino-5-hydroxyoctandioic acid.[11] Another toxic non-protein amino acid has been purified and identified from the same culture filtrates as *trans*-4-amino-D-proline,[12] together with the ascaulitoxin aglycone.[13] When assayed by leaf puncture on several plants, all the three toxins showed modulated phytotoxic activity.

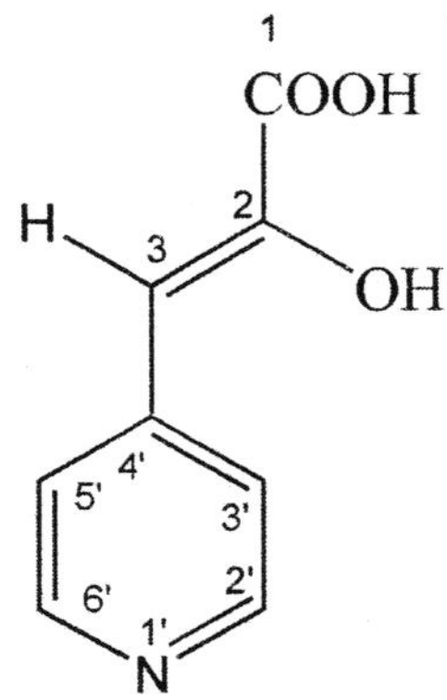

Figure 1. Ascosonchine

Sonchus arvensis is a herbaceous weed occurring through the temperate regions of the world. A major toxic metabolite was purified, isolated, structurally elucidated and biologically characterized from the liquid culture filtrates of *Ascochyta sonchi*, a pathogen responsible for severe necrosis on leaves of this weed. The metabolite, whose structure was determined by spectroscopy, is a novel compound, ascosonchine (Figure 1).[14] The toxin quickly caused the appearance of necrotic circular lesions resembling those caused by the pathogen in the leaf puncture assay. A broad necrosis appeared around 1 mM droplets of toxin. Necrosis was still quite evident at a five times lower concentration. Ascosonchine showed interesting selective properties, being completely ineffective on all the solanaceous species assayed (tomato, eggplant, pepper, potato), slightly active or almost inactive on cucurbitaceous (melon and zucchini) and leguminous (bean and chickpea) plants, but very active causing wide necrosis on many other species, such as *Euphorbia*, *Salvia*, and wheat.[14]

Ascosonchine belongs to the group of α-ketoacid and in particular to that of the heteroarylpyruvic acids. The α-ketoacids, such as phenylpyruvic acid, are biologically important metabolic products.[15] Phenylpyruvic acid is an intermediate in the shikimic acid pathway for the biosynthesis of the aromatic amino acids in plants and bacteria.[16,17]

3.2.1.2. *Toxins from Pathogens of Grass Weeds*

Culture filtrates were produced from pathogenic fungi isolated from grass weeds[18] and applied by infiltration both to the host and non host plant tested. The culture filtrate of one strain of *Drechslera siccans*, isolated from *Lolium perenne* was particularly effective, causing rapid chlorosis in the injected leaf tissues, followed by wide necrosis along the leaves.

Drechslera is a well-known genus producing of phytotoxic metabolites. Most of those pathogens and their toxins have been widely studied due to

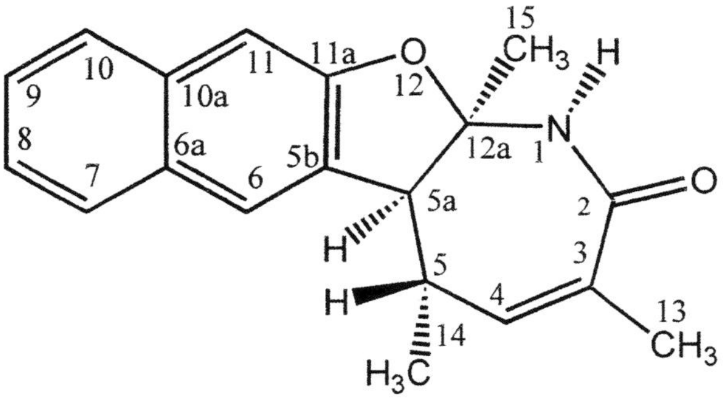

Figure 2. Drazepinone

the very severe disease of cereals.[19–21] Some species were also isolated from grass weeds[22] and their toxins proposed as potential natural herbicides.[23–25]

The main toxin produced in liquid culture by *D. siccans* was identified as a new interesting metabolite, drazepinone (Figure 2).[26] It belongs to a group of naturally occurring compounds that are broadly distributed in nature as plant and marine organism metabolites. Most of them show biological activity.[27,28] Natural compounds containing the naphthoazepin skeleton had not previously been reported, and those having furoazepine rings are only synthetic derivatives with important pharmacological activity.[29] Drazepinone is the first natural compound having both these entities in a new and interesting bioactive fungal metabolite.

The toxin applied to wounded leaves caused necrosis on almost all the species tested. The severity of necrosis ranged from very wide, as in the case of *Urtica dioica*, to small as when applied to *Setaria viridis* and *Lolium perenne* leaves. It caused necrosis of *Euphorbia helioscopia* and *Mercurialis annua* leaves, both Euphorbiaceae, and *Chenopodium album*. *Amaranthus retroflexus* and *Bromus* sp. were completely unaffected by the toxin.[26]

Toxins with structure completely different from drazepinone were previously isolated from other strains of the same fungus, such as de-*O*-methyldiaporthin[30] and siccanol,[31] an isocoumarin and a bicyclic sestert-erpene, respectively. Siccanol completely inhibited the root growth *Lolium multiflorum* seedlings at a level of 100 mg/l.[31] De-*O*-methyldiaporthin was almost inactive when assayed on host plants (*L. perenne* and *Avena sativa*), whereas it was toxic when assayed on corn, soybean, *Echinochloa crus-galli, Amaranthus spinosus* and *Digitaria* sp.,[30] with a toxicity resembling that caused by drazepinone.

Drechslera gigantea is a cosmopolitan fungal pathogen found throughout the world. It causes a zonate eye-spot disease of grasses, banana, and coconut.[32] The leaf spots may coalesce under severe levels of disease, causing leaf lesions and leaf blight. Infected leaves may be killed. Some metabolites were isolated and chemically and biologically characterized from the culture

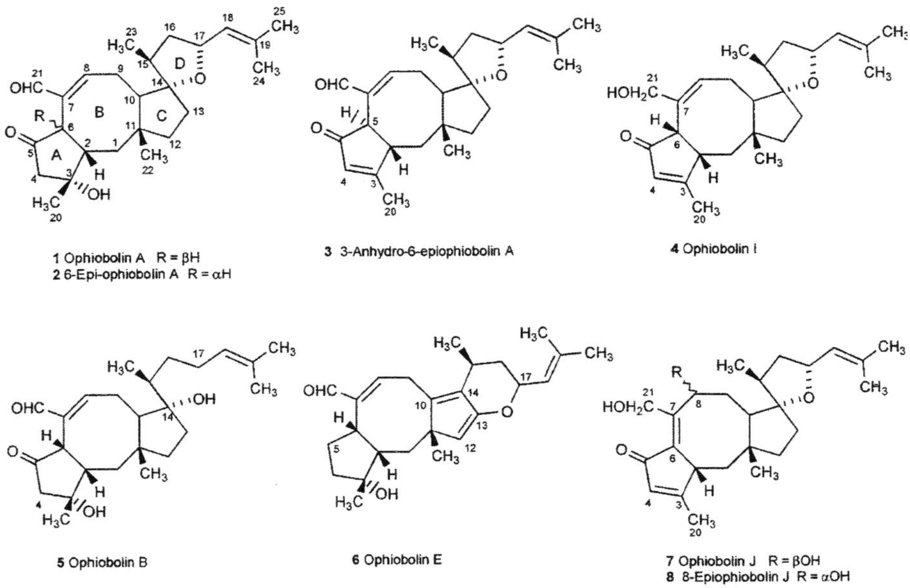

Figure 3. Ophiobolins isolated from *Drechslera gigantea* cultures

extracts of the fungus. The main metabolite, produced at 25 mg/l culture filtrate was identified as ophiobolin A (Figure 3).[33] Three other related compounds, namely 6-*epi*-ophiobolin A, 3-anhydro-6-*epi*-ophiobolin A and ophiobolin I were purified in lesser amounts, together with another new metabolite, named ophiobolin E. The fungus produced polycyclic sesquiterpenoids, ophiobolins B and J, and a new compound identified as 8-*epi*-ophiobolin J when grown in solid media (Figure 3).[34]

Ophiobolin A was highly toxic to almost all the plant species tested, already at the lowest concentration used (0.1 mM). Among dicotyledons, *Sonchus oleraceus* appeared to be particularly sensitive, whereas *Phalaris canariensis* was the most sensitive among monocotyledons. At the highest concentration used, the toxin was almost inactive to *Cynodon dactylon*. Compared to ophiobolin A, 6-*epi*-ophiobolin A had almost the same spectrum of plant sensitivity, but at a lower intensity. 6-*epi*-3-anydro-ophiobolin A was almost inactive on most of the plants tested, with the exception of *Setaria viridis* and rocket. Ophiobolin I was inactive on all the plants tested, even at the highest concentration.[34]

It is interesting to note a certain level of selectivity of the toxins. In fact, on average ophiobolins proved to be more active to grass weeds than to dicotyledonous species.

Although ophiobolins were quite widely studied for their interesting effects on plant physiology and for their biological activities, only limited information is available on their potential herbicidal activity.[3] For example, they can reduce root and coleoptile growth of wheat seedlings, inhibit seed germination, change cell membrane permeability, stimulate leakage of electrolytes and glucose, or cause respiratory changes.[3] In our assays, the necrotic spot lesions on leaves induced by the application of drops of toxins resemble those caused by the pathogen, even if those symptoms are not as specific as the pathogen.

Kenfield et al.[24] had previously studied the metabolites produced by another strain of *D. gigantea* and reported that gigantenone is the main toxin. Although being both terpenoids, gigantenone is a sesquiterpene, whereas ophiobolins are sesterterpenoids.

3.2.1.3. *Toxins from Fungal Pathogens of Parasitic Weeds*

Orobanche ramosa (broomrape) is a widespread parasitic weed of many Solanaceae species, such as tobacco or tomato, and attaches to many other species, including ornamentals and weeds. It causes both qualitative and quantitative damage to crops interfering with water and mineral intake and by affecting photosynthate partitioning. Seeds germinate only by stimulation with host root exudates, and produce a germ tube that, if it attaches to the host root, develops a haustorium penetrating the root and then forms a tubercle. This is followed by the most damaging phase, with the parasitic withdrawal of water, nutrients and photosynthates from the host. Due to the long underground phase, flowering stalk emergence occurs only when most of the damage has already been produced. As stimulated seed germination is a key phase of the parasitic plant life cycle, the search for natural compounds able to inhibit the germination appears to be an attractive and environmentally friendly approach.

Many fungi were isolated from diseased *O. ramosa* plants and some of them proved to be promising potential mycoherbicides for biological control of broomrapes.[35] Fifty-three isolates tested for virulence were also grown in vitro both on liquid and solid media with the main aim to find new metabolites having the ability to inhibit the induced germination of *O. ramosa* seeds.[36] The extracts from the liquid cultures were assayed for the ability to inhibit seed germination. Only the extracts produced by five strains were highly effective, causing the total or nearly complete inhibition of germination and were further considered as sources of new natural compounds. In particular the attention was focused on a strain of *Myrothecium verrucaria* and one of *Fusarium compactum*. Seven compounds were isolated from *M. vurrucaria* culture extracts and identified as verrucarins A, B, M and L acetate, roridin A, isotrichoverrin B and trichoverrol B, whereas the main metabolite produced was verrucarin E, a disubstituted pyrrole not belonging to the trichothecene group (Figure 4). The main metabolite produced by *F. compactum* was

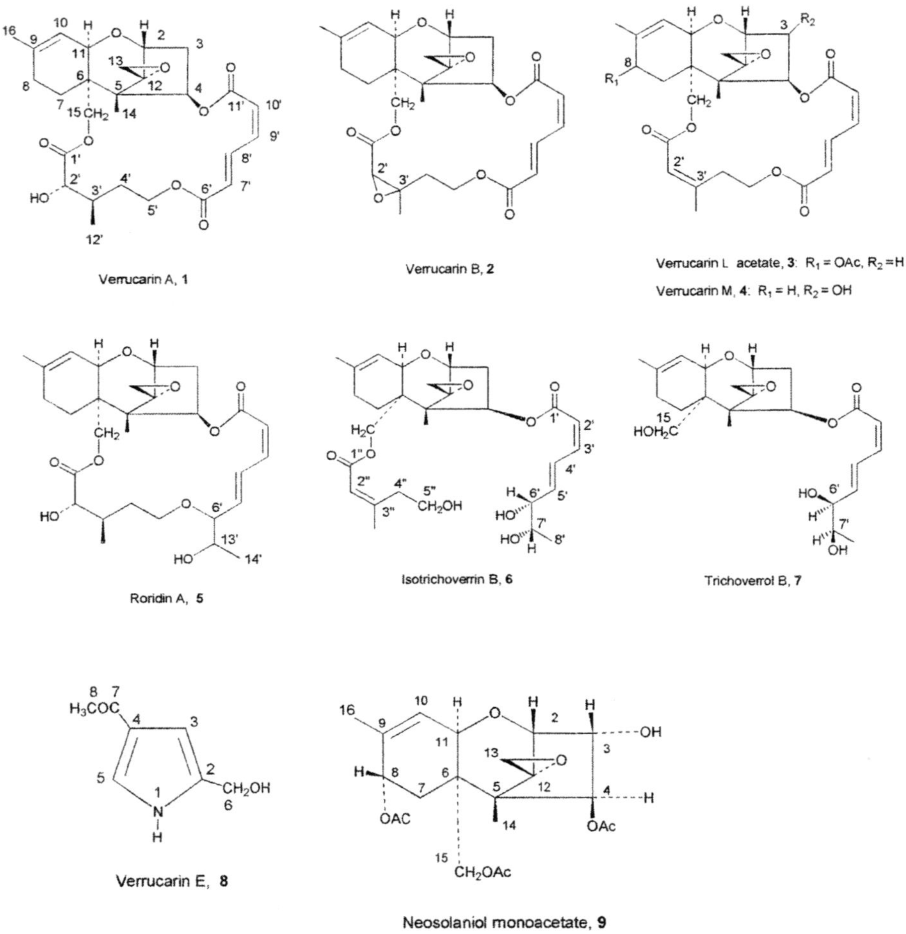

Figure 4. Metabolites isolated from *Myrothecium verrucaria* and *Fusarium compactum* cultures

neosoloaniol monoacetate, a trichothecene (Figure 4). All the trichothecenes were potent inhibitors of *O. ramosa* seed germination, whereas verrucarin E was inactive.[37]

Verrucarins A, B M and L acetate are in a subgroup of macrocyclic trichothecenes having a differently functionalized lactone ring located between C-4 and C-15. This macrocycle was substantially different and open, respectively, in roridin A, isotrichoverrin B and trichoverrol B, which belong to other two subgroups of the macrocyclic trichothecene family. The trichothecenes are a family of tetracyclic sesquiterpenoid substances produced by several fungal species. Macrocyclic trichothecenes have also been reported to cause increased cellular leakage, growth inhibition and chlorophyll loss when tested in *Lemna pausicostata* and *Pueraria lobata*.[38] They all are potent inhibitors of protein synthesis in eucaryotic cells, acting by interfering with peptidyl transferase activity.

Although inhibition of seed germination of many plant species (i.e., broccoli, carrot, radish and turnip) by macrocyclic trichothecenes has been already reported,[39,40] ours was the first report of the inhibitory effect of these metabolites to parasitic plant seeds.[37] Furthermore, this toxic effect on *Orobanche* seeds was observed at 2 μM, much lower than reported for crop seeds. Some seeds (e.g., lettuce, barley, tomato, wheat) were unaffected by 2 μM.[39]

The strong phytotoxicity of the macrocyclic trichothecenes (acting at 0.1 μM) may be due to the presence of an epoxy group, which plays an important role in the biological activity of some classes of naturally occurring compounds.

3.2.1.4. *Toxins Produced by Enthomopathogenic Fungi*

Entomopathogenic fungi produce several secondary metabolites, many of which might be novel sources for natural insecticides.[40] To know which toxins are produced and why they are toxic to the insects is important for a better understanding of the mode of action of entomopathogenic fungi at both the cellular and molecular level.[42] The fungus *Paecilomyces fumosoroseus* (Hyphomycetes) is one of the most infective fungal species on whiteflies *Bemisia tabaci* and *B. argentifolii*, attacking all insect stages.[43] Different metabolites have been isolated from *P. fumosoroseus*, among them beauvericin, beauverolides and 2,6-pyridindicarboxylic acid, also known as dipicolinic acid.[44,45] Beauvericin is a cyclic hexadepsipeptide with insecticidal properties.[46,47] Beauverolides belong to the same family but do not have a direct insecticidal effect. Dipicolinic acid, a pyridine derivative, is also produced and secreted by certain *Penicillium* spp.[48,49] and by several other entomopathogenic fungi, including *Beauveria bassiana, P. farinosus* and *Verticillium lecanii*.[50]

3.2.2. SYNTHESES

The inability of microorganisms to produce large amounts of a toxin or the high costs of purification, represent potential constraints to their practical use as natural pesticides. Their chemical syntheses or the chemical synthesis of the active moiety could overcome those limitations. Unfortunately, the natural compounds often have very unusual and complex structures, and by synthesis only very partial structures can be achieved.

Several fungal pathogens, especially those belonging to the genera *Alternaria* and *Cochliobolus*, produce host-selective toxins that are virulence and/or pathogenicity factors. These compounds are active against the same plant species as the fungal pathogens and low (physiological) concentrations of the toxin are able to reproduce symptoms of the natural infections. These plant-specific metabolites have received attention as models for new herbicides. For example, the synthesis of host-specific toxins has been extensively investigated by Cromby,[51] particularly AK-toxin I and AK-toxin II produced by *Alternaria alternata* (Japanese pear pathotype), which causes disease in pears. The interest in those toxins is due not only to the selectivity, but also because they are active at very low concentrations (5 nM). Another host-specific toxin, the cyclic dehydrodepsipeptide AM-toxin II, produced by *A. alternata*, the fungal agent of apple tree leaf spot disease, was efficiently synthesized using a solid-phase method.[52] It has also been shown that this methodology could be very useful in synthesizing unsaturated compounds using solid-phase chemistry.

Seiridin and its structural isomer isoseiridin are two phytotoxic butenolides produced by three species of *Seiridium*, fungi associated with the canker of cypress trees in the Mediterranean area. Only minute amounts of those compounds were available after a long process of purification of the fungal culture filtrates. The first enantioselective synthesis provided large quantities of the toxin seiridin.[53]

Cyperin is a phytotoxic diphenyl ether natural product. Total synthesis of cyperin has been achieved and its herbicidal activity evaluated.[54] Cyperin inhibited root growth of *Cyperus rotundus* grown on agar; however, root growth of *Cyperus* grown in soil was unaffected. Cyperin also inhibited growth of *Arabidopsis thaliana* and *Agrostis palustris*. The mode of action of cyperin is different from commercial diphenyl ether herbicides that inhibit protoporphyrinogen oxidase.

Dehydrocurvularin and its structural relatives, curvularin and 8-hydroxycurvularin are produced by a number of phytopathogenic fungal species, such as *Curvularia, Penicillium, Cochliobolus and Alternaria*. These metabolites possess biological properties including antifungal, phytotoxic and cytotoxic activities, and are related to octaketide and nonaketide analogs such as lasiodiplodins, resorcyclide, zearalenones and monocillin. The interesting biochemical effects have stimulated studies on their synthesis. The presence of a variety of oxidation states along the backbone of the dehydrocurvularin carbon skeleton has allowed to study its biosynthesis and demonstrate that its assembly by *Alternaria cinerariae* proceeds via a polyketide pathway.[55]

3.2.3. TOXINS AS TEMPLATES FOR NEW PESTICIDES

There are many reasons why natural products might be good sources of molecules or molecular templates for pesticides or at least lead to new targets of action.[4] New mechanisms of action for pesticides are highly desirable to fight the evolution of resistance in the target pests, to create or exploit unique market niches, and to cope with new regulatory legislation. Comparatively little effort has been expended on determination of the sites of action of phytotoxins from natural sources, suggesting that intensive study of these molecules will reveal many more novel mechanisms of action.[56]

Correlations of structure-activity are of utmost importance to the knowledge of the structural characteristics of the fungal metabolites and the determination of their active sites, or to hypothesize their chemical transformation to obtain more active, stable or selective compounds.

As the germination of seeds of parasitic plants depends on the presence of stimulating exudates produced by the roots of the host plant, an alternative approach for the management of parasitic weeds is stimulatory "suicidal germination" by the application of a germination stimulant to the soil, in the absence of the host. The parasite seeds will germinate and die, resulting in a reduction of the seed bank.

The chemical structures of a few *Orobanche* germination stimulants are known, i.e., alectrol and orobanchol. Some natural "strigolactones" (strigol, xenognosin, dihydrosorgoleon, sorgolactone, strigol related compounds) isolated from both hosts and non hosts of *Striga* and *Orobanche* are known. Synthetic analogues of strigolactones named the "GR" family have been developed and tested along with sesquiterpene lactones and their derivatives.[57]

Natural stimulants are mostly unstable in soil, and synthetic ones generally cannot be economically produced at industrial levels. Screening of fungal metabolites with stimulating activity is a very promising strategy to find such compounds. Among several fungal metabolites tested with the aim of finding new natural stimulants, Yoneyama and co-authors[58] reported that fusicoccin and cotylenol at concentrations 10 μM induced seed germination of *S. hermonthica* and *O. minor*.

Fusicoccin (FC) is the major toxic metabolite of *Fusicoccum amygdali*, the causative fungal agent of peach and almond canker.[59] Fusicoccin effectively stimulates seed germination of parasitic plants, and was available in our laboratory with its aglycone, several FC derivatives and natural analogues, and cotylenol. A structure-activity study was carried out using 25 of these compounds, sixteen were glucosides and 9 were aglycones (Table I) of FC.[60] Some natural FC analogues and derivatives showed a higher activity than FC (Table I). This appears to be modulated by chemical modifications, essentially in the functionalities and/or the conformation of the carbotricyclic diterpenoid ring. Noteworthy differences in the activity were noted among the glucosides. Among them, the most active compound was dideactylFC, which could have practical applications because it can be easily prepared from FC in high yields. The importance of the presence of a free primary hydroxy group at C-19 appeared evident. Some FC glucosides, having the acetylation of all hydroxy groups and other significant modifications of functionalities and conformation of the carbotricyclic ring decreased stimulant activity.[60]

TABLE I. Fusicoccin derivatives and analogues stimulate *Orobanche ramosa* seed germination.[60]

#	Compound*	Type	%†
1	Fusicoccin (FC)		0
2		FC d.	16
3	19-DeoxydideacetylFC	FC n.a.	24
4	DideacetylFC	FC n.a.	36
5	19-Dehydroxy-19-fluorodideacetylFC	FC d.	36
6	19-MonoacetyldideacetylFC	FC n.a.	15
7	19-Deoxy-3α- hydroxydideacetylFC	FC n.a.	13
8	3α-HydroxydideacetylFC	FC n.a.	36
9	12-MonoacetyldideacetylFC	FC n.a.	28
10	16-*O*-Demethyl-19-deoxydideactyl- 3-epiFC	FC n.a.	19
11		FC d.	0
12		FC d.	24
13		FC d.	0
14		FC d.	5
15		FC d.	12
16	De-*t*-Pentenyl-16-*O*-demethyl-19-deoxydideacetylFC	FC n.a.	22
17	FC Deacetyl aglycone	FC d.	14
18	FC Aglycone	# 17 d.	6
19	Cotylenol (aglycone of cotylenins)		0
20	Isopropyldene derivative	# 17 d.	54
21	Isopropylidene derivative	# 3 d.	11
22	12-Epi-8,9-isopropyldene	# 3 d.	8
23	12-Oxo-8,9-isopropyldene	# 3 d.	12
24	Isopropyldene derivative	# 17 d.	24
25		# 17 d.	31

LSD (0.05) = 6.9.
n.a. = natural analogue; d. = derivative.
*Compounds 1–16 = glucosides; 17–25 = aglycones.
†Percentage of seed germination—compounds assayed at 10 μM.

Fumonisins A and B are secondary metabolite analogues produced by *Fusarium* spp. The A series fumonisins have an *N*-terminal acetyl group not found on the B. Hydrolytic removal of the propanetricarboxylic acid moieties from fumonisins B1 and B2 yields the aminoalcohols HB1 and HB2, respectively. AAL-toxin is a phytotoxin produced by *Alternaria alternata* chemically related to fumonisins. AAL-toxin and the B series fumonisins at 1 μM caused pronounced cellular leakage of electrolytes and photodegradation of chlorophylls in a duckweed bioassay. These compounds also caused the most marked reductions in duckweed growth. HB1 at 1 μM moderately inhibited growth and caused a low level of cellular leakage. The other compounds were inactive at this concentration. The propanetricarboxylic acid groups of fumonisins and AAL-toxin are necessary for herbicidal activity in this series of compounds, whereas acetylation of the terminal amine group greatly reduced their activity. The structurally related sphingolipids, phytosphingosine and sphingosine, were about two orders of magnitude less phytotoxic than fumonisins and AAL, but the phytotoxicity symptoms were similar.[61]

Maculosin, a host-specific phytotoxin produced by *A. alternata* on *Centaurea maculosa*, is an ideal prototype for creating a safe and an environmentally friendly specific herbicide. A series of 18 maculosin analogs was synthesized and tested in the greenhouse on whole *Centaurea* plants to evaluate this possibility. Many of these maculosin analogs have significant potential as natural herbicides against this weed.[62]

A structure–activity relationship study on ophiobolins showed that ophiobolin A and its 6-epi analog were more phytotoxic than their anhydro derivatives against sorghum, *Senna obtusifolia*, and maize. Epiophiobolin A at a high concentration produced the largest necrosis on leaves of all plants tested except *Ipomoea* sp. The anhydrous derivatives were generally less phytotoxic and not toxic at all to *Ipomoea* leaves, even at concentrations of 2 mg/ml.[63] The results are in agreement with our findings, because also in our assays ophiobolin A proved to be more toxic to almost all the plant species tested, in comparison with 6-epi-ophiobolin A, whereas the 3-anydro compound was much less toxic, being almost inactive to many of the plants tested, even at the highest concentration used.[34]

Structural modification of the light-sensitive polyenic structure of strobilurin A led to the commercialization of the first fungicide of this class of compounds, azoxystrobin, followed by trifloxystrobin and fluoxastrobin, obtained by further chemical transformations.[64]

3.2.4. BIOTRANSFORMATION

Microbial transformation of natural toxins could usefully be utilized in crop protection, facilitating obtaining selected or new metabolites without time-consuming or uncertain chemical synthesis. Microbial transformations of natural or synthetic compounds are mostly used in mammalian drug metabolism studies for pharmacological and toxicological studies. For example, the microbial biotransformation of HR325, a synthetic immunomodulating agent, has been investigated by including it in growth medium of 16 fungal strains. Several fungal strains are able to metabolize this drug in different manners.

A strain of *Beauveria bassiana* produced many unstable products in the first 2–3 days, and then produced two main products after 7 days. Many strains were also unable to metabolize the compound, whereas a strain of *Mortierella isabellina* was best in transforming the drug to a single derivative.[65]

Much work has been done on the biotransformation of the inexpensive hydrocarbon monoterpene limonene, one of the most widely distributed terpene in nature, to obtain novel products having characteristics suitable for cosmetics, by bacterial and fungal conversion.[66] For example, a strain of *Cladosporium* sp. was able to convert the metabolite in trans-limonene-1,2-diol, whereas another transformed it in α-terpineol. In a recent study more than 60 fungal strains were used, and those able to perform transformations belong to genera such as *Aspergillus* and *Penicillium*.[67]

Of several hundred microorganisms randomly selected from the environment, only a fungal isolate identified as *Aspergillus niger* var. *niger* transformed the phytotoxin thaxtomin A to much less toxic metabolites and none more toxic.[68]

A further approach to biotransformation is to use strains having blocked biosynthetic abilities to obtain intermediates. For example, mutant strains of *Fusarium graminearum* obtained by disruption of Tri8, a gene probably encoding an esterase, were able to accumulate 3-acetyl T-2 toxin, 3-acetyl neosolaniol and 3,4,15-triacetoxyscirpenol, rather than T-2 toxin.[69] Such intermediates could have different biological properties. Disruption of *F. sporotrichioides* Tri11, a gene encoding a cytochrome P-450 monooxygenase, led to the accumulation of four trichothecenes not observed in cultures of the parent strain.[70]

Production of metabolites can be further manipulated by the use of specific growth media, to modify the biosynthetic pathways for the production of the compounds. This offers the possibility to obtain "non-natural" natural products, and this could be accomplished by simply adding chemical analogs of key biosynthetic intermediates to the growth medium. These chemicals are recognized by biosynthetic enzymes and enter into the pathway. The end-products are analogues of the normal product or intermediates that are not substrates for subsequent transformations.

Changes in culturing conditions can strongly influence the biosynthetic production of ophiobolins. *Bipolaris maydis* was able to produce ophiobolin A, 3-anhydroophiobolin A, ophiobolin B and ophiobolin L when grown in liquid conditions,[71] whereas it produced ophiobolin M, 6-epiophiobolin M, ophiobolin C, 6-epiophiobolin C, ophiobolin K and 6-epiophiobolin K when grown on solid media.[72]

Phoma exigua var. *heteromorpha* produced very different cytochalasins when grown in different culture conditions. It produced deoxaphomin (a 13-cytochalasan), several 14-cytochalasans (deoxaphomin, cytochalasin A, B, F, T, 7-*O*-Acetyl-CB) and many 24-cytochalasans (cytochalasins Z1–Z5) on solid medium. In liquid culture it produced ascochalasin (13-cytochalasan), deoxaphomin, cytochalasin A and B (all 14-cytochalasans), together with cytochalasin U and V (15- and 16-cytochalasans, respectively). Only three compounds out of fourteen were produced in both cultural conditions.

Three major destruxins A, B, and E produced by *Metarhizium anisopliae* could be detected in liquid medium but not on a solid medium. No toxins could be detected in highly aerated fermentation suggesting that the aeration regime also has a significant impact on destruxin production.[73]

3.2.5. SELECTION OF BIOCONTROL AGENTS OVERPRODUCING TOXINS

The development of pathogens with enhanced biocontrol activity by selection or by the introduction of genes responsible for toxin biosynthesis, seems a reasonable possibility. In fact, several genes in the biosynthetic pathways of fungal toxins have already been identified and cloned.[74,75]

The use of transformed protein toxins is described in Chapter 16 of this book.

3.2.6. SYNERGISTIC USE OF TOXINS WITH BIOCONTROL AGENTS

Toxins can be used to weaken physical and biochemical defenses of the target organism, or to increase the aggressiveness of the pathogen. This approach has been used for example by the application of low doses of the herbicide imazaquin in combination with *Alternaria zinniae*, increasing the efficacy of the fungus to control *Xanthium occidentale* and restricting the plant's ability to recover after fungal application due to the herbicide's ability to interfere with protein synthesis.[77] Application of *Ascochyta caulina* toxins used in combination with the pathogen enhanced disease severity and the speed of symptom appearance. The first symptoms appeared on plants after only 1–2 days in case of simultaneous applications, whereas they appeared more slowly when using the pathogen alone. This faster colonization of plant tissues by the pathogen could also render the pathogen less dependent on environmental conditions that are usually limiting factors to the practical use of mycoherbicides.[78]

3.2.7. TOXINS AS BIOMARKERS

Screening for fungal strains to select the best biological control agents requires time and space consuming experiments. When there is a positive correlation between known toxin production and aggressiveness of the candidate biocontrol agent, analytical methods can be developed to measure the toxins in culture filtrates or partially purified extracts, and choosing the highest toxin producers.

A method of high-performance anion exchange chromatography and pulsed amperometric detection was developed, allowing a quick and simple quantification of three main metabolites produced by *Ascochyta caulina*, the biocontrol agent of *Chenopodium album*, in liquid culture. Preliminary observations seemed to support a positive correlation between toxin production and virulence of the strains.[13] The same approach was unsuccessful in the selection of phytopathogenic strains of *Fusarium oxysporum* for biological control of parasitic weeds. An attempt was made to correlate the virulence of several strains of *F. oxysporum* isolated from *Orobanche ramosa*, and the production of fusaric acid in vitro, but no correlations were observed.[36]

The ascosonchine toxin content in culture filtrates of different strains of *Ascochyta sonchi* was measured, and varied with seven of the nine strains between 0.5 and 2.7 mg/l, whereas two strains did not produce any. No correlation was found between toxin production and strain virulence. Almost all strains, including non-producers, were able to cause leaf disease, regardless their ability to produce the toxin.[79]

Stagonospora convolvuli, a promising biocontrol agent of *Convolvulus arvensis* and *Calystegia sepium* produces the phytotoxins leptosphaerodione and elsinochrome A, whereas another strain produces the toxin cercosporin. Cercosporin and elsinochrome A are closely related photodynamic perylenequinone toxins produced by many *Cercospora* and *Elsinoe* spp., respectively. Thirty isolates of *Stagonospora* sp. were characterised for their aggressiveness on both weed species, and for the production of the three metabolites. Cercosporin producers were less pathogenic on *Convolvolus*. Conversely, there was a positive correlation between elsinochrome A and leptosphaerodione production, and each was positively correlated with aggressiveness of isolates on both *Convolvulus*. Isolates without elsinochrome A were not aggressive.[80]

A significant correlation was found between the titer of destruxin production in vitro by isolates of *Metarhizium anisopliae* var *anisopliae* pathogenic to *Otiorhynchus* and the rapidity of death, suggesting a role for the toxin in isolate virulence. A strong positive correlation was found only between in vitro toxin production and percentage mortality of individuals in which sporulation did not occur on the cadaver. To account for this, it is suggested that if destruxin kills locusts before the fungus has established itself, then the pathogen may not compete effectively with the saprophytic flora and, as a result, fails to sporulate. It is concluded that, in the pathogenesis of *M. anisopliae* var *anisopliae* there is a relationship between the titer of destruxin production of isolates in vitro and the killing power.[81]

3.3. Risks

Increasing public sensitivity to environmental pollution and problems of pest resistance to chemical pesticides has provided the impetus for the development of alternate strategies for pest control. However, consumer concerns regarding mycotoxins entering the food chain has prompted closer scrutiny of the secondary metabolites of all fungal biocontrol agents. Regulatory authorities often require detailed information on "relevant metabolites." It is not clear what constitutes a relevant metabolite when most fungi secrete a disparate array of bioactive compounds with different ones produced under different conditions. Most often the metabolites are secreted in extremely small quantities, even under optimal production systems. Evaluation of the toxicological risks of each secondary metabolite could prove onerous and highly expensive, partly because methods and tools have not been developed for the risk assessment of metabolites of fungal biocontrol agents. Furthermore, no guidelines or simulation models exist to evaluate the fate of secreted fungal metabolites in the environment. Little is known about the full range of metabolites produced and whether they enter the food chain, posing a risk to human and animal health. Often very promising biopesticides have been discarded by final evaluation processes just because in vitro they produce powerful and dangerous toxins. The evaluation of the risk should be ascertained by considering not only the toxicity in vitro of certain known amount of toxins, e.g., lethal or effective doses against chosen organisms, but by evaluating the global environmental impact, e.g., determining the production of those metabolites when fungi are formulated, or when they are applied to, and grown on the target; the toxicity to non-target organisms; the stability of toxins in vivo or the absorption by soil particles; and the risk of water drift.

One of the main problems is to determine the biosynthesis of toxic metabolites by the mycoherbicides and if they are released into the environment. In fact, many fungi are able to produce very high amounts of secondary metabolites when grown for some weeks on solid media where they have at their disposal large amounts of nutrients. The accumulation of metabolites can be different when a fungus is formulated as dried spores or chlamydospores, as biocontrol agents usually are. When distributed in the field, the biocontrol agent is usually applied to young plants or seedlings. The available nutrients are not as plentiful, and usually the fungus is able to cause a high level of disease within a few days, and then disappear together with the diseased target. So, the potential to produce and accumulate high amounts of toxins appears very limited and the likelihood they will arrive in the food chain is exceedingly unlikely. This is also confirmed by the scarcity of information about the detection of phytotoxins in vivo.

For example, *Myrothecium verrucaria* proposed as an agent for the control of *Pueraria montana* var. *lobata* (kudzu), when grown in vitro on both liquid and solid culture, produced a wide range of mycotoxic macrocyclic trichothecenes at concentrations up to milligrams per gram of culture. Conversely, none of those metabolites were detected by HPLC analysis in diseased tissues of *Pueraria* treated with spore suspensions of the fungus.[82]

In a preliminary attempt to determine the levels of elsinochrome A and leptosphaerodione produced by *Stagonospora convolvuli* in diseased *Convolvulus*, none of the toxins were detected in infected leaves.[80] The transformation of the fungal metabolites by microbial or plant metabolism, their immobilization in the soil particles and the physical and chemical changes that can occur, leading to the possible inactivation of the compounds should be considered.

References

1. J. Lacey, *Trichothecenes and Other Mycotoxins* (Wiley, Chichester, UK 1985).
2. M. S. C. Pedras, L. I. Zaharia, and D. E. Ward, The destruxins: Synthesis, biosynthesis, biotransformation, and biological activity, *Phytochemistry* 59, 579–596 (2002).
3. T. K. Au, W. S. H. Chick, and P. C. Leung, The biology of ophiobolins, *Life Sci.* 67, 733–742 (2000).
4. A. M. Rimando and S. O. Duke, Natural products for pest management, in *Natural Products for Pest Management*, edited by A. Rimando and S. Duke (ACS Press, Washington, DC, 2006), pp. 2–21.
5. T. Anke, F. Oberwinkler, W. Steglich, and G. Schramm, The strobilurins—New antifungal antibiotics from the basidiomycete *Strobilurus tenacellus*, *J. Antibiot.* 30, 806–810 (1977).
6. P. R. Graupner, B. C. Gerwick, T. L. Siddall, A. W. Carr, E. Clancy, J. R. Gilbert, K. L. Bailey, and J. Derby, Chlorosis inducing phytotoxic metabolites: New herbicides from *Phoma macrostoma*, in *Natural Products for Pest Management*, edited by A. Rimando and S. Duke (ACS Press, Washington, DC, 2006), pp. 37–47.
7. V. A. Melnik, Taxonomy of the genus *Ascochyta* Lib., *Mikologia Fitopatol.* 5, 15–22 (1971).
8. A. Evidente, R. Capasso, M. Vurro, and A. Bottalico, Ascosalitoxin, a phytotoxic trisubstituted salicylic aldehyde from *Ascochyta pisi*, *Phytochemistry* 34, 995–998 (1993).
9. A. Evidente, R. Lanzetta, R. Capasso, M. Vurro, and A. Bottalico, Pinolidoxin, a phytotoxic nonenolide from *Ascochyta pinodes*, *Phytochemistry* 34, 999–1003 (1993).
10. R. N. Strange, Phytotoxins associated with *Ascochyta* specie, in *Toxins in Plant Disease Development and Evolving Biotecgnology*, edited by R. K. Upadhyay and F. G. Mukerji (Oxford & IBH Publishing Co., New Delhi, 1997), pp. 167–181.
11. A. Evidente, R. Capasso, A. Cutignano, O. Taglialatela-Scafati, M. Vurro, M. C. Zonno, and A. Motta, Ascaulitoxin, a phytotoxic bis-amino acid *N*-glucoside from *Ascochyta caulina*, *Phytochemistry* 48, 1131–1137 (1998).
12. A. Evidente, A. Andolfi, M. Vurro, M. C. Zonno, and A. Motta, Trans-4-aminoproline, a phytotoxic metabolite with herbicidal activity produced by *Ascochyta caulina*, *Phytochemistry* 53, 231–237 (2000).

13. A. Evidente, A. Andolfi, M. Vurro, and M. C. Zonno, Determination of *Ascochyta caulina* phytotoxins by high performance anion exchange chromatography and pulsed amperometric detection, *Phytochem. Anal.* 12(6), 383–387 (2001).
14. A. Evidente, A. Andolfi, M. A. Abouzeid, M. Vurro, M. C. Zonno, and A. Motta, Ascosonchine, the enol tautomer of 4-pyridylpyruvic acid with herbicidal activity produced by *Ascochyta sonchi*, *Phytochemistry* 65(4), 475–480 (2004).
15. A. Meister, *Biochemistry of the Amino Acids,* vol. II (Academic Press, New York, 1965).
16. B. Ganem, From glucose to aromatics: Recent developments in natural products of the shikimic acid path. 4, *Tetrahedron* 34, 3353–3383 (1978).
17. E. Haslam, *Shikimic Acid: Metabolism and Metabolites* (Wiley, Chichester, UK, 1993).
18. M. Fracchiolla, *Biological control of grass weeds by using bioherbicides* (Ph.D. Thesis, 2003).
19. L. A. Tatum, The southern corn leaf blight, *Science* 171, 1113–1115 (1972).
20. S. Y. Padmanabhan, The great Bengal famine, *Annu. Rev. Phytopathol.* 11, 11–26 (1973).
21. G. Strobel, D. Kenfield, and F. Sugawara, The incredible fungal genus *Drechslera* and its phytotoxic ophiobolins, *Phytoparasitica* 16, 145–152 (1988).
22. S. Chandramohan and R. Charudattan, Control of seven grasses with a mixture of three fungal pathogen with restricted host ranges, *Biol. Control* 22, 246–255 (2001).
23. M. A. Kastanias and M. Chrysayi-Tokousbalides, Herbicidal potential of pyrenophorol isolated from a *Drechslera avenae* pathotype, *Pest Manag. Sci.* 56, 227–232 (2000).
24. D. Kenfield, G. Bunkers, Y. U. Wu, G. Strobel, F. Sugawara, Y. Hallock, and J. Clardy, Gigantenone, a novel sesquiterpene phytohormone mimic, *Experientia* 45, 900–902 (1989).
25. D. Kenfield, Y. Hallock, J. Clardy, and G. Strobel, Curvulin and *O*-methylcurvulinic acid: Phytotoxic metabolites of *Drechslera indica* which cause necroses on purslane and spiny amaranth, *Plant Sci.* 60, 123–127 (1989).
26. A. Evidente, A. Andolfi, M. Vurro, M. Frachiolla, M. C. Zonno, and A. Motta, Drazepinone, a trisubstituted tetraidronaphthofuroazepinone with herbicidal activity produced by *Drechslera siccans*. *Phytochemistry* 66, 715–721 (2005).
27. C. A. Mattia, L. Mazzarella, and L. Puliti, 4-(2-Amino-4-oxo-2-imidazolin-5-ylidene)-2-bromo-4,5,6,7-tetrahydropyrrolo[2,3-*c*]-azepin-8-one methanol solvate: A new bromo-compound from the sponge *Acanthella aurantiaca*, *Acta Cryst.* B 38, 2513–2515 (1982).
28. T. Sekine, A. Jiro, S. Kazuki, I. Fumio, O. Siripon, and M. Isamu, (+)-Acacialactam, a new seven-membered lactam from seeds of *Acacia concinna*, *Chem. Pharm. Bull.* 37, 3166–3165 (1989).
29. H. Cho, K. Murakami, H. Nakanishi, A. Fujisawa, H. Isoshima, M. Niwa, K. Hayakawa, Y. Hase, I. Uchida, H. Watanabe, K. Wakitani, and K. Aisaka, Synthesis and structure-activity relationships of 5,6,7,8-tetrahydro-4*H*-thieno[3,2-*b*]azepine derivatives: Novel arginine vasopressin antagonists, *J. Med. Chem.* 47, 101–109 (2004).
30. Y. F. Hallock, J. Clardy, D. S. Kenfield, and G. Strobel, De-*O*-methyldiaporthine, a phytotoxin from *Dreschslera siccans*, *Phytochemistry* 27, 3123–3125 (1988).
31. C. H. Lim, H. Miyagawa, T. Ueno, H. Takenaka, and N. D. Sung, Siccanol: Sesterterpene isolated from pathogenic fungus *Drechslera siccans, Han'guk Nonghwa Hakhoechi* 39, 241–244 (1996).
32. D. F. Farr, G. F. Bills, G. P. Chamuris, and A. Y. Rossman, *Fungi on Plants and Plant Products in the United States* (APS Press, St. Paul, MN, 1989).
33. A. Evidente, A. Andolfi, A. Cimmino, M. Vurro, M. Fracchiolla, and R. Charudattan, Herbicidal potential of ophiobolins produced by *Drechslera gigantea*, *J. Agric. Food Chem.* 54(5), 1779–1783 (2006).

34. A. Evidente, A. Andolfi, A. Cimmino, M. Vurro, M. Fracchiolla, R. Charudattan, and A. Motta, Ophiobolin E and 8-epi-ophiobolin J produced by *Drechslera gigantea*, potential mycoherbicide of weedy grasses, *Phytochemistry* 67, 2281–2287 (2006).
35. A. Boari and M. Vurro, Evaluation of *Fusarium* spp. and other fungi as biological control agents of broomrape (*Orobanche ramosa*), *Biol. Control* 30, 212–219 (2004).
36. M. A. Abouzeid, A. Boari, M. C. Zonno, M. Vurro, and A. Evidente, Toxicity profile of potential agents of *Orobanche ramosa* biocontrol, *Weed Sci.* 52, 326–332 (2004).
37. A. Andolfi, A. Boari, A. Evidente, and M. Vurro, Metabolites inhibiting germination of *Orobanche ramosa* seeds produced by *Myrothecium verrucaria* and *Fusarium compactum*, *J. Agric. Food Chem.* 53, 1598–1603 (2005).
38. H. K. Abbas, B. B. Johnson, W. T. Shier, H. Tak, B. B. Jarvis, and C. D. Boyette, Phytotoxicity and mammalian cytotoxicity of macrocyclic trichothecene mycotoxins from *Myrothecium verrucaria, Phytochemistry* 59, 309–313 (2002).
39. J. O. Kuti, N. Mokhtari, B. B. Jarvis, and G. A. Bean, Allelopathic potential of plant-derived macrocyclic trichothecenes on selected crop species, *Biodeter. Res.* 2, 383–388 (1989).
40. A. Kobayashi, Y. Nakae, T. Kawaski, and K. Kawazu, Fungal trichothecenes which promote callus initiation from the alfalfa cotyledon, *Agric. Biol. Chem.* 53, 585–589 (1989).
41. P. Mollier, J. Lagnel, B. Fournet, A. Aïoun, and G. Riba, A glycoprotein highly toxic for *Galleria mellonela* larvae secreted by the entomopathogenic fungus *Beauveria sulfurecens*, *J. Invertebr. Pathol.* 64, 200–207 (1994).
42. A. Vey, J. M. Quiot, I. Mazet, and C. W. McCoy, Toxicity and pathology of crude broth filtrate produced by *Hirsutella thompsonii* var. *thompsonii* in shake culture, *J. Invetebr. Pathol.* 61, 131–137 (1993).
43. A. Lacey, A. Kirk, L. Millar, G. Mercadier, and C. Vidal, Ovicidal and larvicidal of conidia and blastospores of *Paecilomyces fumosoroseus* (Deuteromycotina: Hyphomicetes) against *Bemisia argentifolii* (Homoptera: Aleyrodidae) with a description of a bioassay system allowing prolonged survival of control insects, *Biocontrol Sci. Technol.* 9, 9–18 (1999).
44. M. Bernardini, A. Carilli, G. Pacioni, and B. Santurbano, Isolation of beauvericin from *Paecilomyces fumosoroseus*, *Phytochemistry* 14, 1865 (1975).
45. A. Jegorov, P. Sedmera, V. Matha, P. Simek, H. Zahradnickova, Z. Landa, and J. Eyal, Beauverolides L and La from *Beauveria tenella* and *Paecilomyces fumosoroseus*, *Phytochemistry* 37, 1301–1303 (1994).
46. J. F. Grove and M. Pople, The insecticidal activity of beauvericin and the enniatin complex, *Mycopathologia* 70, 103–105 (1980).
47. S. Gupta, S. B. Krasnoff, N. L. Underwood, J. A. A. Renwick, and D. W. Roberts, Isolation of beauvericin as an insect toxin from *Fusarium semitectum* and *Fusarium moniliforme* var. *subglutinans*, *Mycopathologia* 115, 185–189 (1991).
48. J. Oyama, Biosynthesis of dipicolinic acid by molds, II: Dipicolinic acid producers, *Rep. Ferment. Res. Inst.* 20, 105–112 (1961).
49. J. Oyama, N. Nakamura and O. Tanabe, Biosynthesis of dipicolinic acid by molds. I. Isolation and indentification of dipicolinic acid from the culture filtration of a *Penicillium* sp., *Rep. Ferment. Res. Inst.* 19, 5–81 (1961).
50. N. Claydon and J. Grove, Insecticidal secondary metabolitic products from the entomogenous fungus *Verticilliun lecanii*, *J. Invertebr. Pathol.* 40, 413–418 (1982).
51. L. Cromby, Natural product chemistry and its part in the defense against insects and fungi in agriculture, *Pest. Sci.* 55, 761–774 (1999).
52. E. Horikawa, M. Kodaka, Y. Nakahara, H. Okuno, and K. Nakamura, Solid-phase synthesis of dehydropeptide, AM-toxin II, using a novel selenyl linker by side-chain tethered strategy, *Tetrahedron Lett.* 42, 8337–8339 (2001).

53. C. Bonini, L. Chiummiento, A. Evidente, and M. Funicello, First enantioselective synthesis of (–)-seiridin the major phytotoxic metabolite of *Seiridium* species pathogenic for cypress, *Tetrahedron Lett.* 36, 7285–7286 (1995).
54. P. M. Harrington, B. K. Singh, I. T. Szamosi, and J. H. Birk, Synthesis and herbicidal activity of cyperin, *J. Agric. Food Chem.* 43, 804–808 (1995).
55. Y. Liu, Z. Li, and J. C. Vedras, Biosynthetic incorporation of advanced precursors into dehydrocurvularin, a poliketide phytotoxin from *Alternaria alternate*, *Tetrahedron* 54, 15,937—15,958 (1998).
56. S. O. Duke, J. G. Romagni, and F. E. Dayan, Natural products as sources for new mechanisms of herbicidal action, *Crop Prot.* 19, 583–589 (2000).
57. K. Yoneyama, D. Sato, Y. Takeuchi, H. Sekimoto, T. Yokota, and T. Sassa, Search for germination stimulants and inhibitors for root parasitic weeds, in *Natural Products for Pest Management*, edited by A. Rimando and S. Duke (ACS Press, Washington, DC, 2006), pp. 88–98.
58. K. Yoneyama, Y. Takeuchi, M. Ogasawara, M. Konnai, Y. Sugimoto, and T. Sassa, Cotylenins and fusicoccins stimulate seed germination of *Striga hermonthica* (Del.) Benth and *Orobanche minor* Smith, *J. Agric. Food Chem.* 46, 1583–1586 (1998).
59. A. Ballio, E. B. Chain, P. De Leo, B. F. Erlanger, M. Mauri, and A. Tonolo, Fusiccocin: A new wilting toxin produced by *Fusicocum amygdali* Del. *Nature* 203, 297 (1964).
60. A. Evidente, A. Andolfi, M. Fiore, A. Boari, and M. Vurro, Stimulation of *Orobanche ramosa* seed germination by fusicoccin derivatives: A structure-activity relationship study, *Phytochemistry* 67, 19–26 (2006).
61. T. Tanaka, H. K. Abbas, and S. O. Duke, Structure-dependent phytotoxicity of fumonisins and related compounds in a duckweed bioassay, *Phytochemistry* 33, 779–785 (1993).
62. M. M. Bobylev, L. I. Bobyleva, and G. A. Strobel, Synthesis and bioactivity of analogs of maculosin, a host-specific phytotoxin produced by *Alternaria alternata* on spotted knapweed (*Centaurea maculosa*), *J. Agric. Food Chem.* 44, 3960–3964 (1996).
63. L. M. Pena-Rodriguez and W. S. Chilton, 3-Anhydroophiobolin A and 3-anhydro-6-epi-ophiobolin A, phytotoxic metabolites of the johnson grass pathogen *Bipolaris sorghicola*, *J. Nat. Prod.* 52, 1170–1172 (1989).
64. P. Jeschke, F. Lieb, R. Velten, and W. B. Wiese, Natural products and their role in the design of active ingredients for modern crop protection, in *Natural Products for Pest Management*, edited by A. Rimando and S. Duke (ACS Press, Washington DC, 2006), pp. 128–141.
65. I. Lacroix, J. Biton, and R. Azerad, Microbial biotransformations of a synthetic immunomodulatin agent, HR325, *Bioorg. Med. Chem.* 5, 1369–1380 (1997).
66. J. Demyttenaere, M. del Carmen Herrera, and N. De Kimpe, Biotransformation of geraniol, nerol and citral by sporulated surface cultures of *Aspergillus niger* and *Penicillium* sp., *Phytochemistry* 55, 363–373 (2000).
67. J. Demyttenaere, K. Van Belleghem, and N. De Kimpe, Biotransformation of (*R*)-(+)-and (*S*)-(–)-limonene by fungi and the use of solid phase microextraction for screening, *Phytochemistry* 57, 199–208 (2001).
68. G. Lazarovits, J. Hill, R. R. King, and L. A. Calhoun, Biotransformation of the *Streptomyces scabies* phytotoxin thaxtomin A by the fungus *Aspergillus niger*, *Can. J. Microbiol.* 50, 121–126 (2004).
69. S. P. McCormick and N. J. Alexander, Fusarium Tri8 encodes a trichothecene C-3 esterase, *Appl. Environ. Microb.* 68, 2959–2964 (2002).
70. S. P. McCormick and T. M. Hohn, Accumulation of trichothecenes in liquid cultures of a *Fusarium sporotrichioides* mutant lacking a functional trichothecene C-15 hydroxylase, *Appl. Environ. Microb.* 63, 1685–1688 (1997).

71. E. Li, A. M. Clark, D. P. Rotella, and C. D. Hufford, Microbial metabolites of ophiobolin A and antimicrobial evaluation of ophiobolins, *J. Nat. Prod.* 58, 74–81 (1995).
72. A. Tsipouras, A. A. Adefarati, J. S. Tkacz, E. G. Frazier, S. P. Rohrer, E. Birzin, A. Rosegay, D. L. Zink, M. A. Geotz, S. B. Singh, and J. M. Schaeffer, Ophiobolin M and analogues, noncompetitive inhibitors of ivermectin binding with nematocidal activity, *Bioorg. Med. Chem.* 4, 531–536 (1996).
73. C. Wang, A. Skrobek, and T. M. Butt, Investigations on the destruxin production of the entomopathogenic fungus *Metarhizium anisopliae, J. Inv. Pathol.* 85, 168–174 (2004).
74. J. Yu, D. Bhatnagar, and T. E. Cleveland, Completed sequence of aflatoxin pathway gene cluster in *Aspergillus parasiticus*, *FEBS Lett.* 564, 126–130 (2004).
75. J. Seo, R. H. Proctor, and R. D. Plattner, Characterization of four clustered and coregulated genes associated with fumonisin biosynthesis in *Fusarium verticillioides*, *Fungal Genet. Biol.* 34, 155–165 (2001).
76. Z. Amsellem, B. Cohen, and J. Gressel, Engineering hypervirulence in a mycoherbicidal fungus for efficient weed control, *Nat. Biotechnol.* 20, 1035–1039 (2002).
77. B. A. Auld, H. E. Smith, and S. Qiang, Control of cocklebur with a combination of *Alternaria zinniae* and low rates of imaziquin, in *Proceedings of the 16th Asian-Pacific Weed Science Society Conference,* Asian-Pacific Weed Sci. Soc., edited by Anonymous (Kuala Lampur, Malaysia, 1997) pp. 345–347.
78. M. Vurro, M. C. Zonno, A. Evidente, A. Andolfi, and P. Montemurro, Enhancement of efficacy of *Ascochyta caulina* to control *Chenopodium album* by use of phytotoxins and reduced rates of herbicides, *Biol. Cont.* 21, 182–190 (2001).
79. A. Evidente, A. Berestetskiy, A. Andolfi, M. C. Zonno, A. Cimmino, and M. Vurro, Relation between in vitro production of ascosonchine and virulence of strains of the potential mycoherbicide *Ascochyta sonchi*: A method for its quantification in complex samples. *Phytochem. Anal.* 17, 357–364 (2006).
80. M. O. Ahonsi, M. Maurhofer, D. Boss, and G. Defago, Relationship between aggressiveness of *Stagonospora* sp. isolates on field and hedge bindweeds, and in vitro production of fungal metabolites cercosporin, elsinochrome A and leptosphaerodione, *Eur. J. Plant Pathol.* 111, 203–215 (2005).
81. M. J. Kershaw, E. R. Moorhouse, R. Bateman, S. E. Reynolds, and A. K. Charnley, The role of destruxins in the pathogenicity of *Metarhizium anisopliae* for three species of insect, *J. Invertebr. Pathol.* 74, 213–223 (1999).
82. H. K. Abbas, H. Tak, C. D. Boyette, W. T. Shier, and B. B. Jarvis, Macrocyclic trichothecenes are undetectable in kudzu (*Pueraria montana*) plants treated with a high-producing isolate of *Myrothecium verrucaria*, *Phytochemistry* 58, 269–276 (2001).

4. BIOCONTROL OF WEEDS WITH ALLELOPATHY: CONVENTIONAL AND TRANSGENIC APPROACHES

Stephen O. Duke,[1*] Scott R. Baerson,[1] Agnes M. Rimando,[1] Zhiqiang Pan,[1] Franck E. Dayan,[1] and Regina G. Belz[2]
[1]*USDA, ARS, Natural Products Utilization Research Unit, P. O. Box 8048, University, MS 38677, USA*
[2]*Institute of Phytomedicine 360, Department of Weed Science, University of Hohenheim, 70593 Stuttgart, Germany*

Abstract. Growing highly allelopathic crops has the potential to significantly reduce our reliance on synthetic herbicides for weed management. Specific phytotoxins have been found in allelopathic rice, wheat, and rye varieties, but this information has not been used in breeding varieties that can be marketed on the basis of their weed management properties. Although such a conventional approach is viable, transgenic strategies may be better. For example, genes encoding enzymes of the highly potent phytotoxin sorgoleone in *Sorghum* spp. might be transgenically manipulated to enhance the allelopathic properties of sorghum crops. This potent phytotoxin is exclusively synthesized and secreted by root hairs. The sorgoleone pathway has been elucidated and putative genes encoding them have been identified and partially verified.

Keywords: allelopathy; rice; sorghum; sorgoleone; transgene; weed; wheat

4.1. Introduction

Allelopathy is the chemical warfare component of interference between plants. Allelopathy became an "in vogue" area of research in the 1960s and 70s, after which the importance and even the existence of allelopathy was questioned by prestigious ecologists such as Harper.[1] The status of allelopathy research was further diminished by the poor quality of much of the research that was being conducted.

In the past decade, more rigorous research in this area with more powerful techniques has demonstrated unequivocally that allelopathy can be a powerful influence on plant/plant interactions. This has led to the hope that this phenomenon can be utilized to manage weeds with less dependence on

*To whom correspondence should be addressed, e-mail: sduke@olemiss.edu

M. Vurro and J. Gressel (eds.), Novel Biotechnologies for Biocontrol Agent Enhancement and Management, 75–85.

TABLE I. Some crops that have surveyed for allelopathic potential, along with allelochemicals associated with the crops

Crop	Allelochemical	References
Barley (*Hordeum vulgare*)	Hordenine, benzoxazinones	5
Cucumber (*Cucumis sativa*)	Phenolic acids, *p*-thiocyanatophenol	6
Oats (*Avena sativa*)	Phenolic acids	7
Rice (*Oryza sativa*)	Momilactone B	8, 9
Rye (*Secale cereale*)	Benzoxazinones	10
Sorghum spp.	Sorgoleone, dhurrin	2, 11
Sunflower (*Helianthus* spp.)	Various sesquiterpenes	12
Wheat (*Triticum aestivum*)	Benzoxazinones	13, 14

synthetic herbicides. This chapter will discuss non-transgenic and transgenic approaches to this goal.

The potential for use of allelopathic crops to reduce synthetic herbicide inputs has been discussed in many reviews.[2–4] Despite surveys of the allelopathic potential of several crops (Table I), no crop variety is being sold on the basis of its ability to suppress weeds by allelopathy. Significant recent progress has been made in identification of potentially important allelochemicals produced by crops. The identification of genes involved in biochemical pathways that produce allelochemicals is now relatively straightforward, providing the potential to enhance allelopathy by molecular breeding or by transgene technology. This review is meant to provide a brief update on the most recent advances in crop allelopathy in the crops that we think offer the best opportunity for improving crop allelopathy using new technologies. All of these crops exude allelochemicals from their roots, something that we consider essential to the practical success of allelopathy in an annual crop.

4.2. Conventional Allelopathy

4.2.1. RICE

Rice is perhaps the most intensively studied case of crop allelopathy. An international effort to generate allelopathic rice varieties has been underway for more than a decade.[15] Thousands of varieties have been screened for allelopathic potential.[8,9] Although weeds such as *Echinochloa crus-galli*[8,16]and *Cyperus difformis*[17] are suppressed by some rice varieties, the level of weed management does not reach that obtained with herbicides. Still, substantially reduced herbicide use rates can provide excellent weed control with such varieties.[18]

Several phytotoxic compounds are found in root exudates of allelopathic rice varieties. They include momilactone B[19,20]; glucosides of two resorcinols,

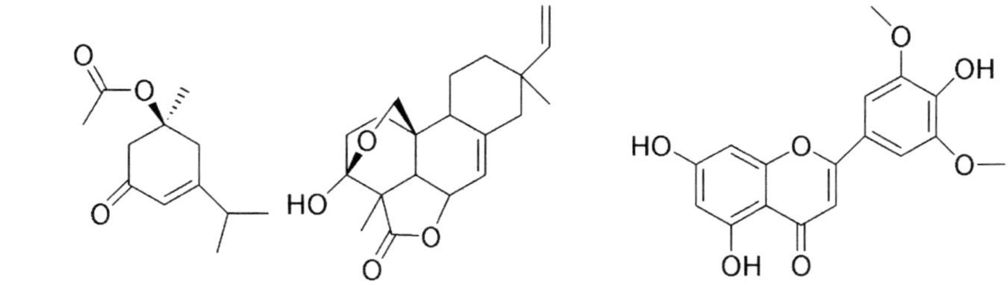

2-Isopropyl-5-acetoxy-cyclohexene-2-one-1   Momilactone B   3,7,4-Trihydroxy-3',5'-dimethoxyflavone

Figure 1. Structures of some of rice allelochemicals mentioned in the text

a glucoside of a flavone, and glucosides of two benzoxazinoids[9]; and a cyclohexenone[17] (Figure 1). Thus, more than one type of phytotoxin may play a role in fighting weeds in the most allelopathic varieties of rice. Whether these compounds act synergistically has not been determined. Momilactone B is released from roots throughout all growth stages of rice, increasing up until flowering.[21] Another phytotoxic compound, lanast-7,0(11)diene-3α, 15α-diol-3α-D-glucofuranoside, was isolated from rice seed hulls,[22] but no evidence of exudation from roots was presented. Similarly, momilactone A was found to be in higher concentrations than momilactone B in rice hulls and was shown to be more phytotoxic than momilactone B to some weed species.[23] However, without root exudation, this compound could not be very useful in weed management in this annual crop.

The synthesis of two compounds that are phytotoxic to *Echinochloa crus-galli*, a flavone and a cyclohexenone (Figure 1), are induced in rice plants by the presence of this weed.[24] Constitutive versus inducible allelochemical production should be considered when altering allelochemical production of a crop, because induction of allelochemical synthesis by a chemical clue provided by a competing species may be more energy efficient.

Others have developed genetic information related to the allelopathy of rice, such as quantitative trait loci mapping of allelopathic traits,[16] but no direct link between this genetic information to production of any particular allelochemical has yet been demonstrated. To exploit the newfound knowledge of root-exuded allelochemicals in rice, the discovery of the genes involved in the production of the more important compounds is needed. The availability of the complete rice genome should make this feasible in the near future. For example, this genomic information was used by Xu et al.[25] to find that the gene for *syn*-copalyl diphosphate synthase, which plays a regulatory role in the synthesis of the momilactones and structurally related phytoalexins. The laboratory of Reuben Peters is identifying and characterizing the genes and enzymes required for momilactone B biosynthesis in rice.[26] Kato-Noguchi and Ino[27] found momilactone B in eight rice cultivars, suggesting that the genes for production are present in all rice cultivars, with production levels dependent on other factors.

4.2.2. WHEAT

Extensive laboratory evaluations of wheat cultivars for their allelopathic potential have been conducted.[13,14,28] Related wheat species such as *Triticum durum*, *Triticum spelta*, and *Triticum speltoides*, as well as rye, have been less intensively evaluated as possible sources of allelopathic germplasm. Some cultivars of these species have high allelopathic potential.[28–30] Wheat allelochemicals have recently been the focus of a large, multi country chemical ecology effort (the FATEALLCHEM project) funded by the European Commission.[31]

Allelopathy of both wheat and rye has been attributed to root secretions of phytotoxins.[13,14,29] This information has yet to be used in any strategy to genetically improve allelopathy of these crops. Laboratory studies have shown genetic variability of allelopathic properties among wheat cultivars, indicating that breeding for allelopathy may be promising.[13,14] Two major quantitative trait loci associated with allelopathy in wheat have been detected.[32] However, both loci accounted for only a small portion of the phenotypic variation, and whether they are directly linked to any allelochemical involved is unknown. Rigorous studies demonstrating that indications of high allelopathic activity of wheat cultivars in the laboratory translate to significant allelopathy under field conditions have not been published.

The allelopathic effects of wheat cannot be accounted for by a single chemical class of allelochemicals. Allelopathic effects are apparently due to a fluctuating mixture of two categories of phytotoxins, phenolic acids and natural benzoxazinoids, whose contribution may vary according to genotype, developmental stage, and environmental factors. Seven phenolic acids (e.g., *p*-hydroxybenzoic acid, *trans*-ferulic acid, vanillic acid) and two benzoxazinoids (2,4-dihydroxy-2*H*-benzoxazin-3(4*H*)-one (DIBOA) and 2,4-dihydroxy-7-methoxy-2*H*-benzoxazin-3(4*H*)-one (DIMBOA)) (Figure 2) are reliable biochemical markers for the allelopathic potential of wheat cultivars in bioassays, with an estimated contribution to overall allelopathy of more than 90%.[28,33,34]

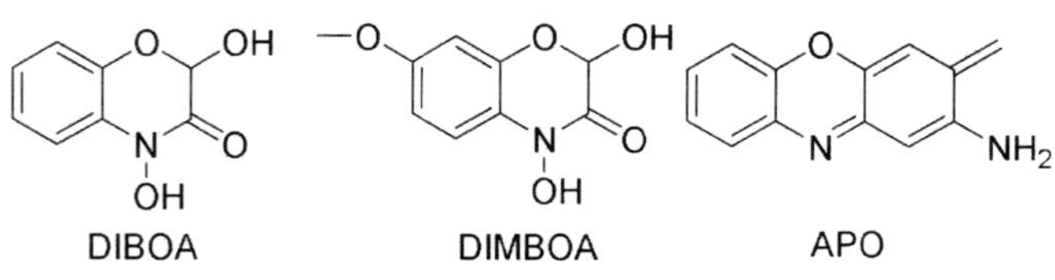

DIBOA DIMBOA APO

Figure 2. Structures of some of wheat and rye allelochemicals mentioned in the text

These weakly phytotoxic compounds apparently do not act synergistically as phytotoxins in binary nor ternary mixtures.[35] Although synergism as frequently been invoked to explain significant effects of mixtures of weak phytotoxins in the allelopathy literature, it has never been proven in properly conducted studies. However, proper studies have not been conducted compounds likely to have different modes of action. Additive and antagonistic interactions have been reported in properly conducted studies of mixtures of similar compounds.

Several laboratory findings suggest that allelopathy in wheat and rye is related to root exudation of benzoxazinoids. Recent studies indicate that the contribution of these unstable compounds to allelopathy in the soil may be facilitated by their conversion to more phytotoxoic metabolites by soil microbes. For example, DIBOA is converted to 2-aminophneoxazine-3-one (APO) (Figure 2) in soil.[36] APO is much more phytotoxic than DIBOA.[37]

The gene sequences encoding the five homologous enzymes for the biosynthesis of DIBOA, beginning with indole-3-glycerol phosphate, in wheat and rye are largely identified.[38,39] The relatively short biosynthetic pathway of the benzoxazinoids should facilitate genetic engineering.[40]

4.3. Transgenic Approaches

Transgenic technologies with crops are providing alternatives to synthetic insecticides and antimicrobials, but have not been used to provide an alternative to herbicides.

Enhancement of allelopathy in crops via transgene technology may be a viable approach to managing weeds with reduced levels of synthetic herbicides. Ideally, the allelochemical should be highly potent and be produced and exuded by roots only. A potent compound produced by a limited portion of the plant would use fewer of the resources of the plant. Likewise, induction of synthesis or of increased synthesis of the allelochemical by the presence of weed species should further reduce the metabolic cost to the crop.

We chose sorghum to try this approach. Many species of *Sorghum* produce a group of root-exuded hydrophobic compounds, cumulatively called sorgoleone. Sorgoleone was identified as the leading source of the allelopathy properties of sorghums.[41] The term sorgoleone is also used to specifically describe 2-hydroxy-5-methoxy-3-[(8′Z, 11′Z)-8′, 11′, 14′ pentadecatriene]-*p*-benzoquinone (structure shown in Figure 3), the most abundant metabolite in this hydrophobic exudate.

The biosynthesis of sorgoleone involves the convergence of the fatty acid and polyketide pathways (Figure 3).[42,43] The hydrophobic tail is derived from

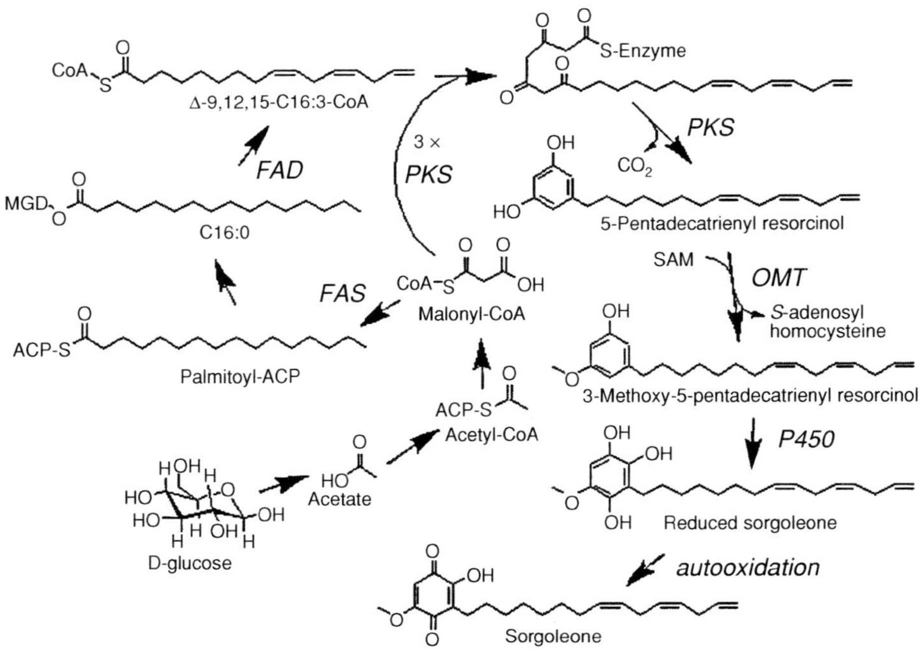

Figure 3. Biosynthetic pathway of sorgoleone

a 16:3 fatty acid intermediate synthesized by the combined action of fatty acid synthase and desaturases. The ring is derived from action of a polyketide synthase that produces 5-pentadecatriene resorcinol as an intermediate. This lipid resorcinol intermediate is then methylated by a SAM-dependent *O*-methyltransferase and dihydroxylated by a P450 monooxygenase to yield the reduced form of sorgoleone. The reduced form of the molecule is probably oxidized to the active quinone after being secreted by the root hair.

Sorgoleone is secreted only from root hairs in droplets[44] containing 90% sorgoleone and its 1,4-hydroquinone form. The droplets also contain several minor congeners varying in the substitutions in the aromatic ring, and/or in number of carbon and the level of unsaturation in the tail.[42,45,46] All of these compounds appear to be derived from the same biosynthetic pathway and contribute to the overall allelopathic potential of sorghum.[45]

Sorgoleone inhibits growth of many weeds.[41,46,47] It is a strong inhibitor of in vitro PSII activity.[46,48,49] Sorgoleone also inhibits mitochondrial functions[50] as well as the enzyme *p*-hydroxyphenylpyruvate dioxygenase (HPPD).[51] Having multiple modes of action is desirable from the standpoint of slowing evolution of resistance in target species. Research is underway by our group to determine the how each of these target sites contribute to the mode of action of sorgoleone in whole plants.

Extensive genomic resources such as those developed for rice are unlikely to become available for other allelopathic crops such as sorghum, rye, and wheat in the near future. Thus, researchers working with these species will have to generate sequence data to meet specific objectives. Expressed sequence tag (EST) analysis, the generation of single-pass DNA sequence data sets from randomly selected cDNA library clones has recently emerged as a highly effective approach for identifying genes involved in secondary metabolic pathways, particularly in cases where the pathway of interest is highly expressed and restricted to a specific cell type or developmental stage.[52,53] EST analysis has also proven useful for identifying genes potentially involved in the biosynthesis of the allelochemical sorgoleone,[54] which, due to its high levels of biosynthesis, specifically in root hair cells of sorghum,[42] is well-suited to this approach. An annotated sorghum EST data set containing approximately 5,500 sequences generated from a root hair-specific cDNA library was analyzed.

Highly expressed candidate sequences were found representing all of the enzymes expected to be involved in the final steps of sorgoleone biosynthesis. Functional analysis of some of these genes has led to the characterization of a resorcinol-specific fatty acid desaturases, *O*-methyltransferases, and polyketide synthases likely to be involved in sorgoleone biosynthesis.[54] Upon completion of the characterization of the genes and their products involved in sorgoleone synthesis, manipulation of the pathway in sorghum, or transferring all or part of the pathway to selected sorghum cultivars may result in crops with enhanced allelopathy. Rice has a 5-heptadenyl resorcinol pathway that would require only the last two enzymes of the sorgoleone pathway to produce a compound identical to sorgoleone, except for the tail length and desaturation pattern.[55] Such a compound is likely to have similar biological activity to sorgoleone, as small variations in the tail of sorgoleone-type compounds produced by sorghum have little influence on biological activity.[45,46]

DNA microarrays represent another potentially important tool for gene discovery research in the field of allelopathy. Starting with only knowledge about the pattern of accumulation for a given allelochemical, correlations with the expression patterns of specific genes can quickly be discerned, thus narrowing the list of candidate enzyme sequences required for subsequent biochemical screening. This approach has been successfully applied in both plant and non-plant systems for the identification of genes encoding metabolic enzymes involved in various pathways.[56,57] The recent commercial release of DNA microarrays for rice and wheat should accelerate discovery efforts for genes involved in allelochemical biosynthesis for these two species.

The potential environmental and social benefits of success in creating highly allelopathic crops are great. However, crops with enhanced allelopathy via molecular breeding or by transgenes present potential environmental and toxicological hazards that must be studied and evaluated. New weed problems could be created by gene flow to weedy relatives or by the crop itself in a feral form. Fail-safe methods to eliminate gene flow could mitigate the first of these potential problems.[58] If managed carefully, we believe that the benefits of such crops would substantially outweigh the risks.

References

1. J. L. Harper, *Population Biology in Plants* (Academic Press, London, 1977).
2. L. A. Weston, Utilization of allelopathy for weed management in agroecosystems, *Agron. J.* 88, 860–866 (1996).
3. L. A. Weston and S. O. Duke, Weed and crop allelopathy, *Crit. Rev. Plant Sci.* 22, 367–389 (2003).
4. S. O. Duke, R. G. Belz, S. R. Baerson, Z. Pan, D. D. Cook, and F. E. Dayan, The potential for advances in crop allelopathy. *Outlook Pest Manag.* 16, 64–68 (2005).
5. J. V. Lovett, A. H. C. Hoult, and O. Christen, Biologically active secondary metabolites of barley, IV: Hordenine production by different barley lines. *J. Chem. Ecol.* 20, 1945–1954 (1994).
6. A. R. Putnam and W. B. Duke, Biological suppression of weeds: Evidence for allelopathy in accessions of cucumber. *Science* 185, 370–373 (1974).
7. P. K. Fay and W. B. Duke, An assessment of allelopathic potential in *Avena* germplasm. *Weed Sci.* 25, 224–228 (1977).
8. R. H. Dilday, J. D. Mattice, K. A. Moldenhauer, and W. Yan, Allelopathic potential in rice germplasm against ducksalad, redstem and barnyardgrass, *J. Crop Prod.* 4, 287–301 (2001).
9. C. Kong, X. Xu, F. Hu, X. Chen, B. Ling, and Z. Tan, Using specific secondary metabolites as markers to evaluate allelopathic potentials of rice varieties and individual plants. *Chin. Sci. Bull.* 47, 839–843 (2002).
10. F. J. Pérez and J. Ormeño-Núñez, Difference in hydroxyamic acid content in roots and root exudates of wheat (*Triticum aestivum* L.) and rye (*Secale cereale* L.): Possible role in allelopathy. *J. Chem. Ecol.* 17, 1037–1043 (1991).
11. M. A. Czarnota, A. M. Rimando, and L. A. Weston, Evaluation of root exudates of seven sorghum accessions, *J. Chem. Ecol.* 29, 2073–2083 (2003).
12. F. A. Macías, R. M. Oliva, R. M. Varela, A. Torres, and J. M. G. Molinollo, Allelopathic studies in cultivar species, 14: Allelochemicals from sunflower leaves cv. Peredovick. *Phytochemistry* 52, 613–621 (1999).
13. H. Wu, J. Pratley, D. Lemerle, and T. Haig, Evaluation of seedling allelopathy in 453 wheat (*Triticum aestivum*) accessions against annual ryegrass (*Lolium rigidum*) by the equal-compartment-agar-method. *Aust. J. Exp. Agric.* 51, 937–944 (2000).
14. H. Wu, J. Pratley, D. Lemerle, and T. Haig, Laboratory screening for allelopathic potential of wheat (*Triticum aestivum*) accessions agains annual ryegrass (*Lolium rigidum*). *Aust. J. Exp. Agric.* 51, 259–266 (2000).
15. M. Olofsdotter (Ed.), *Allelopathy in Rice* (International Rice Research Institute, Manila, Philippines, 1998).
16. L. B. Jenson, B. Courtois, L. Shen, Z. Li, M. Olofsdotter, and R. P. Mauleon, Locating genes controlling allelopathic effects against barnyardgrass in upland rice, *Agron. J.* 93, 21–26 (2001).

17. C. Kong, W. Liang, X. Xu, F. Hu, and Y. Jiang, Release and activity of allelochemicals from allelopathic rice seedlings. *J. Agric Food Chem.* 19, 2861–2865 (2004).
18. R. S. C. Chavez, D. R. Gealy, and H. L. Black, Reduced propanil rates and naturally suppressive cultivars for barnyardgrass control in drill-seeded rice. In B. R. Wells Rice Res. Studies–1998. Series 468 (Arkansas Agricultural Experimental Station, University of Arkansas, Fayetteville, AR, USA, 1999), pp. 43–50.
19. H. Kato-Noguchi and T. Ino, Release of momilactone B from rice plants. *Plant Product Sci.* 7, 189–190 (2004)
20. H. Kato-Noguchi, Allelopathic substance in rice root exudates: Rediscovery of momilactone B as an allelochemical, *J. Plant Physiol.* 161, 271–276 (2004).
21. H. Kato-Noguchi, T. Ino, and M. Ichii, Changes in release of momilactone B into the environment from rice throughout its life cycle, *Funct. Plant Biol.* 30, 995–997 (2003).
22. I. M. Chung, M. Ali, A. Ahmad, J. D. Lim, C. Y. Yu, and J. S. Kim, Chemical constituents of rice (*Oryza sativa*) hulls and their herbicidal activity against duckweed (*Lemna paucicostata* Hegelm 381), *Phytochem. Anal.* 17, 36–45 (2006).
23. I. M. Chung, J. T. Jung, and S.-H. Kim, Evaluation of allelochemical potential and quantification of momilacton A, B from rice hull extracts and assessment of inhibitory bioactivity on paddy field weeds, *J. Agric. Food Chem.* 54, 2527–2536 (2006).
24. C. Kong, X. Xu, B. Zhou, F. Hu, and C. Zhang, Two compounds from allelopathic rice asccession and their inhibitory activity on weeds and fungal pathogens, *Phytochemistry* 65, 1123–1128 (2004).
25. M. Xu, M. L. Hillwig, S. Prisic, R. M. Coates, and R. J. Peters, Functional identification of rice *syn*-copalyl diphosphate synthase and its role in initiating biosynthesis of diterpenoid phytoalexin/allelopathic products. *Plant J.* 39, 309–318
26. M. Xu, S. Prisic, P. R. Wilderman, Y. Jin, R. M. Coates, and R. J. Peters, Elucidating biosynthesis of the rice allelochemical/phytoalexin momilacton B, in *Proceedings of the 4th World Congress on Allellopathy* (Regional Institute Ltd., Gosford, Australia), pp. 218–222 (2005).
27. H. Kato-Noguchi and T. Ino, Concentration and release level of momilacton B in the seedlings of eight rice cultivars, *J. Plant Physiol.* 162, 965–969 (2005).
28. R. G. Belz and K. Hurle, Differential exudation of two benzoxazinoids—One of the determining factors for seedling allelopathy of Triticeae species, *J. Agric. Food Chem.* 53, 250–261 (2005).
29. F. J Pérez and J. Ormeño-Núñez, Difference in hydroxamic acid content in roots and root exudates of wheat (*Triticum aestivum* L.) and rye (*Secale cereale* L.): Possible role in allelopathy, *J. Chem. Ecol.* 17, 1037–1043 (1991).
30. M. Quader, G. Daggard, R. Barrow, S. Walker, and M.W. Sutherland, Allelopathy, DIMBOA production and genetic variability in accessions of *Triticum speltoides*, *J. Chem. Ecol.* 27, 747–760 (2001).
31. I. S. Fomsgaard, Chemical ecology in wheat plant—Pest interactions. How the use of modern techniques and a multidisciplinary approach can throw new light on a well-known phenomenon: Allelopathy, *J. Agric. Food Chem.* 54, 987–990 (2006).
32. H. Wu, J. Pratley, W. Ma, and T. Haig, Quantitative trait loci and molecular markers associated with wheat allelopathy, *Theor. Appl. Genet.* 107, 1477–1481 (2003).
33. H. Wu, J. Pratley, D. Lemerle, and M. An, Biochemical basis for wheat seedling allelopathy on the suppression of annual ryegrass, (*Lolium rigidum*), *J. Agric. Food Chem.* 50, 4567–4571 (2002).
34. Z. Huang, T. Haig, H. Wu, M. An, and J. Pratley, Correlation between phytotoxicity on annual ryegrass (*Lolium rigidum*) and production dynamics of allelochemicals within root exudates of an allelopathic wheat, *J. Chem. Ecol.* 29, 2263–2279 (2003).

35. J. Chunghong, P. Kudsk, and S. K. Mathiassen, Joint action of benzoxazinone derivatives and phenolic acids, *J. Agric. Food Chem.* 54, 1049–1057 (2006).
36. R. W. Gagliardo and W. S. Chilton, Soil transformation of 2(3*H*)-benzoxazolone of rye into phytotoxic 2-amino-3*H*-phenoxazin-3-one, *J. Chem. Ecol.* 18, 1683–1691 (1992).
37. F. A. Macías, D. Marín, A. Oliveros-Bastidas, D. Castellano, A. M. Simonet, and J. M. G. Molinollo, Structure-activity relationship (SAR studies of benzazinones, their degradation products, and analogues). Phytoxicity on problematic weeds *Avena fatua* L. and *Lolium regidum* Gaud., *J. Agric. Food Chem.* 54, 1040–1048 (2006).
38. T. Nomura, A. Ishihara, H. Imaishi, T. R. Endo, H. Ohkawa, and H. Iwamura, Molecular characterization and chromosomal location of cytochrome P450 genes involved in the biosythesis of cyclic hydroxyamic acids in hexaploid wheat, *Molec. Genet. Genomics* 267, 210–217.
39. T. Nomura, A. Ishihara, H. Imaishi, H. Ohkawa, T. R. Endo, and H. Iwamura, Rearrangement of the genes for the biosynthesis of benzoxazinones in the evolution of Triticeae species, *Planta* 217, 776–782.
40. S. O. Duke, S. R. Baerson, F. E. Dayan, I. A. Kagan, A. Michel, and B. E. Scheffler, Biocontrol of weeds without the biocontrol agent, in *Enhancing Biocontrol Agents and Handling Risks*, edited by M. Vurro, J. Gressel, T. Butt, G. E. Harmon, A. Pilgeram, R. J. St. Leger, and D. L. Nuss (IOS Press, Amsterdam, 2001), pp. 96–105.
41. D. H. Netzly and L. G. Butler, Roots of sorghum exude hydrophobic droplets containing biologically active components. *Crop Sci.* 26, 775–778 (1986).
42. G. D. Fate and D. G. Lynn, Xenognosin methylation is critical in defining the chemical potential gradient that regulates the spatial distribution in striga pathogenesis. *J. Am. Chem. Soc.* 118, 11369–11376 (1996).
43. F. E. Dayan, I. A. Kagan, and A. M. Rimando, Elucidation of the biosynthetic pathway of the allelochemical sorgoleone using retrobiosynthetic NMR analysis, *J. Biol. Chem.* 278, 28607–28611 (2003)
44. M. A. Czarnota, R. N. Paul, L. A. Weston, and S. O. Duke, Anatomy of sorgoleone-secreting root hairs of *Sorghum* species. *Int. J. Plant Sci.* 164, 861–866 (2003)
45. I. A. Kagan, A. M. Rimando, and F. E. Dayan, Chromatographic separation and in vitro activity of sorgoleone congeners from the roots of *Sorghum bicolor*, *J. Agric. Food Chem.* 51, 7589–7595.
46. A. M. Rimando, F. E. Dayan, M. A. Czarnota, L. A. Weston, and S. O. Duke, A new photosystem II electron transfer inhibitor from *Sorghum bicolor*, *J. Nat. Prod.* 61, 972–930 (1998).
47. F. A. Einhellig and I. F. Souza, Phytotoxicity of sorgoleone found in grain sorghum root exudates, *J. Chem. Ecol.* 18, 1–11 (1992).
48. F. A. Einhellig, J. A. Rasmussen, A. M. Hejl, and I. F. Souza, Effects of root exudate sorgoleone on photosynthesis, *J. Chem.Ecol.* 19, 369–375 (1993).
49. V. M. Gonzalez, J. Kazmir, C. Nimbal, L. A. Weston, and G. M. Cheniae, Inhibition of photosystem II electron transfer reaction by the natural product sorgoleone, *J. Agric. Food Chem.* 45, 1415–1421 (1997).
50. J. A. Rasmussen, A. M. Hejl, F. A. Einhellig, and J. A. Thomas, Sorgoleone from root exudates inhibits mitochondrial functions, *J. Chem. Ecol.* 18, 197–207 (1992).
51. G. Meazza, B. E. Scheffler, M. R. Tellez, A. M. Rimando, N. P. D. Nanayakkara, I. A. Khan, E. A. Abourashed, J. G. Romagni, S. O. Duke, and F. E. Dayan, The inhibitory activity of natural products on plant *p*-hydroxyphenylpyruvate dioxygenase, *Phytochemistry* 59, 281–288 (2002).

52. I. Guterman, M. Shalit, N. Menda, D. Piestun, M. Dafny-Yelin, G. Shalev, E. Bar, O. Davydov, M. Ovadis, M. Emanuel, J. Wang, Z. Adam, E. Pichersky, E. Lewinsohn, D. Zamir, A. Vainstein, and D. Weiss, Rose scent genomics approach to discovering novel floral fragrance-related genes, *Plant Cell* 14, 2325–2338 2002).
53. B. M. Lange, M. R. Wildung, E. J. Stauber, C. Sanchez, D. Pouchnik, and R. Croteau, Probing essential oil biosynthesis and secretion by functional evaluation of expressed sequesnce tags from mint trichomes, *Proc. Natl. Acad. Sci. USA* 97, 2934–2939.
54. S. R. Baerson, F. E. Dayan, A. M. Rimando, Z. Pan, D. Cook, N. P. D. Nanayakkara, and S. O. Duke, A functional genomics approach for the identification of genes involved in the bioysnthesis of the allelochemical sorgoleone, *Am. Chem. Soc. Symp. Ser.* 927, 265–276 (2006).
55. Dayan, F. E., D. Cook, S. R. Baerson, and A. M. Rimando, Manipulating the lipid resorcinol pathway to enhance allelopathy in rice, in *Proceedings of the 4th World Congress on Allelopathy* (Regional Institute Ltd., Gosford, Australia, 2005), pp. 96–105.
56. P. Mercke, I. F. Kappers, F. W. Verstappen, O. Vorst, M. Dicke, and H. J. Bouwmeester, Combined transcript and metabolite analysis reveals genes involved in spider mite induced volatile formation in cucumber plants. *Plant Physiol.* 135, 2012–2024 (2004).
57. R. Niwa, T. Matsuda, T. Yoshiyama, T. Namiki, K. Mita, Y. Fujimoto, and H. Kataoka, CYP306A1, a cytochrome P450 enzyme, is essential for ecdysteroid biosynthesis in the prothoracic glands of *Bombyx* and *Drosophila*. *J. Biol. Chem.* 279, 35942–35949 (2004)
58. J. Gressel, *Molecular Biology of Weed Control* (Taylor & Francis, London, 2002), 504 pp.

5. SELECTING, MONITORING, AND ENHANCING THE PERFORMANCE OF BACTERIAL BIOCONTROL AGENTS: PRINCIPLES, PITFALLS, AND PROGRESS

Linda S. Thomashow,[1*] David M. Weller,[1] Olga V. Mavrodi,[2] and Dmitri V. Mavrodi[2]
[1] *USDA-ARS, Root Disease and Biological Control Research Unit, Pullman, WA, USA*
[2] *Department of Plant Pathology, Washington State University, Pullman, WA 99164-6430, USA*

Abstract. Genetic resistance to root diseases of plants is rare, and agriculture controls these diseases through practices such as crop rotation and soil fumigation. However, plants have evolved a strategy of stimulating and supporting specific groups of antagonistic rhizosphere microorganisms as a defense against diseases caused by soilborne pathogens. Antibiotic production has a significant role in plant defense by many of these rhizobacteria. Information now is available about the genetics, biochemistry, and regulation of synthesis of some of the most commonly-produced antibiotics. Similarly, many genes that contribute to the ability of these bacteria to colonize roots have been identified. Studies of naturally suppressive soils have provided evidence of preferential interactions between plant hosts and protective populations, revealing the existence of functional diversity among otherwise almost indistinguishable strains. Here, we consider how this knowledge can be applied to aid in the selection of more effective biological control agents and the development of recombinant strains that may overcome impediments to inoculum preparation, formulation, and cost that currently limit commercial acceptance of highly promising candidate strains.

Keywords: antibiotics; 2,4-diacetylphloroglucinol; *Pseudomonas;* rhizobacteria; real-time PCR; root colonization

5.1. Introduction

Whereas genetic resistance has long been the method of choice for the control of foliar plant diseases, resistance to common soilborne pathogens such as

*To whom correspondence should be addressed, e-mail: Thomashow@wsu.edu

M. Vurro and J. Gressel (eds.), Novel Biotechnologies for Biocontrol Agent Enhancement and Management, 87–105.

species of *Pythium*, *Rhizoctonia*, and *Fusarium* has remained elusive. These pathogens typically are controlled in agricultural systems through practices including tillage, crop rotation, and the use of chemical pesticides. In undisturbed ecosystems, plants depend on a more ancient mechanism provided by root-colonizing microorganisms supported by rhizodeposition, the release of organic materials from roots as they grow through soil. These microorganisms collectively provide a basal level of biological buffering and general disease suppression due to their metabolic activity. However, many rhizosphere isolates can actively antagonize soilborne pathogens, and this ability has been a major driving force in bacterial biological control research over the past 50 years. Hundreds, if not thousands, of antagonistic strains have been described, representing diverse genera and with the potential for use as bacterial biological control agents. Remarkably few of these actually have attained commercial status. Here, we draw from studies on fluorescent *Pseudomonas* species to illustrate lessons learned, pitfalls revealed, and progress towards exploiting these underutilized bacteria to better serve the needs of contemporary agriculture.

5.2. The Quest for New Agents—Brute Force or a More Targeted Approach?

The search for new biocontrol agents typically begins with screening procedures that are inherently laborious because effective isolates are only a minor component of the large and phylogenetically diverse populations indigenous to soil and the rhizosphere. While introduced strains must both compete as rhizosphere colonizers and suppress disease, the importance of colonization depends in part on the amount and mode of delivery of the biocontrol inoculum and the duration required for protection. Many strains have the capacity to colonize roots for days or weeks in relatively undisturbed systems, with genes related to such diverse properties as bacterial cell surface structures (flagella, fimbriae, and lipopolysaccharides), catabolic activity, and global regulation of gene expression[1] implicated in the process. Considering the redistribution of rhizobacteria that occurs after watering,[2] however, it is likely that for relatively short-term threats such as those associated with seedling pre- and post-emergence damping-off diseases, inundative applications of highly antagonistic strains may eliminate the need for aggressive colonization and persistence on the roots. Forage and field crops, in contrast, may require more sustained protection. In these cases it is important to consider the source and method of strain selection as well as the factors that contribute to the establishment and long-term maintenance of biocontrol populations.

5.2.1. SELECTION BASED ON ROOT COLONIZATION

It generally is thought that biocontrol agents will be adapted to the pathosystem or environment from which they were obtained. Thus, candidate strains often are isolated from the intended area of use. Such soils contain far too many rhizobacteria to evaluate every isolate in the greenhouse or field, however, and those chosen at random, whether or not they produce antagonistic metabolites, seldom have proven effective in practice. In contrast, exceptionally competitive rhizosphere colonizers were recovered from roots when long-term monoculture or suppressive soils first were subjected to a process of selective enrichment in which the monocultured crop was sown for several consecutive cycles of growth.[3–5] These strains consistently established protective population densities in excess of 10^5 per gram on roots of the crop from which they originally were isolated. They persisted over extended periods of time, out-competing strains recovered by cycling other crops and revealing the existence of functional diversity among very closely related strains.[4–6] All of the strains in these cycling studies produce the antibiotic 2,4-diacetylphloroglucinol (DAPG). While DAPG itself does not appear to be responsible for strain competitiveness,[7] restriction fragment length polymorphisms (RFLPs) within the DAPG biosynthetic locus are predictive of the rhizosphere competence of a strain on at least some hosts. These RFLPs are correlated with broader differences among strains revealed by a variety of genomic fingerprinting techniques,[8] enabling isolates to be assigned to one of 22 currently recognized genotypes. They serve as a convenient indicator and the first genetic marker associated with a capacity for extended root colonization. More generally, the results from these and other studies[9–13] support the concept that plants select over time for bacterial populations adapted to their particular rhizosphere conditions. This highlights the need to identify and understand not only the microbial factors, but also those of the host, that contribute to the establishment and long-term maintenance of protective rhizosphere populations.

Enrichment also is the basis for the selection of enhanced colonizers recovered from root tips after cycling tomato or cucumber in sterile quartz sand[14] or stonewool[15] (rockwool), a popular industrial substrate for hydroponically-grown vegetables. The latter procedure is thought to select for strains that utilize citrate, a major carbon source in root exudates,[16] and to suppress tomato foot and root rot caused by *Fusarium oxysporum* f. sp. *radidicis-lyscopersici* by competition for nutrients and niches, rather than through the production of antifungal metabolites.[14,15] It remains to be determined whether strains selected in this way would be protective in the physically, chemically, and microbiologically more challenging environment of natural soil, and if differences in exudate composition and catabolism can account for the preferential interactions between some strains and host crops. However, it must noted that few differences have been found among the carbon utilization profiles of genetically-distinct isolates of DAPG producers that differ markedly in their colonization properties and affinity for host crops.[3]

5.2.2. SELECTION BASED ON INHIBITORY ACTIVITY IN VITRO

There is no general relationship between the ability of a rhizosphere isolate to inhibit a pathogen in vitro and to suppress disease caused by that pathogen on roots.[17] Nonetheless, inhibition in vitro often is used as a first screen of potential biocontrol agents, even though it eliminates nonproducing candidates that may control pathogens via mechanisms such as niche exclusion[14,15] or induced systemic resistance.[18] Pathogen inhibition in vitro results from the synthesis of one or, more likely, a mixture of metabolites and is strongly influenced by cultural conditions. This may account for the failure of some antagonistic strains to function effectively in the rhizosphere.

5.2.3. MOLECULAR SCREENING—A NEW APPROACH TO STRAIN SELECTION?

Considerable research in recent years has focused on four antibiotics frequently produced by antagonistic fluorescent *Pseudomonas* spp. isolated from the rhizosphere. These low molecular weight organic compounds typically include phenazines, DAPG, pyrrolnitrin, and pyoluteorin. They have broad-spectrum activity against plant pathogens and are produced by diverse strains isolated from the roots of a wide variety of crops grown worldwide.[19] They are fairly simple in structure and *Pseudomonas* spp. are amenable to genetic manipulation. Thus, much is now known about the genetics, biochemistry, and regulation of synthesis of these compounds. The biosynthesis genes are chromosomally encoded, well-conserved, and mostly clustered into operons. Not surprisingly, their transcription and translation are controlled by global mechanisms responsible for the overall response of microorganisms to their environment as well as by specific, genetically linked regulatory elements.[20] The biosynthesis operons responsible for the synthesis of these antibiotics have been cloned and sequenced, opening new possibilities for directed strain selection and enhancement of activity. Molecular identification of locally-adapted strains capable of producing metabolites of known efficacy can streamline screening approaches. It also eliminates the possibility that such strains will be overlooked in inhibition assays conducted on media or under conditions unfavorable for antibiotic production.

The probes and primers[21–23] available for the major antibiotics produced by fluorescent *Pseudomonas* spp. have in most cases been used to confirm the presence of known antibiotic biosynthesis genes in rhizosphere isolates that exhibit antifungal activity in vitro. However, techniques are readily available for applying the same molecular tools to rapidly screen locally-adapted populations for strains with the capacity to produce these well-characterized compounds. Thus, Raaijmakers et al.[21] determined the frequency of phenazine- and phloroglucinol-producing fluorescent pseudomonads on roots of wheat grown in seven natural soils by colony hybridization followed by confirmatory PCR with primers specific for the biosynthesis genes. de Souza and Raaijmakers[23] used the same approach to show that *Pseudomonas* and *Burkholderia* spp. harboring pyrrolnitrin and pyoluteorin genes were not present at detectable levels in five Dutch agricultural soils. The use of dilute plating media was important in these assays to reduce the colony size and the amount of polysaccharide produced, thereby enhancing the effectiveness of the hybridization technique. Likewise, gene-specific primers can be used in PCR reactions to screen for the presence of target antibiotic genes in pools of 100 or more isolates prior to localizing a positive signal or inhibitory activity to an individual strain. This approach has not been applied to mixed populations obtained by dilution-plating soil or rhizosphere samples, but it is useful in screening large genomic libraries for target genes.[24]

The detection of biosynthesis genes indicates that an isolate has the potential to produce antibiotics, but synthesis must be confirmed to rule out regulatory or other mutations that occur at low frequency in natural populations and render strains ineffective. The simple demonstration of antagonistic activity in vitro is not sufficiently specific for this purpose but a variety of chromatographic techniques[25] are available for the detection of these antibiotics.

5.3. Field Performance: More Than Just a Numbers Game

5.3.1. CULTURE-DEPENDENT QUANTIFICATION OF RHIZOSPHERE POPULATIONS

The rhizosphere population density of introduced *Pseudomonas* strains is an important determinant of their ability to suppress disease. Threshold population densities of approximately 10^5 CFU per gram of root are necessary for significant disease control by introduced agents expressing mechanisms ranging from siderophore-mediated competition for iron(III) and induced systemic resistance[26] to antifungal activity mediated by DAPG.[27] Introduced strains traditionally are quantified by dilution plating after having been made antibiotic-resistant to facilitate recovery from soil or rhizosphere samples. This approach is labor-intensive and inadequate, even with the use of semi-selective media,[28] to quantify functionally distinct communities within indigenous populations of *Pseudomonas*. Raaijmakers et al.[21,27] overcame these limitations by using colony hybridization and confirmatory PCR with antibiotic-specific probes and primers to enumerate indigenous DAPG-producing strains in soils naturally suppressive or conducive to take-all, an important root disease of cereal crops. Later, McSpadden-Gardener et al.[22] developed a PCR-based dilution-endpoint assay for quantifying DAPG producers that includes an enrichment step comprised of incubating serially-diluted root washes in media selective for fluorescent pseudomonads. PCR is used to detect the DAPG biosynthesis gene, and analysis of restriction fragment polymorphisms in the PCR product enables determination of the genotype of the dominant DAPG producer in the population. Dilution plating, colony hybridization, and the PCR-based dilution-endpoint assay with the enrichment step all are suitable for monitoring the population dynamics of antibiotic-producing *Pseudomonas* strains introduced into the rhizosphere. All three detect similar population densities, but the latter method allows much more rapid sample processing and is less sensitive to operator error.[29] Primers can be developed to detect almost any target gene or even alleles of a target, as recently was done to quantify individual strains in mixed populations of DAPG producers in the rhizosphere of wheat.[30]

5.3.2. CULTURE-INDEPENDENT QUANTIFICATION OF RHIZOSPHERE POPULATIONS: REAL-TIME PCR

The practical detection limit for the PCR-based dilution-endpoint method is $\geq \log 3.1$ cells per rhizosphere after the initial round of selective enrichment but declines to about log 5.6 cells per rhizosphere without enrichment,[22] making it unsuitable for quantifying *Pseudomonas* strains directly from soil or rhizosphere samples. This dependence on culturing extends the turnaround time for assays to about 5 days and introduces a degree of uncertainty as to whether the populations detected after enrichment accurately reflect the *Pseudomonas* community structure in situ because of the potential for inhibitory interactions among strains during growth.[31] Further, when strains of different genotypes are present in the same sample, the dominant genotype is readily detected but population sizes of subdominant genotypes are difficult to estimate. We have developed a culture-independent quantitative real-time PCR technique[32] to overcome these limitations. It has a detection limit comparable to those of culture-based methods, is capable of detecting both introduced and indigenous strains of DAPG producers, and of distinguishing among genotypes of these strains. It reduces the turnaround time of assays to about 2 days.

Both real-time and standard PCR depend on the same principles governing sensitivity, specificity, and primer design. However, data collection and analysis occur in real-time PCR as the reaction proceeds in the instrument, making the technique much faster and less prone to contamination than standard PCR methods requiring post-PCR processing such as gel electrophoresis. Amplification in real-time PCR is detected as an increase in fluorescence emitted by a dye present either in the reaction mix or incorporated into a primer or probe. Regardless of how it is introduced and detected (which are largely determined by the thermocycler itself), this dye will fluoresce above a background level only after amplification has resulted in de novo synthesis of double-stranded DNA. The method is inherently quantitative because the cycle with the first significant increase in fluorescence above the background (Ct, the threshold cycle) is correlated with the initial amount of target template. For measurements to be meaningful, the reactions must be highly optimized with regard to amplification conditions (annealing temperature and $MgCl_2$ concentration), amplification efficiency, and primer concentration and specificity. Standard curves must be developed over a range of DNA concentrations and depending on the thermocycler and software, DNA concentration standards may need to be included in each run with unknown samples. In addition, the genome size and copy number of the template gene must be known in order to relate template DNA concentration to the population size of the bacteria of interest. We estimated a genome size of approximately 7 Mb, comparable to the recently sequenced[33] genome of *P. fluorescens* Pf-5 for our *P. fluorescens* strains, which contain a single copy of the DAPG biosynthesis operon.

The real-time system we use detects fluorescence emitted from SYBR green, which increases as the dye, initially present in the reaction mix, is bound to the accumulating double-stranded DNA amplification product. Emission also will occur upon binding to non-specific amplification products, however, and must be minimized through primer design and optimization of annealing conditions. In addition, the melting temperature and a melting curve must be determined empirically for each target amplicon in order to distinguish it from non-specific amplification products. Melting curve analysis of unknown samples is qualitatively analogous to the analysis of reaction products from a standard PCR reaction by gel electrophoresis in that aberrant melting profiles and the appearance of unexpected bands both are indicative of non-specific amplification. The two differ, however, in that melting curve analysis prevents overestimation of the DNA concentration and hence, the population size of a target organism in a sample. This is not the case in standard PCR.

The procedure for recovering DNA from bulk soil or rhizosphere samples also must be optimized. Recoveries may vary for rhizosphere and soil samples differing in their physical and chemical properties, requiring that recovery values from each sample matrix be determined separately. In our system, root washes are processed using the UltraClean™ Soil DNA Isolation kit (MO BIO Labs, Carlsbad, California) by a modification of the alternative protocol for wet soil samples. DNA recovery was approximately 10% as determined by adding known amounts of bacteria to wheat roots suspended in a wash solution, shaking, and then extracting DNA from the wash solution.

The goals of our initial studies with real-time PCR were to determine whether strains of different genotypes could be detected in a single rhizosphere sample and to compare the sensitivity of real-time PCR with that of the PCR-based terminal dilution endpoint assay. To these ends we developed primer sets that amplified unique fragments, differing slightly in size and melting temperature, from the *phlD* gene of the DAPG operon in strains of *P. fluorescens* representative of four different genotypes. Following optimization, PCR efficiencies for DNA extracted from inoculated root washes ranged from 80 to 98%, depending on the strain, and the amplification products could readily be distinguished from one another by their melting curves. Detection limits also varied among strains but averaged approximately 1,000 CFU per rhizosphere, approximately the same as detected by the terminal dilution endpoint assay. Side-by-side comparisons of population sizes determined by both methods on roots colonized by each of the four strains also indicated that population densities determined by the real-time PCR assay are comparable to those determined by the end-point dilution assay.[32]

5.3.3. ANTIBIOTIC SYNTHESIS, ACTIVITY, AND DETECTION IN SITU

Factors such as temperature, aeration, and the quantity and quality of minerals and carbon and nitrogen nutrients available strongly affect antibiotic synthesis by fluorescent *Pseudomonas* spp. in vitro. The same is true in the rhizosphere, where antibiotic concentrations are influenced not only by abiotic factors including the soil matrix, but also by the population density of the strain.[34–38] Other biotic factors include the species, cultivar, and age of the host plant[39] and the presence of other microorganisms including pathogens[40–42] and the indigenous microflora.[43–45] Thus, questions arise as to the relationship between the population size of a biocontrol agent, its ability to synthesize antibiotics in situ, and whether population size can be considered indicative of antibiotic synthesis and accumulation in amounts sufficient to suppress pathogens. This issue has been indirectly addressed through the use of reporter gene constructs, and directly, by isolating and quantifying antibiotics from the rhizosphere. The two approaches are complementary; they have different advantages and limitations, and both suffer from the difficulties inherent in working with soil systems.

5.3.3.1. *Antibiotic Gene Expression in Situ*

Transcriptional analyses of antibiotic gene expression are a sensitive and convenient alternative to the isolation and quantification of antibiotics produced in situ, particularly when the objective is to monitor synthesis over time or in response to environmental conditions. Such studies typically employ strains expressing a reporter gene product that can readily be monitored and is not naturally present in the rhizosphere. The reporter gene is placed under transcriptional control of a promoter regulating expression of the antibiotic biosynthesis genes. The speed and sensitivity with which reporters such as the green fluorescent protein *gfp* or the ice nucleation gene *inaZ* can be assayed facilitate the use of samples as small as single seeds or seedlings. This allows sufficient replication to detect significant differences among treatments despite the sample-to-sample variation typical in such studies.[25,46]

Reporter gene expression provides evidence that antibiotic synthesis can occur under prevailing environmental conditions. However, the presence of antibiotics also must also be determined empirically; otherwise, expression levels cannot be considered proportional to the actual amounts of antibiotics present in the rhizosphere. This is partially because transcriptional activity is measured relative to the total population size, but the sampled population is physiologically heterogeneous, having been recovered from a variety of different microhabitats on the roots. A further confounding factor relates to turnover rates of antibiotics and reporter gene products. Antibiotics produced in situ can become biologically unavailable over time[25] as they rapidly adsorb to organic matter and to charged groups on the surface of soil particles. Degradation by the producer strain itself[47] or by the indigenous microflora also may occur. Some reporters, and especially green fluorescent protein, are relatively stable and may more accurately reflect cumulative gene expression than instantaneous transcription rates. Finally, the complex autoregulatory circuitry and posttranscriptional control mechanisms involved in antibiotic synthesis[20,36] and the nature of the reporter gene construct itself[48] can potentially influence the relationship between reporter gene expression and amount of antibiotic actually synthesized. Reporter gene expression was an accurate indicator of antibiotic accumulation in a gnotobiotic system,[39] but it is not known whether this also would be true in studies conducted in natural soils.

5.3.3.2. *Extraction and Analysis: The Direct Approach*

The simple isolation and identification of an antibiotic from the rhizosphere provides incontrovertible evidence that the genetic and physiological potentials for its synthesis have been met. Still, quantitative data are needed to relate the presence of biocontrol strains to antibiotic synthesis and disease suppression in the rhizosphere. The high-performance liquid chromatography systems required to detect and quantify antibiotics in rhizosphere samples have become less expensive and more widely available over the past two decades. Detection limits remain a major limiting factor in the design of most studies. Efficient and reproducible extraction protocols also must be developed, taking into account the chemical and physical properties of a substance and its probable interactions with soil constituents. Analytical methods must be validated with the use of authentic standards, some of which are not readily available through commercial sources. Constraints to detection necessitate the use of fairly large samples, and samples must be adequately replicated to compensate for inherent variability due to the variable nature of root colonization. The lower limit of sample size is determined by the efficiency of extraction and the sensitivity of detection, whereas the upper limit is set by how much material can conveniently be processed. Soil sample sizes of one gram or larger, and root systems from 50 to 200 seedlings or 25–30 g of roots with adhering soil are typical.[25] These limitations make studies labor-intensive, technically demanding, and expensive. Large sample sizes also preclude direct analyses in the spatially restricted sites where bacterial populations are localized and antibiotic concentrations are likely to be higher than elsewhere on the root surface. Accordingly, antibiotic concentrations are typically expressed as average values per entire root system or gram (fresh weight) of root tissue, or in relation to the population size of antibiotic-producing bacteria on the roots. Thus, in one study,[49] the total amount of DAPG produced on roots of wheat by *P. fluorescens* Q2-87 was proportional to the rhizosphere population density over a range of 10^5 to 10^7 CFU per gram root, and DAPG production per population unit was a constant (0.62 ng per 10^5 CFU). The results indicate a clear relationship between population size and the accumulation of DAPG over a range typical of introduced rhizosphere populations and suggest that populations within this range are not limited by the resources needed to produce DAPG.

One procedure[50] suitable for the extraction of many of the antibiotics produced by fluorescent *Pseudomonas* spp. involves shaking samples of up to 30 g of roots for 2 h in 40 ml of 80% acetone acidified to pH 2.0 with trifluoroacetic acid (TFA), followed by filtration to remove plant material, and centrifugation at 4°C to remove residual soil particles. The solvent is then evaporated to 8 ml, again acidified to pH 2.0, extracted twice with 10 ml volumes of ethyl acetate, and evaporated to dryness. The dried extracts can be stored frozen in the dark, preferably under nitrogen. Recoveries of DAPG isolated by this method averaged 60% for replicated controls of roots grown in soil not inoculated with antibiotic-producing bacteria and amended with the purified antibiotic in quantities sufficient to span the range of concentrations expected in unknown samples. The least amount of DAPG extractable under these conditions was 200 ng.[49] An internal standard can be added to all samples prior to extraction as an additional control. Such standards should have chemical properties similar to those of the antibiotic of interest, should not occur naturally in the sample matrix, and must not interfere with subsequent analyses. Commercially available phenazine is suitable for DAPG extractions.

Prior to analysis, dried samples are suspended in 1 ml of 35% acetonitrile-0.1% TFA, centrifuged at maximum speed for 20 min at 4°C, and subjected to further clean-up as necessary to remove soil residues that can foul chromatography equipment and interfere with UV detection. For example, some organic contaminants can be sedimented by centrifugation after freezing solutions of DAPG in acidified 35% acetonitrile at −20°C,[50] but other antibiotics may not remain soluble under these conditions. Phenazine-1-carboxylic acid and other antibiotics with ionizable residues can be partitioned away from salts and other impurities and into organic solvents by exploiting the pH-dependent differential solubility of the neutral and charged forms.

The versatility, resolving capability, and accuracy of HPLC have made it the preferred method for analyzing antibiotics produced in situ. Chromatographic systems for antibiotic analysis typically employ reversed-phase columns and a variety of mobile (solvent) phases and elution profiles[25] optimized to permit rapid resolution and quantification of the antibiotic of interest. Detection is usually by UV absorbance, with photodiode array detectors preferable to fixed wavelength instruments because each compound within a mixture can be monitored at its own spectral maximum, increasing sensitivity. Compounds typically are identified based their retention time and spectral properties compared to those of known standards. When additional sensitivity and resolution are required, as when working at near-baseline detection levels, HPLC can readily be coupled with mass spectrometry[49] or other techniques.

5.4. Engineered Strains: Are They Better? Are They Safe?

The application of bacteria to seeds or soil to control plant diseases or to improve plant growth has been studied since the early 1900s, but the use of beneficial microorganisms in agriculture has remained primarily an academic exercise for most of the last century. Only in the last two decades has there been a scientific consensus that microbial inoculants have a role in commercial agriculture, and research in the field has increased dramatically. Several biocontrol and growth-promoting agents are now sold commercially worldwide, but their use remains miniscule compared to the use of synthetic chemical pesticides. Perhaps the most important event to alter perceptions about the utility of biocontrol and growth-promoting agents has been the emergence of modern molecular biology. The tools of molecular biology have facilitated the identification of fundamental mechanisms of biocontrol and growth promotion, revealed constraints to the performance of existing strains, and made possible the engineering of novel agents that perform more consistently and with broader activity spectra than their wild-type counterparts. These and other, still-to-be developed recombinant strains, or even transgenic plants engineered to express bacterial antifungal genes in their roots, have the potential to overcome the obstacles associated with inoculum production, formulation, and cost that have until now impeded commercial acceptance of *Pseudomonas* biocontrol agents.

5.4.1. ACTIVITY IS ENHANCED IN ENGINEERED STRAINS

Single genes such as *chiA*, encoding a chitinase enzyme, and *acdS*, encoding 1-aminocyclopropane-1-carboxylic acid deamidase, which interferes with the synthesis of the plant growth regulator ethylene, were among the first to be transferred and expressed in heterologous bacteria with the goal of enhancing their ability to provide biocontrol or promote plant growth.[51–55] However, the most compelling support for the capacity of recombinant DNA technology to enhance strain performance has come from studies involving antibiotic biosynthesis in fluorescent *Pseudomonas* spp. These genes are expressed from operons and can be transferred as intact functional units to enhance or confer new biosynthetic capabilities to biocontrol agents. Thus, constitutive expression of plasmid-borne copies of the pyrrolnitrin biosynthesis operon *prnABCD* in *P. fluorescens* BL915 (from which the genes were cloned) resulted in fourfold more pyrrolnitrin produced, and the modified strain protected cucumber and impatiens against damping-off disease caused by *R. solani* as well as 10-fold higher doses of the parental strain BL915. Modified strains producing elevated levels of pyrrolnitrin also provided significantly better disease control than the parental strain in the field. Control with the enhanced strain was not significantly different from a chemical treatment or the healthy control.[56] These results suggest that the wild-type strain does not provide the threshold antibiotic level needed for maximum disease suppression in this system.

We, with collaborators, have introduced the biosynthesis operon for phenazine-1-carboxylic acid (PCA) under control of the consititutive tac promoter into random sites in the genome of *P. fluorescens* 54/96. This approach provides genetic stability and effective gene containment, which are important in minimizing the potential for horizontal gene transfer in the rhizosphere. The PCA-producing derivatives reduced damping-off disease caused by *P. ultimum* on pea seedlings significantly more than did strain 54/96. The efficacy and persistence of the bacteria correlated with the level of PCA produced, and pretreatment of the soil with the modified strain effectively decontaminated it, reducing disease incidence.[57] A similar strategy was used to generate modified derivatives of another strain of *P. fluorescens* enhanced in DAPG production, and these too provided increased control of *P. ultimum*.[58]

The PCA cassette also has been used to extend the range of diseases controlled by *P. fluorescens* Q8r1-96, which produces DAPG and is highly effective against take-all disease of wheat but less effective against root rot caused by *R. solani*. The recombinant strains produced more DAPG than did wild-type Q8r1-96 and more PCA than did *P. fluorescens* 2-79 (the source of the cloned genes) in vitro and in the wheat rhizosphere. In the greenhouse, PCA-producing strains suppressed *R. solani* root rot at only 100 CFU per seed, a dose one to two orders of magnitude less than the dose of wild-type Q8r1-96 required for comparable control.[59] Wheat treated with the PCA- and DAPG-producing recombinant derivatives of strain Q8r1-96 consistently had yields 8–20% greater than those from treatments with Q8r1-96 in three years of field trials.[60]

5.4.2. RECOMBINANT STRAINS AND RHIZOSPHERE FITNESS

The ecological fitness of biocontrol agents, whether genetically modified or not, is a key factor in evaluating risks associated with their release into the environment. Using isogenic derivatives of *P. fluorescens* SBW25 tagged with different marker genes, De Leij et al.[61] detected no effect of metabolic burden in the rhizosphere of pea or wheat. The question of fitness also has been addressed in wild-type Q8r1-96 and its PCA-producing derivatives. Because PCA contributes to the competitiveness of *Pseudomonas* strains,[62] it is conceivable that Q8r1-96 constitutively producing PCA would be more competitive than the wild-type. Conversely, PCA synthesis is energetically costly, suggesting that such a strain might be *less* fit due to the metabolic burden imposed by expressing the introduced genes. To distinguish between these possibilities, the persistence on wheat of Q8r1-96 and its PCA-producing derivative were compared under controlled conditions and in the field. No consistent strain-specific differences in rhizosphere competence were observed in either case. For three years after being introduced into the field, both strains established population densities on roots sufficient to control take-all when wheat was again sown. When the strains were co-inoculated to provide the most intense competitive pressure, the wild-type displaced the recombinant strain, which declined to nondetectable population levels.[63] Collectively, these data suggest that any benefit of PCA synthesis ultimately was overridden by its metabolic cost to the recombinant strain. Apparently that cost was not enough to impact on biocontrol activity, even over three field seasons. The burden imposed by other genes, in strains that differ in competitiveness from Q8r1-96, and on other plant hosts, remains to be determined.

5.4.3. NON-TARGET EFFECTS OF WILD-TYPE AND RECOMBINANT BIOCONTROL AGENTS

Considerable research has been conducted on the non-target effects of antibiotic-producing and non-producing biocontrol rhizobacteria introduced into the rhizosphere, as recently reviewed by Winding et al.[64] The work with fluorescent pseudomonads including *P. fluorescens* strains F113,[65] CHA0,[66] SBW25,[61,67] DR54,[68] and *P. putida* WCS358r[69] is of particular interest. These studies have considered non-target effects on the abundance and community structure of microorganisms that are closely related or not related to the introduced rhizobacteria, on soil enzyme activities and available nutrients, on microbial indicators such as rhizobia, on protozoa and nematodes, and on the plant.[64] One of the most thorough studies to date of the population dynamics and non-target effects of recombinant rhizobacteria has been conducted with *P. putida* WCS358r, modified to produce either PCA or 2,4-DAPG.[69–71] PCA was produced by the recombinant strain in the rhizosphere of plants grown in the field, and both cultivation-dependent and cultivation-independent methods were employed to quantify non-target effects. The wild-type and recombinant strains both had transient effects on the composition of the rhizosphere fungal and bacterial microflora of wheat, and the effects of the recombinant strains sometimes were longer-lasting. The impact of the recombinant strains differed from year to year and study to study. These results, which mirror those of other studies conducted under controlled and field conditions, are perhaps not surprising given that WCS358r and the rhizobacteria tested in other studies generally establish very high population sizes in the rhizosphere or soil immediately after inoculation. Then the densities decline (sometimes precipitously) over time and distance from the inoculum source. In addition, introduced rhizobacteria do not become uniformly distributed throughout the rhizosphere or among roots of the same or different plants. Collectively, studies of the non-target effects of wild-type and recombinant biocontrol rhizobacteria indicate that the bacteria have definite impacts on non-target bacterial, fungal and protozoan populations. The effects vary from study to study, often are less than those associated with routine agronomic practices, and are transient.

Acknowledgments

This work was supported by the U. S. Department of Agriculture, National Research Initiative, Competitive Grants Program (grant 2003-35319-13800).

References

1. B. J. J. Lugtenberg, L. Dekkers, and G. Bloemberg, Molecular determinants of rhizosphere colonization by *Pseudomonas*, *Annu. Rev. Phytopathol.* 39, 461–490 (2001).
2. C. M. Liddell and J. L. Parke, Enhanced colonization of pea taproots by a fluorescent pseudomonad biocontrol agent by water infiltration into soil, *Phytopathology* 79, 1327–1332 (1989).

3. J. M. Raaijmakers and D. M. Weller, Exploiting genotypic diversity of 2,4-diacetylphloroglucinol-producing *Pseudomonas* spp.: Characterization of superior root-colonizing *P. fluorescens* strain Q8r1-96, *Appl. Environ. Microbiol.* 67, 2545–2554 (2001).
4. B. B. Landa, O. V. Mavrodi, J. M. Raaijmakers, B. B. McSpadden-Gardener, L. S. Thomashow, and D. M. Weller, Differential ability of genotypes of 2,4-diacetylphloroglucinol-producing *Pseudomonas fluorescens* to colonize the roots of pea, *Appl. Environ. Microbiol.* 68, 3226–3237 (2002).
5. B. B. Landa, O. V. Mavrodi, K. L. Schroeder, R. Allende-Molar, and D. M. Weller, Enrichment and genotypic diversity of *phlD*-containing fluorescent *Pseudomonas* spp. in two soils after a century of wheat and flax monoculture, *FEMS Microbiol. Ecol.* 55, 351–368 (2006).
6. B. B. Landa, D. V. Mavrodi, L. S. Thomashow, and D. M. Weller, Interactions between strains of 2,4-diacetylphloroglucinol-producing *Pseudomonas fluorescens* in the rhizosphere of wheat, *Phytopathology* 93, 982–994 (2003).
7. J. T. de Souza, D. M. Weller, and J. M. Raaijmakers, Frequency, diversity, and activity of 2,4-diacetylphloroglucinol-producing fluorescent *Pseudomonas* spp. in Dutch take-all decline soils, *Phytopathology* 93, 54–63 (2003).
8. D. M. Weller, B. B. Landa, O. V. Mavrodi, K. L. Schroeder, L. De La Fuente, S. Blouin Bankhead, R. Allende Molar, R. F. Bonsall, D. V. Mavrodi, and L. S. Thomashow, Role of 2,4-diacetylphloroglucinol-producing fluorescent *Pseudomonas* spp. in the defense of plant roots, *Plant Biol.*, in press.
9. P. Lemanceau, T. Corberand, L. Gardan, X. Latour, G. Laguerre, J. M. Boeufgras, and C. Alabouvette, Effect of two plant species, flax (*Linum usitatissimum* L.) and tomato (*Lycopersicon esculentum* Mill.), on the diversity of soilborne populations of fluorescent pseudomonads, *Appl. Environ. Microbiol.* 61, 1004–1012 (1995).
10. M. Bergsma-Vlami, M. E. Prins, and J. M. Raaijmakers, Influence of plant species on population dynamics, genotypic diversity and antibiotic production in the rhizosphere by indigenous *Pseudomonas* spp., *FEMS Microbiol. Ecol.* 52, 59–69 (2005).
11. C. Picard, E. Frascarioli, and M. Bosco, Frequency and biodiversity of 2,4-diacetylphloroglucinol-producing rhizobacteria are differentially affected by the genotype of two maize inbred lines and their hybrid, *FEMS Microbiol Ecol.* 49, 207–215 (2004).
12. L. De La Fuente, B. B. Landa, and D. M. Weller, Host crop affects rhizosphere colonization and competitiveness of 2,4-diacetylphloroglucinol-producing *Pseudomonas fluorescens*, *Phytopathology* 96, 751–762 (2006).
13. R. Costa, J. Falcão Salles, G. Berg, and K. Smalla, Cultivation-independent analysis of *Pseudomonas* species in soil and in the rhizosphere of field-grown *Verticillium dahliae* host plants, *Environ. Microbiol.* 8, 2136–2149 (2006).
14. F. Kamilova, S. Validov, T. Azarova, I. Mulders, and B. Lugtenberg, Enrichment for enhanced competitive plant root tip colonizers selects for a new class of biocontrol bacteria, *Environ. Microbiol.* 7, 1809–1817 (2005).
15. S. Validov, F. Kamilova, S. Qi, D. Stephan, J. J. Wang, N. Makarova, and B. Lugtenberg, Selection of bacteria able to control *Fusarium oxysporum* f. sp. *radicis-lycopersici* in stonewool substrate, *J. Appl. Microbiol.*, in press.
16. F. Kamilova, L. V. Kravchenko, A. Shaposhnikov, T. Azarova, N. Makarova, and B. Lugtenberg, Organic acids, sugars, and L-tryptophane in exudates of vegetables growing on stonewool and their effects on activities of rhizosphere bacteria, *Mol. Plant-Microbe Interact.* 19, 250–256 (2006).
17. D. M. Weller, Biological control of soilborne plant pathogens in the rhizosphere with bacteria, *Annu. Rev. Phytopathol.* 26, 379–407 (1988).

18. C. M. J. Pieterse, S. C. M. Van Wees, J. Ton, J. A. Van Pelt, and L. C. Van Loon, Signalling in rhizobacteria-induced systemic resistance in *Arabidopsis thaliana*, *Plant Biol.* 4, 535–544 (2002).
19. J. M. Raaijmakers, M. Vlami, and J. T. de Souza, Antibiotic production by bacterial biocontrol agents, *Antonie van Leeuwenhoek* 81, 537–547 (2002).
20. D. Haas and C. Keel, Regulation of antibiotic production in root-colonizing *Pseudomonas* spp. and relevance for biological control of plant disease, *Annu. Rev. Phytopathol.* 41, 117–153 (2003).
21. J. M. Raaijmakers, D. M. Weller, and L. S. Thomashow, Frequency of antibiotic-producing *Pseudomonas* spp. in natural environments. *Appl. Environ. Microbiol.* 63, 881–887 (1997).
22. B. B. McSpadden Gardener, D. V. Mavrodi, L. S. Thomashow, and D. M. Weller, A rapid polymerase chain reaction-based assay characterizing rhizosphere populations of 2,4-diacetylphloroglucinol-producing bacteria, *Phytopathology* 91, 44–54 (2001).
23. J. T. De Souza and J. M. Raaijmakers, Polymorphisms within the *prnD* and *pltC* genes from pyrrolnitrin- and pyoluteorin-producing *Pseudomonas* and *Burkholderia* spp., *FEMS Microbiol. Ecol.* 43, 21–34 (2003).
24. O. V. Mavrodi, D. V. Mavrodi, D. M. Weller, and L. S. Thomashow, The role of *ptsP*, *orfT*, and *sss* recombinase in root colonization by *Pseudomonas fluorescens* Q8r1-96, *Appl. Environ. Microbiol.* 72, 7111–7122 (2006).
25. L. S. Thomashow, R. F. Bonsall, and D. M. Weller, Antibiotic production by soil and rhizosphere microbes in situ, in *Manual of Environmental Microbiology*, edited by C. J. Hurst (ASM Press, Washington, 2002), pp. 638–647.
26. J. M. Raaijmakers, M. Leeman M. M. P. Van Oorschot, I. Van der Sluis, B. Schippers, and P. A. H. M. Bakker, Dose-response relationships in biological control of Fusarium wilt of radish by *Pseudomonas* spp., *Phytopathology* 85, 1075–1081 (1995).
27. J. M. Raaijmakers and D. M. Weller, Natural plant protection by 2,4-diacetylphloroglucinol-producing *Pseudomonas* in take-all decline soils, *Mol. Plant-Microbe Interact.* 11, 144–152 (1998).
28. A. Simon and E. H. Ridge, The use of ampicillin in a simplified selective medium for the isolation of fluorescent pseudomonads, *J. Appl. Bacteriol.* 37, 459–460 (1974).
29. B. B. Landa, H. A. E. de Werd, B. B. McSpadden-Gardener, and D. M. Weller, Comparison of three methods for monitoring populations of different genotypes of 2,4-diacetylphloroglucinol-producing *Pseudomonas fluorescens* in the rhizosphere, *Phytopathology* 92, 129–137 (2002).
30. L. De La Fuente, D. V. Mavrodi, B. B. Landa, L. S. Thomashow, and D. M. Weller, *phlD*-based genetic diversity and detection of genotypes of 2,4-diacetylpholorogucinol-producing *Pseudomonas fluorescens*, *FEMS Microbiol. Ecol.* 56, 64–78 (2006).
31. S. Validov, O. Mavrodi, L. De La Fuente, A. Boronin, D. Weller, L. Thomashow, and D. Mavrodi, Antagonistic activity among 2,4-diacetylphloroglucinol-producing fluorescent *Pseudomonas* spp. *FEMS Microbiol. Lett.* 242, 249–256 (2005).
32. O. V. Mavrodi, D. V. Mavrodi, L. S. Thomashow, and D. M. Weller, Application of real-time PCR quantification of 2,4-diacetylphloroglucinol-producing *Pseudomonas fluorescens* in the plant rhizosphere, in *International Plant Growth Promoting Rhizobacteria Workshop* (May 20, 2006); available at http://www.ars.usda.gov/research/publications/ publications. htm?seq_ no_115=199460.
33. I. T. Paulsen, C. Press, J. Ravel, D. Y. Kobayashi, G. S. A. Myers, D. V. Mavrodi, R. T. DeBoy, R. Seshadri, Q. Ren, R. Madupu, R. J. Dodson, A. S. Durkin, A. M. Brinkac, S. C. Daugherty, S. A. Sullivan, M. J. Rosovitz, M. L. Gwinn, L. Zhou, D. J. Schneider, S. W. Cartinhour, W. C. Nelson, J. Weidman, K. Watkins, K. Tran, H. Khouri, E. A. Pierson,

L. S. Pierson, L. S. Thomashow, and J. E. Loper, Complete genome sequence of the plant commensal *Pseudomonas fluorescens* Pf-5: Insights into the biological control of plant disease, *Nature Biotechnol.* 23, 873–878 (2005).

34. L. S. Pierson III, V. D. Keppenne, and D. W. Wood, Phenazine antibiotic biosynthesis in *Pseudomonas aureofaciens* 30-84 is regulated by PhzR in response to cell density, *J. Bacteriol.* 176, 3966–3974 (1994).
35. D. W. Wood, F. Gong, M. M. Aykin, P. Williams, and L. S. Pierson III, *N*-Acyl-homoserine lactone-mediated regulation of phenazine gene expression by*Pseudomonas aureofaciens* 30-84 in the wheat rhizosphere, *J. Bacteriol.*179, 7663–7670 (1997).
36. D. Haas, C. Blumer, and C. Keel, Biocontrol ability of fluorescent pseudomonads genetically dissected: Importance of positive feedback regulation. *Curr. Opin. Biotechnol.* 11, 290–297 (2000).
37. U. Schnider-Keel, A. Seematter, M. Maurhofer, C. Blumer, B. Duffy, C. Gigot-Bonnefoy, C. Reimmann, R. Notz, G. Défago, D. Haas, and C. Keel, Autoinduction of 2,4-diacetylphloroglucinol biosynthesis in the biocontrol agent *Pseudomonas fluorescens* CHA0 and repression by the bacterial metabolites salicylate and pyoluteorin, *J. Bacteriol.* 182, 1215–1225 (2000).
38. A. Abbas, J. P. Morrissey, P. Carnicero Marquez, M. M. Sheehan, I. R. Delany, and F. O'Gara, Characterization of interactions between the transcriptional repressor PhlF and its binding site at the phlA promoter in *Pseudomonas fluorescens* F113, *J. Bacteriol.* 184, 3008–3016 (2002).
39. R. Notz, M. Maurhofer, S. Schnider-Keel, B. Duffy, D. Haas, and G. Défago, Biotic factors affecting expression of the 2,4-diacetylphloroglucinol biosynthetic gene *phlA* in *Pseudomonas fluorescens* biocontrol strain CHA0 in the rhizosphere, *Phytopathology* 91, 873–881 (2001).
40. B. K. Duffy and G. Défago, Zinc improves biocontrol of Fusarium crown and root rot of tomato by *Pseudomonas fluorescens* and represses the production of pathogen metabolites inhibitory to bacterial antibiotic synthesis, *Phytopathology* 87, 1250–1257 (1997).
41. R. Notz, M. Maurhofer, H. Dubach, D. Haas, and G. Défago, Fusaric acid-producing strains of *Fusarium oxysporum* alter 2,4-diacetylphloroglucinol biosynthetic gene expression in *Pseudomonas fluorescens* CHA0 in vitro and in the rhizosphere of wheat, *Appl. Environ. Microbiol.* 68, 2229–2235 (2002).
42. M. P. Lutz, S. Wenger, M. Maurhofer, G. Défago, and B. Duffy, Signaling between bacterial and fungal biocontrol agents in a strain mixture, *FEMS Microbiol. Ecol.* 48, 447–455 (2004).
43. M. Maurhofer, E. Baehler, R. Notz, V. Martinez, and C. Keel, Cross talk between 2,4-diacetylphloroglucinol-producing biocontrol pseudomonads on wheat roots, *Appl. Environ. Microbiol.* 70, 1990–1998 (2004).
44. E. A. Pierson, D. W. Wood, J. A. Cannon, F. M. Blachere, and L. S. Pierson III, Interpopulation signaling via *N*-acyl-homoserine lactones among bacteria in the wheat rhizosphere, *Mol. Plant-Microbe Interact.* 11, 1078–1084 (1998).
45. J. E. Morello, E. A. Pierson, and L. S. Pierson III, Negative cross-communication among wheat rhizosphere bacteria: Effect on antibiotic production by the biological control bacterium *Pseudomonas aureofaciens* 30-84, *Appl. Environ.Microbiol.* 70, 3103–3109 (2004).
46. J. E. Loper and S. E. Lindow, Reporter gene systems useful in evaluating in situ gene expression by soil- and plant-associated bacteria, in *Manual of Environmental Microbiology*, edited by C. J. Hurst (ASM Press, Washington, DC, 2002), pp. 627–637.

47. M. Bottiglieri and C. Keel, Characterization of PhlG, a hydrolase that specifically degrades the antifungal compound 2,4-diacetylphloroglucinol, *Appl. Environ. Microbiol.* 72, 418–427 (2006).
48. G. Pessi, C. Blumer, and D. Haas, *lacZ* fusions report gene expression, don't they?, *Microbiology* 147, 1993–1995 (2002).
49. J. M. Raaijmakers, R. F. Bonsall, and D. M. Weller, Effect of population density of *Pseudomonas fluorescens* on production of 2,4-diacetylphloroglucinol in the rhizosphere of wheat, *Phytopathology* 89, 470–475 (1999).
50. R. F. Bonsall, D. M. Weller, and L. S. Thomashow, Quantification of 2,4-diacetylphloroglucinol produced by fluorescent *Pseudomonas* spp. in vitro and in the rhizosphere of wheat, *Appl. Environ. Microbiol.* 63, 951–955 (1997).
51. I. Chet, Cloning chitinases for plant protection: Hope or hazard for the environment, in *Chitin Enzymology,* edited by R. A. A. Muzzarelli (Plenum Press, New York, 1998), pp. 25–30.
52. K. Downing and J. A. Thomson, Introduction of the *Serratia marcescens chiA* gene into an endophytic *Pseudomonas fluorescens* for the biocontrol of phytopathogenic fungi, *Can. J. Microbiol.* 46, 363–369 (2000).
53. L. Sundheim, A. R. Poplawsky, and A. H. Ellingboe, Molecular cloning of two chitinase genes from *Serratia marcescens* and their expression in *Pseudomonas* species. *Physiol. Molec. Plant Pathol.* 33, 483–491 (1998).
54. S. Koby, H. Schickler, I. Chet, and A. B. Oppenheim, The chitinase encoding Tn7-based *chiA* gene endows *Pseudomonas fluorescens* with the capacity to control plant pathogens in soil, *Gene* 147, 81–83 (1994).
55. C. X. Wang, E. Knill, B. R. Glick, and G. Défago, Effect of transferring 1-aminocyclopropane-1-carboxylic acid (ACC) deaminase genes into *Pseudomonas fluorescens* strain CHA0 and its *gacA* derivative CHA96 on their growth-promoting and disease-suppressive capacities, *Can. J. Microbiol.* 46, 898–907 (2000).
56. J. M. Ligon, D. S. Hill, P. E. Hammer, N. R. Torkewitz, D. Hofmann, H. J. Kempf, and K. H. van Pee, Natural products with antifungal activity from *Pseudomonas* biocontrol bacteria, *Pest Management Sci.* 56, 688–695 (2000).
57. T. M. Timms-Wilson, R. J. Ellis, A. Renwick, D. J. Rhodes, D. V. Mavrodi, D. M. Weller, L. S. Thomashow, and M. M. Bailey, Chromosomal insertion of the phenazine biosynthetic pathway (*phzABCDEFG*) enhances the efficacy of damping off disease control by *Pseudomonas fluorescens* 54/96, *Mol. Plant-Microbe Interact.* 13, 1293–1300 (2000).
58. B. W. Alsanius, M. Hultberg, and J.-E. Englund, Effect of lacZY-marking of the 2,4-diaceylphloroglucinol producing *Pseudomonas fluorescens*-strain 5-2/4 on its physiological performance and root colonization ability, *Microbiol. Res.* 157, 39–45 (2002).
59. Z. Huang, R. F. Bonsall, D. V. Mavrodi, D. M. Weller, and L. S. Thomashow, Transformation of *Pseudomonas fluorescens* with genes for biosynthesis of phenazine-1-carboxylic acid improves control of Rhizoctonia root rot and in situ antibiotic production, *FEMS Microbiol. Ecol.* 49, 243–251 (2004).
60. D. M. Weller, unpublished data.
61. F. A. A. M. De Leij, E. J. Sutton, J. M. Whipps, J. S. Fenlon, and J. M. Lynch, Impact of a field release of genetically modified *Pseudomonas fluorescens* on indigenous microbial populations of wheat, *Appl. Environ. Microbiol.* 61, 3443–3453 (1995).
62. M. Mazzola, R. J. Cook, L. S. Thomashow, D. M. Weller, and L. S. Pierson III, Contribution of phenazine antibiotic biosynthesis to the ecological competence of fluorescent pseudomonads in soil habitats, *Appl. Environ. Microbiol.* 58, 2616–2624 (1992).

63. S. Blouin-Bankhead, B. Landa, E. Lutton, D. M. Weller, and B. B. McSpadden Gardener, Minimal changes in rhizobacterial population structure following root colonization by wild type and transgenic biocontrol strains, *FEMS Microbiol. Ecol.* 49, 307–318 (2004).
64. A. Winding, S. J. Binnerup, and H. Pritchard, Non-target effects of bacterial biological control agents suppressing root pathogenic fungi, *FEMS Microbiol. Ecol.* 47, 129–141 (2004).
65. Y. Moënne-Loccoz, H.-V. Tichy, A. O'Donnell, R. Simon, and F. O'Gara, Impact of 2,4-diacetylphloroglucinol-producing biocontrol strain *Pseudomonas fluorescens* F113 on intraspecific diversity of resident culturable fluorescent pseudomonads associated with the roots of field-grown sugar beet seedlings, *Appl. Environ. Microbiol.* 67, 3418–3425 (2001).
66. A. Natsch, C. Keel, N. Hebecker, E. Laasik, and G. Défago, Impact of *Pseudomonas fluorescens* strain CHA0 and a derivative with improved biocontrol activity on the culturable resident bacterial community on cucumber roots, *FEMS Microbiol. Ecol.* 27, 365–380 (1998).
67. T. M. Timms-Wilson, K. Kilshaw, and M. J. Bailey, Risk assessment for engineered bacteria used in biocontrol of fungal disease in agricultural crops, *Plant and Soil* 266, 57–67 (2004).
68. L. Thirup, K. Johnsen, and A. Winding, Succession of indigenous *Pseudomonas* spp. and actinomycetes on barley roots affected by the antagonistic strain *Pseudomonas fluorescens* DR54 and the fungicide imazalil, *Appl. Environ. Microbiol.* 67, 1147–1153 (2001).
69. D. C. M. Glandorf, P. Verheggen, T. Jansen, J.-W. Jorritsma, E. Smit, P. Leeflang, K. Wernars, L. S. Thomashow, E. Laureijs, J. E. Thomas-Oates, P. A. H. M. Bakker, and L. C. van Loon, Effect of genetically modified *Pseudomonas putida* WCS358r on the fungal rhizosphere microflora of fieldgrown wheat, *Appl. Environ. Microbiol.* 67, 3371–3378 (2001).
70. P. Leeflang, E. Smit, D. C. M. Glandorf, E. J. van Hannen, and K. Wernars, Effects of *Pseudomonas putida* WCS358r and its genetically modified phenazine producing derivative on the *Fusarium* population in a field experiment, as determined by 18S rDNA analysis, *Soil Biol. Biochem.* 34, 1021–1025 (2002).
71. M. Viebahn, D. C. M. Glandorf, T. W. M. Ouwens, E. Smit, P. Leeflang, K. Wernars, L. S. Thomashow, L. C. van Loon, and P. A. H. M. Bakker, Repeated introduction of genetically modified *Pseudomonas putida* WCS358r without intensified effects on the indigenous microflora of field-grown wheat, *Appl. Environ. Microbiol.* 69:3110–3118 (2003).

6. EXPLOITING THE INTERACTIONS BETWEEN FUNGAL ANTAGONISTS, PATHOGENS AND THE PLANT FOR BIOCONTROL

Sheridan L. Woo and Matteo Lorito*
Department of Arboriculture, Botany and Plant Pathology, Plant Pathology Section, University of Naples "Federico II", 80055 Portici (NA), Italy

Abstract. The soil community supports an enormous variety of biological interactions among its living inhabitants, which include those occurring between animals, insects, microorganisms and plants. Some of the most commonly found soil microbes belong to different species of *Trichoderma* and function as antagonists of phytopathogens, thus protecting plants and reducing disease incidence in many different soil types. Together with other species, such as *Pseudomonas* spp., *Bacillus* spp., *Coniothyrium* spp., *Pythium* spp. etc., these highly interacting microbes have been extensively studied and commercially marketed mainly as biopesticides/biofertilizers and soil amendments, all containing live cells. *Trichoderma* spp. are also known to produce many different bioactive compounds, including dozens of cell wall degrading enzymes, and thus their biodegradation by-products, hundreds of antibiotics and many others still uncharacterized but highly reactive molecules. In fact, the broad spectrum biological effects of the relative fungal extracts suggest their use as alternatives to or additives with live microbes in diverse agriculture and industrial applications, including plant/fruit protection and food processing. These mixtures of fungal compounds can be easily produced at an industrial level and effectively applied to enhance antimicrobial activity of common fungicidal compounds, as well as activate or stimulate biocontrol agents and plant resistance to pathogen attack. Studies into the complex three-way relationship that *Trichoderma* establishes with the plant and pathogen are revealing mechanisms involved in partner recognition and the molecular cross-talk used to maintain the stable association that provides benefits both for the fungal antagonist and the plant.

Keywords: *Trichoderma*, biocontrol, cell wall degrading enzymes, plant elicitors, fungal-plant interaction, induced systemic resistance

---

*To whom correspondence should be addressed, e-mail: lorito@unina.it

M. Vurro and J. Gressel (eds.), Novel Biotechnologies for Biocontrol Agent Enhancement and Management, 107–130.

6.1. Introduction

Agricultural research has been oriented more and more towards finding biological control and integrated pest management techniques for plant disease control that use compounds that are non-toxic to man and the environment, with the goal of reducing the dose of pesticides used.[1–4] The most promising microorganisms are antagonists of important plant pathogens, including bacteria such as *Bacillus, Pseudomonas* and *Enterobacter,* numerous yeasts such as *Pichia guillermondii, Candida sake, C. pulcherrima, Cryptococcus laurentii* and *C. flavus,* and fungi including *Acremonium breve, Sepedonium* spp., *Trichoderma* spp., and *Gliocladium* spp. (Chapters 5 and 12).[2,5–9] These biocontrol agents have been largely used in single or combined applications of the whole microorganism to the plant or product, in the field or during storage, to control disease (Chapter 7). However, a possible negative side-effect of utilizing combinations of actual microbes in treatments may be that they are antagonistic not only to the disease causing agents, but also to one another, thus reducing control efficacy.

The application of antimicrobial compounds that these microorganisms produce is an alternative to the direct use of live antagonists; such formulations include mixtures of lytic enzymes (such as endochitinase, exochitinase, glucanase) that degrade the fungal cell wall or antibiotics that are toxic or inhibit the pathogen. In general, many of the molecules that are secreted into the growth medium by the antagonist can negatively affect the target pathogen. Conditions can be selected for the production of substances with high biological activity, and these compounds can be made in diverse commercial formulations (i.e., powder, granules, dip, drench), and applied directly to vegetation in the field or greenhouse, or to produce in storage, in a manner similar to chemical pesticides.

There are several advantages to using natural compounds rather than live microorganisms: they have the intrinsic characteristic of wide spectrum antimicrobial inhibitory activity that can be exploited; their production can be readily manipulated and regulated at an industrial level; and the final product is stable, easy to store and transport. All these conditions are much less restrictive than the use of the whole organism for commercialization and use.[6] In addition, these compounds can potentially augment the biological activity of a biocontrol agent with synergistic effects. The combined use of the different enzymes, for example, with themselves, with antibiotics or with synthetic pesticides could provide a high level of synergism. These compounds could also act as inducers of the antagonistic mechanisms of the whole organism, or function as elicitors of the plant defense system against pathogen attack.

6.2. Trichoderma—A Fungal Antagonist

Fungal biological control agents such as *Trichoderma* species have been studied extensively in the past 70 years and are commercially marketed worldwide as biofungicides, plant biostimulants, and biological soil amendments (Chapter 7).[2,9–13] A search on the Internet reveals more than 50 different *Trichoderma*-based agricultural products, many of which are registered in various countries across five continents; sold and applied for the protection and yield improvement of horticultural, ornamental, fruit, recreational, spice etc., for plantings in the field, nursery, orchard, garden, and greenhouse.[9] *Trichoderma* is listed both in Europe and USA as an active principal ingredient permitted for use in organic farming for plant disease control.

Trichoderma spp., along with *Fusarium* spp., are among the fungi most frequently isolated from the soil. Many *Trichoderma* species are strong opportunistic invaders, fast growing, producers of enormous quantities of spores and powerful antibiotics, all properties that make these fungi ecologically successful.[10] They are resistant to many natural and man-made chemicals, and able to effectively degrade compounds such as hydrocarbons, chlorophenols, polysaccharides and synthetic pesticides. *Trichoderma* alone or in combination with beneficial plants is being used for bioremediation in the recovery of contaminated sites.[10,14]

Many *Trichoderma* species utilize highly effective antagonistic mechanisms to survive and colonize the competitive environment of the rhizosphere, phyllosphere and spermosphere. The biocontrol ability of *Trichoderma* can be attributed to numerous modes of antagonism in confrontation with various disease causing agents, as well as with the plant. *Trichoderma* uses: parasitism, in the case of fungi, mycoparasitism, to directly attack the pathogen; its ability to colonize a niche or to compete for nutrients thus excluding a pathogen from the plant roots or exudates; the production of secondary metabolites that have a toxic or inhibitory effect on the plant pathogen; ability to create a suppressive environment by its interactions in the soil community to produce unfavorable ecological conditions that limit the development or multiplication of pathogenic populations; the secretion of numerous compounds that induce plant resistance mechanisms to pathogen attack.[2–4,7,8,11,13,15,16]

Although fungi such as *Trichoderma* have potential for a variety of applications, there are some associated difficulties that limit the development and application of these antagonists as biopesticides. The main factor restricting product development is the lack of both broad spectrum effective strains and successful commercial formulations. A greater understanding is required of the mechanisms of action employed by *Trichoderma* spp. during biocontrol to aid in the selection of potential strains, determine the biological, agricultural and/or biotechnological aspects to develop, and broaden the application spectrum of these biopesticides in different locations, various conditions, on diverse crops etc.

Traditional methods of pathogen biocontrol using fungal agents have mainly involved the selection of antagonistic strains from in vitro plate assays with plant pathogens, the production of conidia, and the innudative application of the whole organism to target plants for control or to provide protection from numerous pathogens without considering the mechanisms involved.[4] However, research conducted in the last two decades has produced a completely new perspective into the manner by which these fungi interact with other microbes, as well as with plants and soil components, and their potential for agricultural and biotechnological applications (Chapter 7).[14,17–22]

6.3. *Trichoderma* Interactions in the Soil Community

It is necessary to step back and obtain a holistic view of this antagonistic fungus in its natural ecosystem to appreciate the biological role of *Trichoderma*. *Trichoderma* spp. are saprophytic, filamentous fungi, ubiquitous to the soil community. They have the incredible ability to interact both as parasites and as symbionts with different living organisms (Chapter 7) and different substrates by establishing neutral, antagonistic or beneficial interactions with microbes, animals and plants[10] (Figure 1). They utilize various mechanisms including nutrient competition, antibiosis, antagonism, inhibition of pathogen or plant

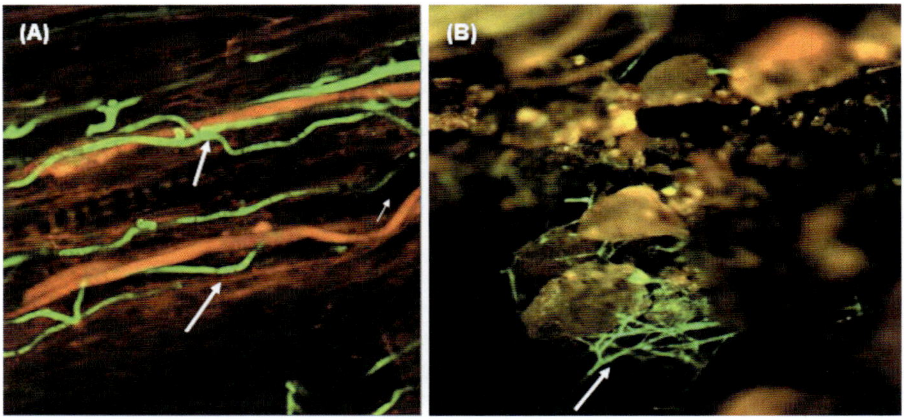

Figure 1. Colonization of cucumber roots (A) and a Pythium-infested soil (B) by a GFP transformant of Trichoderma atroviride strain P1 (light gray mycelia marked by arrows), photographed by a confocal laser microscope (Lu Z. et al., *Appl. Environ. Microbiol.* 70, 3073–3081)

enzymes; processes of biodegradation, carbon and nitrogen cycling; complex interactions with plants in the root zone of the rhizosphere, which involve various processes such as colonization, plant growth stimulation, biocontrol of diverse plant pathogens, decomposition of organic matter, symbiosis, and nutrient exchange.[6,7,10] There are strains that are strong antagonists towards phytopathogenic fungi, others that are effective soil colonizers and biodegraders, and others that are producers of important metabolites.[10] Recent studies indicate that these fungi can induce systemic resistance in plants, thus increasing the plant defense response to diverse pathogen attack.[15,16,18]

Research interest recently has focused on the complex multiple interactions involving antagonistic *Trichoderma* species (mainly *T. atroviride*, *T. asperellum* and *T. harzianum*, used as models), crop plants and plant fungal pathogens. The use of relatively new techniques to study these complex processes, such as proteomic analysis, the use of gene expression reporter systems and high throughput methods to study gene function have indicated that an intricate molecular cross-talk occurs between *Trichoderma*, the plant, and the pathogen.[16,18,23,24] Some molecules may act as hormones to stimulate plant growth and development, while others may function as inducers of *Trichoderma* antagonism or as elicitors that activate plant disease resistance to pathogen attack.

6.3.1. TRICHODERMA–FUNGUS INTERACTIONS

The main biocontrol mechanism that *Trichoderma* utilizes in direct confrontation with fungal pathogens is mycoparasitism.[7,10,13] This mechanism relies on the recognition, binding and enzymatic disruption of the host fungus cell wall. The *Trichoderma* antifungal system consists of numerous genes encoding for an great variety of secreted lytic enzymes such as endochitinases, *N*-acetyl-β-glucosaminidases (exochitinases), proteases, endo- and exo-glucan β-1,3-glucosidases, endoglucan β-1,6-glucosidases, lipases, xylanases, mannanases, pectinases, amylases, phospholipases, RNases, DNases, etc.[6] Chitinolytic and glucanolytic enzymes are particularly useful for biocontrol applications because of their ability to efficiently degrade the cell wall of plant pathogenic fungi by hydrolyzing biopolymers not present in plant tissues.[10,25] We have demonstrated the role of specific cell wall degrading enzymes (CWDEs), an endochitinase (CHIT42), produced by *T. atroviride* during biocontrol with different fungal pathogens.[23,26,27] Each class of enzymes contains a number of proteins with different enzyme activities, many of which have been purified and characterized and their genes cloned.[25,28] Most enzymes tested as purified proteins have very strong antifungal activity, especially when assayed in combinations, against a variety of fungi, i.e., *Rhizoctonia, Alternaria, Pythium, Phytophthora, Colletotrichum*, and in particular

Botrytis.[29–31] A substantial amount of work performed mainly in the past five years has indicated that cell wall degrading enzymes from *Trichoderma* strains have great potential for agriculture as active components in new fungicidal formulations.

6.3.1.1. *Application of Extracts from Beneficial Fungi As an Alternative to the Use of the Living Microbes*

The use of anti-microbial compounds produced by fungal biocontrol agents has numerous advantages over the use of the whole "live" organisms in all aspects related to industrial production, commercialization and application. Major concerns are eliminated about: the production of a high quantity of pure propagules; the ability to survive downstream manufacturing processes (drying or formulation); storage stability and sufficient shelf-life; resistance to variable environmental conditions in the field (temperature, water, pH, light etc.).[32,33] The induction of specific active compounds can be selectively induced and enhanced by controlled variation of the culture growth conditions, i.e., substrate components, pH, temperature etc.[35]

Trichoderma spp. produce large quantities of lytic enzymes, many which have proven antifungal activity. The production of cell wall degrading enzymes can be induced by the addition of various carbon sources to the growth medium such as different sugars, colloidal chitin, purified fungal cell walls or fungal biomass, both live and killed.[7,34] Generally, these enzymes are stable at room temperature, having efficacy levels similar to commercial fungicides, and retain their biological activity even when externally applied to plants in the greenhouse or to produce in post-harvest storage.[21,35] Combinations of these fungal enzymes with different classes of synthetic fungicides, in particular azole and other cell membrane-affecting compounds, have a strong synergistic effect on the inhibition of pathogens.[25,36,37]

The use of cell wall degrading enzymes from *Trichoderma* in biocontrol have many positive attributes that support their application for plant disease control. A comparison between chitinases and glucanases produced by *Trichoderma* and similar enzymes produced by plants indicates that the fungal lytic enzymes are more potent against fungal pathogens.[6,25,28,29] They are able to degrade not only the "tender" immature cell wall found at the hyphal tips, but also the strong chitin–glucan complexes of mature cell walls and dormant sclerotia, chlamydospores and spores. Therefore, they are effective in both reducing disease development, pathogen infection and inoculum distribution. Purified cell wall degrading enzymes originating from different *Trichoderma* strains are capable of inhibiting the spore germination and mycelia growth of a broad range of pathogens, not only in chitin-containing fungi including *Botrytis, Rhizoctonia, Fusarium, Alternaria, Ustilago, Venturia* and *Colletotrichum*, but also in fungus-like organisms such as the Oomycetes *Pythium* and

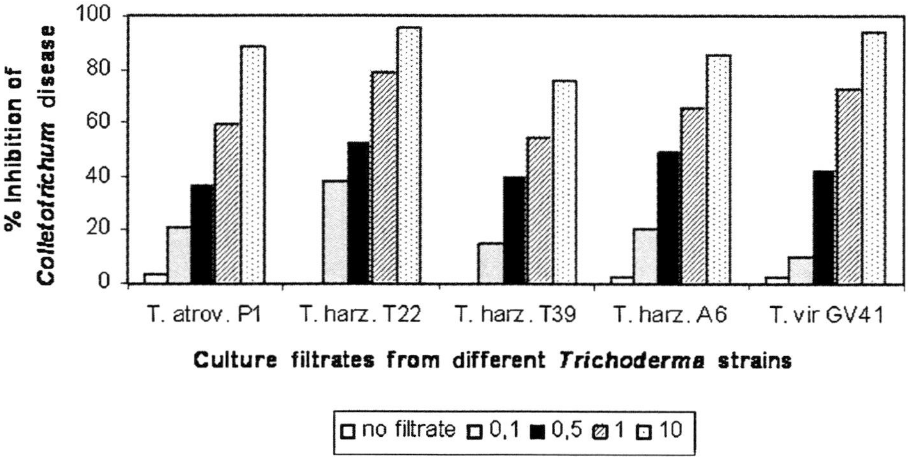

Figure 2. Application of culture filtrates from different *Trichoderma* species/strains (*T. atroviride, T. harzianum, T. virens*), containing a mixture of lytic enzymes (no enzyme, 0.1, 0.5, 1 and 10 μl applied at the infection site), inhibits disease development caused by *Colletotrichum acutatum* on strawberry

Phytophthora that lack chitin in their cell walls.[26,38,39] It is not necessary to apply the purified enzymes to obtain good disease control. It is possible to apply *Trichoderma* culture filtrates produced under different inducing conditions directly to the plant or plant products to attain good fungal pathogen control (Figure 2).

As mentioned above, the antifungal activity of *Trichoderma* cell wall degrading enzymes can be synergistically enhanced by combining enzymes with different lytic activities. For example, a treatment containing a combination of purified *T. harzianum* P1 (= *T. atroviride*) endochitinase, exochitinase and β-1,3-glucanase to *Botrytis* spores results in an ED_{50} dose of about 1 mg/l, an effective dose that is comparable to that produced by most chemical fungicides.[30,36,37] *Trichoderma* cell wall degrading enzymes combined with bacterial metabolites, such as lipodepsipeptides from *Pseudomonas*,[40] synergistically increased the antifungal activity to different plant pathogens (Chapter 5). The antagonistic ability of the biocontrol agent *Enterobacter cloacae* against spores of *Botrytis*, *Fusarium* and *Uncinula* phytopathogens was synergistically increased when a combination of *Trichoderma* cell wall degrading enzymes and the bacterial culture filtrate was applied to the live bacteria.[41] Further, the addition of *Trichoderma* chitinolytic enzymes, in particular the endochitinase, to cultures of *E. cloacae* stimulated the growth of the bacteria. The fact that the cell wall degrading enzymes can be effectively combined with the whole, live microorganisms, *Trichoderma* itself and/or microbial biocontrol agents, enhances the possibilities for improving disease control activity.[26,29,41]

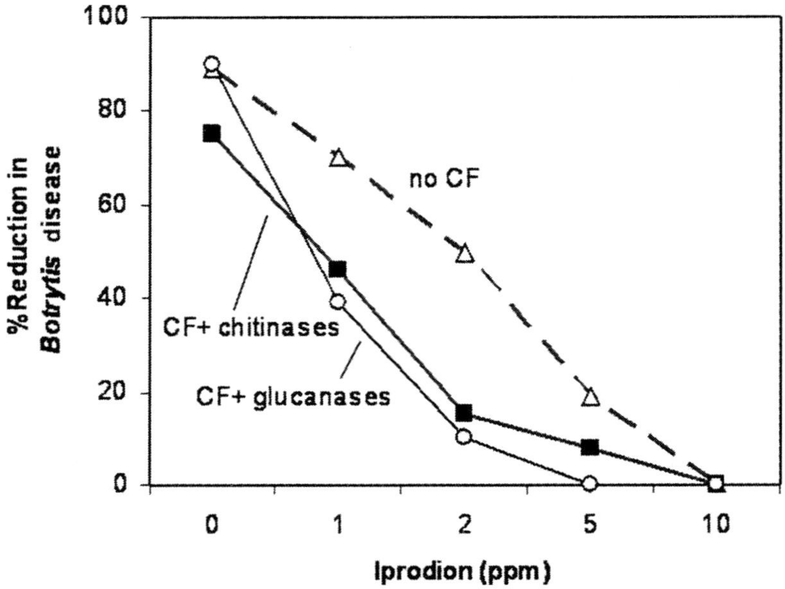

Figure 3. Chitinases or glucanases secreted by a *Trichoderma* strain (crude preparations from liquid cultures were used) enhance the effect of the fungicide iprodion on *Botrytis cinerea* attacking strawberry fruits. CF = fungal culture filtrate containing mainly chitinases or glucanases

Trichoderma spp. also produce a large number of antibiotics, including acetaldehydes gliotoxin and viridin, alpha-pyrones, terpenes, polyketides, isocyanide derivatives, piperacines, and complex families of peptaibols.[10,43,44] Many of these antibiotics are synergistic when combined with various cell wall degrading enzymes originating from *Trichoderma* and other microbial sources, thus producing a strong inhibitory effect on many plant pathogens.[37,40,44]

The inhibitory activity of chemical fungicides applied to *Botrytis* (Figure 3) and other plant pathogens can be greatly enhanced by the addition of minute quantities (10–20 ppm) of *Trichoderma* cell wall degrading enzymes. For example, azole compounds showed an up to tenfold increase in antifungal effect with the addition of *T. harzianum* endochitinase to the treatment.[36] Other commonly used commercial fungicides synergistic with the *Trichoderma* cell wall degrading enzymes include those containing principal ingredients of benzimidazole, dicarboximide and pyrimidine.[36] Importantly, cell wall degrading enzymes are not harmful to humans and animals, and they readily degrade into environmentally friendly residues, as determined by EPA tests conducted for the registration of two *Trichoderma* strains as biocontrol agents in the U.S.A.

6.3.1.2. *Molecules Involved in the Activation and Stimulation of Biocontrol Processes in Trichoderma and Other Beneficial Fungi*

How does a microorganism such as *Trichoderma* react in the vast, variable environment of the soil microbe community? It can be hypothesized that as a *Trichoderma* sp. grows, it constitutively secretes various lytic enzymes, including fungal cell wall degrading enzymes such as endo- and exo-chitinases, plant cell wall degrading enzymes such as xylanases and cellulases, as well as antibiotics into the surroundings. Contact of the enzymes with appropriate substrates releases the breakdown products of the lysis into the environment. Filamentous hyphal growth advances in a chemotactic manner "testing" the compounds released.[6,7,27] The response of the fungus may be considered analogous to the process employed by bats, which use sonar to determine the position of insect prey in their spatial environment. Once a "desired" compound is encountered, for example, a nutrient such as the chitin breakdown products liberated from the cell wall of the plant pathogen *Rhizoctonia solani*, *Trichoderma* will react by growing towards the chemical signal emitted by the fungal pathogen. The sensing and recognition of this compound will stimulate *Trichoderma* to augment secretion of chitinolytic enzymes.[28] Further, its movement will be directed towards the target organism, as determined by the detection of an increased concentration gradient of the released molecules. As *Trichoderma* gets physically closer to *Rhizoctonia*, there will be full initiation of the mycoparasitism process, activation of gene expression and secretion of specific cell wall degrading enzymes such as the endochitinase CHIT42 in *T. atroviride* strain P1,[16,23,28] plus antibiotics such as peptaibols.[42,44] The combination of the cell wall degrading enzymes and the antibiotics may result in a synergistic antifungal effect on *Rhizoctonia*, whereby, the breakdown of the cell wall by the enzymes may aid the penetration of the antibiotics. In turn, these affect the function of the cell membrane and the subsequent associated processes such as cell wall synthesis at this location.[17] Once contact occurs, there is full activation of the mechanisms and processes involved in mycoparasitism, resulting in biological control.[23,34] *Trichoderma* similarly senses plant structures such as roots in the soil.[45] The secretion of fungal enzymes causes the release of plant cell wall constituents that may indicate to *Trichoderma* the location for colonization, the presence of nutrients, i.e., root exudates or the presence of other potential microbial food sources, resulting in mutual benefits for both antagonist and plant.[15,46,47]

The obvious key stimulus for inducing mycoparasitism in *Trichoderma* are the chitin breakdown products liberated from the fungal cell wall, the molecules released from the pathogen by the *Trichoderma* cell wall degrading enzymes. These molecules not only stimulate the biocontrol fungus and its antagonistic activity by activating the mycoparasitic gene expression cascade,[48,49] but they also act as elicitors of the plant defense system in plant cells exposed to them or when injected to root and leaves.[16,18] This topic will be discussed in more detail below.

We used mutants of *T. atroviride* strain P1 containing the gene encoding for the green fluorescent protein (GFP) or the glucose oxidase (GOX) protein in a reporter system utilizing inducible promoters (exochitinase *nag*1 or endochitinase *ech*42), known to be actively involved in biocontrol by P1, to determine the compounds that stimulate the mycoparasitic response.[23,26,34] Purified P1 lytic enzymes, various *Trichoderma* culture filtrates, and pathogen culture filtrates were used to digest whole intact fungal biomass, purified fungal cell walls, and colloidal crab shell chitin. Extracts from cucumber leaves, stems or roots were also tested. The numerous digestion products were assayed singly and in various combinations to determine those that activated the biocontrol gene expression cascade in the antagonist. Various digestion products produced by the treatment of fungal cell walls and colloidal chitin with the purified enzymes or fungal culture filtrates induced the strongest mycoparasitism.[48,49] Interestingly, the use of the culture filtrates from the *ech*42 knock-out mutants for digesting, and digestion of the purified cell walls from the Oomycete *Pythium* with CHIT42 or chitinase containing filtrates, were less active in producing the stimulus for mycoparasitism. This indicates that, in the case of *Trichoderma* P1, the endochitinase and different chitin containing substrates, such as the fungal cell wall, play an important role in the mechanism of biocontrol.[23,26,34] Further, different phytopathogens, such as Oomycetes that do not contain chitin, activate diverse mechanisms for mycoparasitism (Figure 4), as noted by biocontrol of *Pythium* attack to beans.[26]

The products from the digestion of fungal cell walls with purified hydrolytic enzymes were separated to determine the size range of the compounds that were able to activate biocontrol in *Trichoderma*. Micromolecules less than 3000 Da triggered mycoparasitism gene expression before physical contact with the host pathogen.[48,49] Applications of these low molecular weight compounds to the antagonist stimulated mycelial growth and rate of spore germination. These host-derived compounds were separated by HPLC and the fractions were tested in vivo to determine the highest anti-fungal activity to pathogens. The selected inducers stimulated both the production of endochitinases and exochitinases in vitro, even under repressing conditions in the presence of glucose.[48,49] Furthermore, in vitro, these inducers stimulated the biological activity of P1 in the presence of the host fungus. The development of disease symptoms on bean leaves inoculated with both *Botrytis* and *Trichoderma* spores was clearly reduced by the addition of the inducers in comparison to treatments not containing the inducers or to treatments using the specifically inactivated molecules (Figure 5). The addition of the purified inducers to liquid cultures of *T. atroviride* P1 stimulated the production of

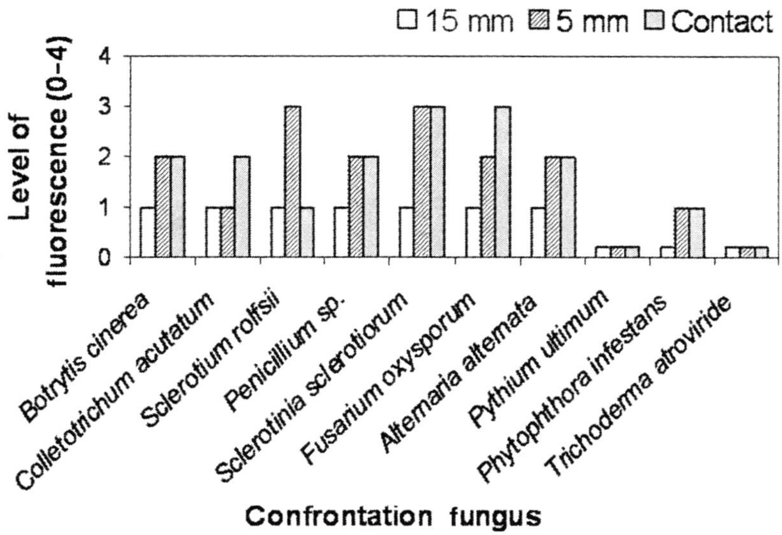

Figure 4. Induction of the biocontrol-related gene *ech42* (endochitinase-encoding) of *T. atroviride* strain P1 during confrontation in vitro with different plant pathogenic fungi (*T. atroviride* self-confrontation used as a control), measured by quantifying the level of fluorescence of a GFP-expressing transformant of PI with the *ech42* promoter (*ech42*::gfp). Fluorescence intensity (scale from 0 = no fluorescence to 4 = maximum fluorescence) was monitored when the two fungal colonies were at a distance of 15 mm, 5 mm and in physical contact

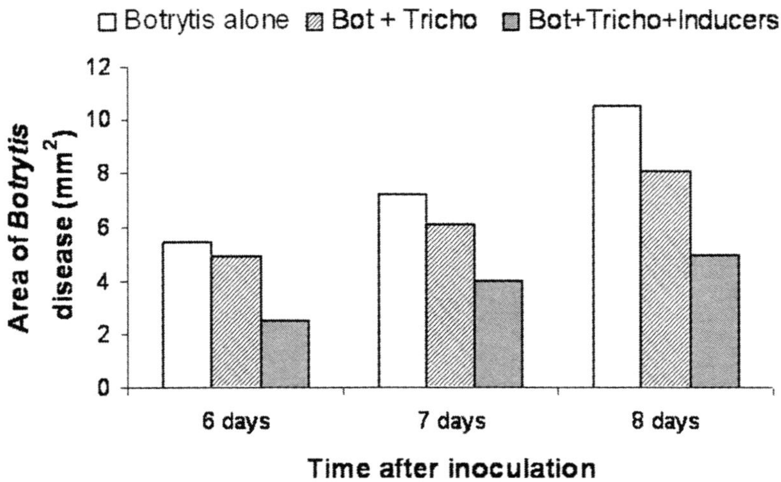

Figure 5. Suppression of *Botrytis* disease symptoms on bean leaves by *Trichoderma atroviride* strain P1 is enhanced with the addition of "biocontrol inducers" obtained by digesting pathogen cell walls with *Trichoderma* enzymes. Inoculation was done with spores of *B. cinerea* alone, *B. cinerea* + *T. atroviride* P1, *B. cinerea* + *T. atroviride* P1 + inducers, and averaged area of the necrotic spots was measured at 6, 7 and 8 days after infection. Inducer mixtures used here contained several bioactive compounds such as identified small oligosaccharides made of two different sugar monomer types and one amino acid

antibiotics and/or other secondary metabolites, which had an inhibitory effect on the spore germination of *Botrytis*. Mass spectrometry analysis (ESI-MS) was used to purify and characterize these novel mycoparasiticsm-related inducers and plant elicitors. These compounds were comprised of short oligosaccharides made of two types of monomers, one with and another without an amino acid residue. Further analysis by MS/MS involving selective fragmentation of peaks in the spectrum, demonstrated the presence of at least three distinct biologically active compounds (S. L. Woo and M. Lorito, unpublished data).

6.3.2. TRICHODERMA–PLANT INTERACTIONS

Recently, much emphasis has been placed on studies investigating the little known interactions between *Trichoderma* and the plant.[7,18,50,51] It was always assumed that the beneficial effects of *Trichoderma* to the plant were limited to the biocontrol of the pathogens causing disease, particularly in the rhizosphere. It has now also been demonstrated that *Trichoderma* is able to systemically activate resistance mechanisms of the plant to pathogen attack.[18,52–54] Different monocotyledonous and dicotyledonous crops, including Gramineae, Solanaceae and Cucurbitaceae infected with diverse fungi (*Rhizoctonia solani*, *Botrytis cinerea*, *Colletotrichum* spp., *Magnaporthe grisea*, *Phytophthora* spp., *Alternaria* spp. etc.), bacteria (*Xanthomonas* spp., *Pseudomonas syringae*, etc.) as well as viruses (cucumber mosaic virus), were more resistant to disease development when the plants were treated with *Trichoderma* prior to pathogen attack.[18] Plant colonization by certain *Trichoderma* spp., reduced disease symptoms caused by one or two different pathogens even when the biocontrol fungus was inoculated at a different time and at a different location on the plant than the pathogen. Further, this effect to the plant occurred at a molecular level; whereby extracts from plants subjected to *Trichoderma* root treatments had more anti-microbial activity than extracts derived from untreated plants. This inhibitory effect corresponded to the up-regulation of different endogenous pathogenesis-related (PR) and defense-related proteins (chitinases, glucanases, peroxidases and specific phytoalexins) and enzyme activities (HPL, PAL1 etc.) in the plant; production that typically augments when the plant incurs pathogen attack.[50,51]

The reaction of the plant to *Trichoderma* is similar to the induced systemic resistance (ISR) elicited by the interaction of the plant with rhizobacteria (RISR).[52,55,56] The ISR mechanism activated by *Trichoderma* may play a greater role in plant protection than the mechanisms of biocontrol. The ISR effect appears to be strongly dependent on the strain of the antagonist and species/cultivar of the plant used in the combination. Similarly, the growth promotion effect has been observed with some, but not all, cultivars of

Figure 6. Differences in the plant growth-promotion effect of two *Trichoderma* species (*T. atroviride* strain P1 and *T. harzianum* strain T22) relatively to an untreated control (C), on four different tomato commercial cultivars. Lines indicate average height of untreated plants. *Trichoderma* spores were applied once as a seed treatment (Ruocco et al., unpublished data)

tomato (Figure 6, M. Ruocco et al., unpublished data) and maize[18] (Chapter 7) in the presence of *Trichoderma*. Nevertheless, these findings could potentially enlarge the commercial market of *Trichoderma* as a biological control agent not only for biofungicide applications, but as a general broad spectrum protectant that stimulates plant disease resistance and enhances plant growth.

6.3.2.1. Trichoderma *the Plant Pathogen versus* Trichoderma *the Plant Symbiont*

The interaction between *Trichoderma* and the plant had been assumed to be superficial and limited to the rhizosphere region external to the plant. It is now known that the hyphae of *Trichoderma* penetrate the root cortex and fungal colonization is limited only to the first few cell layers of plant tissue, probably due to the deposition of callose barriers by the surrounding plant tissues.[50,57] The situation between the fungus and the plant appears to stabilize; that is the fungus does not continue to penetrate and the plant does not proceed to destroy the intruder. However, a very active, direct interaction occurs at the molecular level between the fungus and the plant, that activates the expression of numerous defense genes in the plant and "biocontrol" genes in the fungus.[53,56,57,58]

What are the factors determining that *Trichoderma* does not develop as a pathogen? *Trichoderma* has the weaponry to act as phytopathogen: producing a variety of plant-degrading enzymes, due to its saprophytic lifestyle, as well as other hydrolytic enzymes, and over 200 antibiotics highly toxic to cells of many macro- and microorganisms.[10] *Trichoderma* culture filtrates contain macromolecules and low molecular weight compounds capable of inducing a strong peak of calcium uptake in isolated plant cell cultures and cause apoptosis (Programmed Cell Death).[59] Therefore, *Trichoderma* species have an intrinsic ability to behave as a plant pathogen, but they have developed as a symbiont that has mutualistic interactions with the plant: benefits for the plant by stimulating resistance to pathogen attack; and benefits for itself by limiting niche colonization to competing microbes, stimulating plant growth and root system development that increases the colonization zone and production of root exudates that provide a nutrient basis for the fungus in the rhizosphere.[18,46,47,54] The actual mechanisms utilized in the symbiosis process by *Trichoderma* are not yet fully understood, but probably they include its ability to activate plant defense by the production of avirulence (avr)-like compounds and other forms of elicitors.[23,60,61]

Recent investigations, in collaboration with Pierre de Wit (University of Wageningen, the Netherlands), provide an indication to how one *Trichoderma* strain could elicit plant defenses. *T. atroviride* strain P1 was transformed with the avirulence gene encoding for the Avr4 avirulence protein of the tomato pathogen *Cladosporium fulvum*,[62] under the regulation of either a strong constitutive or an inducible promoter. Transgenic lines of *Trichoderma* overexpressing Avr4 were used to treat seeds of tomato cultivars with and without the corresponding *Cf4* resistance gene. Seeds from the *Cf4*-containing cultivar exhibited lower germination, emerging plants were stunted and generally less healthy in appearance than the plants emerging from the non-*Cf4*-containing cultivar, which were similar to the untreated controls.[18] Further, *Trichoderma*-avr treatments to the roots of mature *Cf4* tomato plants caused the rapid appearance of many hypersensitive response (HR)-like necrotic zones to the root and leaf surface (M. Ruocco, and M. Lorito, unpublished data). These results suggest that not only is a molecular cross-talk established between *Trichoderma* and the colonized plant, but that the fungus can transfer to a receptive plant, molecules that are recognized by the plant. These molecules play an important role in activating the defense system in the plant to various biotic factors (micro-, macrophagous pathogens, viruses), and/or abiotic factors (drought, pH, elemental deficiency).[18,63,64] Therefore, *Trichoderma* strains have the potential to be used as "vectors" of beneficial molecule transfer to the plant, and the application of constitutive or selected inducible promoters allows the targeting of gene expression products at the specific moment of interaction with the plant[15,58] or the pathogen.[23,34,57,58]

6.3.2.2. *The Language Used in* Trichoderma-*plant Molecular Cross-talk*
What molecules does *Trichoderma* use in its exchange with the plant that can significantly alter the plant proteome and that can strongly activate the plant resistance mechanisms? One kind of elicitors produced by *Trichoderma* includes peptides and proteins ranging from 6 to 40 kDa, such as a serine proteinase, a xylanase, an endopolygalacturonase, and a chitin deacetylase.[6,60,61] Other proteins that activate a defense response in the plant include the lytic enzymes that *Trichoderma* secretes during its mycoparasitic and antagonistic activities.[23,26–28,39] These enzymes plus their degradation by-products are probably detected by specific receptors in the plant. This may signal that there is a pathogen risk in the vicinity and that plant defense mechanisms should be activated either preventively before damage is incurred or immediately upon attack. Increased disease resistance in the plant suggests that *Trichoderma* sensitizes or pre-activates the defense system instead of causing the production and accumulation of defense proteins by long term expression of the encoding genes. Applications of *Trichoderma* metabolites, produced by the antagonist alone or in the presence of the foliar pathogen *Botrytis cinerea*, to aequorin-expressing soy bean cell suspension cultures were differentially perceived by the cells, activating Ca^{2+}-mediated signaling and typical plant cell-related responses[59] (Navazio et al., unpublished data). Processes downstream of the Ca^{2+} signal, which can be effected by metabolites secreted by biocontrol agents and phytopathogens, include various differential cell reactions: reactive oxygen species (ROS) accumulation, reduced cell viability, programmed cell death (PCD) vs. necrosis (induction of caspase 3-like activity, chromatin condensation and other morphological cell alterations).[65–67] Comparative testing of material produced by wild type and *ech*42-disruptant strains of *Trichoderma* P1 demonstrated that the secreted endochitinase has an important role in determining a reaction by the plant and in regulating its response to the biocontrol fungus (Navazio et al., unpublished data).

A second type of elicitor used in *Trichoderma*-plant molecular cross-talk may involve avirulence-like proteins similar to those found in avirulent races of pathogens that function as elicitors of related responses in plants containing the corresponding resistance gene,[62,63] and the new avr proteins typical of *Trichoderma* itself. Avr homologues of *avr4* and *avr9* from *C. fulvum* were found in *T. harzianum* and *T. atroviride* by sequence hybridization to labeled probes[18] (Ruocco et al, unpublished data). In addition to these homologues, the search for specific avr genes in *Trichoderma* has produced several putative proteins that are being isolated and investigated further. These *avr* genes may be used transgenically to stimulate related reaction in a variety of plant species, given the avirulence effect of these fungi on many different crops.[50,52,53,56]

A third type of elicitor used in *Trichoderma*-plant communication consists of the breakdown products released from the pathogen and the plant cell walls by digestion with the different *Trichoderma* lytic enzymes, as described in Section 6.3.1.2. These molecules not only stimulate biological and antagonistic activity of the fungus by but they also elicit the resistance mechanisms when plant cells are exposed to them or when they are injected to the root or leaves.[18,48,49,68,69] Some of these low molecular weight mycoparasitic-related inducers and plant elicitors are comprised of short oligosaccharides (S. L. Woo and M. Lorito, unpublished data). These <3 kDa fractions had a greater effect on soybean cell suspension cultures than fractions from the higher molecular weight fractions, in stimulating a related response reflected as an increase in $[Ca^{2+}]_{cyt}$.[59] Other possible compounds in the low MW portion include antibiotics such as peptaibols that affect the membrane permeability of fungi and plant cells, resulting in leakage of cytoplasmic material and cell death.[44] Therefore, the plant, may respond to these compounds by increasing $[Ca^{2+}]_{cyt}$, defending itself and reducing subsequent plant cell damage.[59,68]

The fourth type of *Trichoderma* elicitors, those that produce beneficial effects to the plant by inducing resistance as well as stimulating plant growth, are determined by the substrates that the fungus metabolizes from its surrounding environment, i.e., various nutrient sources. We have found that applications of *Trichoderma* culture filtrates, produced in the presence of glycerol, chitin or *Botrytis* cell walls, to tomato seeds germinated on water agar stimulated plant growth both in root length and secondary root branching, as well as in the overall stem elongation (Ambrosino et al. unpublished data). Plant growth promotion was also clearly noted when seeds of lettuce were treated with the *Trichoderma* culture filtrates produced not only with glycerol and chitin, but especially when produced in the presence of barley fiber, residues from milling (Figure 7). Post-harvest treatments of fruit with culture filtrates of *Trichoderma* grown with barley fiber exhibited the greatest biocontrol effect on pathogens, whereas the *Trichoderma* culture filtrates grown with *Botrytis* cell walls produced the greatest ISR effect when used as pre-pathogen treatments to the plant.

6.3.2.3. *Trichoderma a Member the Multi-component Community*

We are using proteome, micro- and macro-array or functional analyses to determine the identity of the many compounds that are produced during the various interactions between *Trichoderma* with different pathogens and with different plant species. This permits us to analyze the antagonist–pathogen–plant relationships at all levels, each single component alone, each component in all two-way combinations and finally in a three-way interaction. In order to understand this complex relationship, we use subtractive analysis of different proteomes separated by 2-D gel electrophoresis to reveal differential protein spots. In brief: (1) analysis of the plant proteome indicates that the plant produces a very different set of proteins when it is colonized by the biocontrol

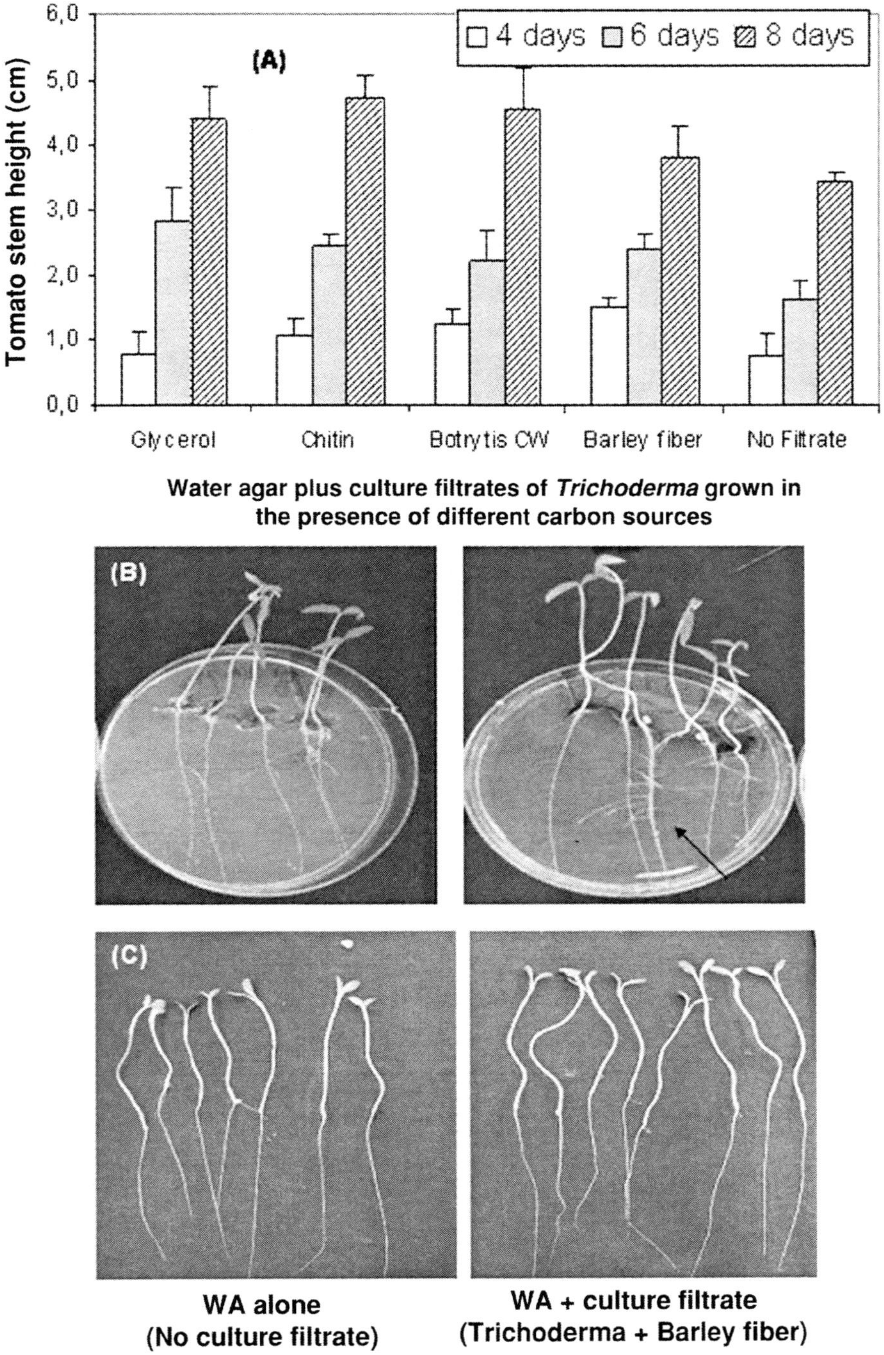

Figure 7. Plant growth-promotion effect obtained *in vitro* by applying crude culture filtrates of *Trichoderma* instead of the living fungus. Filtrates were obtained by growing *T. harzianum* strain T22 in a minimal, salt-based medium supplemented with different carbon sources, applied to seeds of tomato (A, B) and lettuce (C) placed on water agar, and the effect evaluated 4, 6 and 8 days after germination. Stimulation of secondary roots in tomato (indicated by arrows, control at right) (B) and of overall growth of lettuce seedlings (control at left) (C)

fungus *Trichoderma* rather than the foliar pathogen *Botrytis*.[70] The presence of the antagonist in the three-way interaction between the fungal pathogen and the plant results in a strong reduction in the number and the level of intensity of plant proteins produced in comparison to the simpler two-way plant-*R. solani* interaction, which produced the highest number of novel and increased differential spots versus the plant alone. This indicates that the presence of *Trichoderma* greatly modifies the way the plant responds to, and interacts with the pathogen. (2) Analysis of the *Trichoderma* proteome found that *Trichoderma* alone compared to *Trichoderma* with the plant, or *Trichoderma* with the plant and pathogen produced about 270 differential spots.[70] A comparison between *Trichoderma*-bean roots-*Rhizoctonia* and *Trichoderma*-plant, revealed more than 230 differential spots accumulated in the antagonist proteome. This indicates that the presence of a pathogen causes major changes in the *Trichoderma* proteome when *Trichoderma* is colonizing the plant. (3) Analysis of different pathogen proteomes showed that differential proteins were produced by *B. cinerea* in the presence of the plant alone or plant with *Trichoderma*, as compared to the control of *Botrytis* alone. The *Trichoderma*—plant-*Botrytis* produced 204 differential spots in respect to the *Botrytis*-plant interaction.[70] This result indicates that the presence of *Trichoderma* induces major changes in the *Botrytis* proteome when the pathogen is in contact with the plant.

6.4. Application of *Trichoderma* Interaction Products for Improvement of Biocontrol

The interactions of *Trichoderma* in the soil microbial community are indeed intricate. As a result, the fungus produces a multitude of biological products that have great potential applications.[6,11,14,15,20,22] Prospective applications of the "mycoparasitism" inducers in agricultural and industrial production are numerous. They can be used as stimulators of biocontrol in the native, existing, fungal antagonist populations of the soil community. Timed applications of the compounds may be conducted preventively in order to avoid the initiation of disease from infective material in the environment, using spring applications to activate soil antagonists for control of pathogen spores that have overwintered on plant debris, or sclerotia and other resistant structures in the soil (Chapter 12). Alternatively, inducers can be used as curative treatments, applied after plant disease is detected or reaches an economic threshold of crop damage.

These low MW compounds also have good perspectives as growth enhancers in the industrial production of fungal biomass for commercial biopesticides. The greatest potential of these inducers derives from their capability to activate and stimulate the production of fungal enzymes, in particular cell wall degrading enzymes, and antibiotics.[6] The agricultural-food industry has a notable commercial interest in many of the cell wall degrading enzymes produced by *Trichoderma*. The β-(1,4)-endoglucanases produced by *T. longibrachiatum* and *T. reesei* are used to clear the turbidity caused by β-glucans in beer production. *Trichoderma* cellulases and hemicellulases are regularly added to chicken feed to improve the digestibility of the feed and facilitate faecal movement in the fowl. These inducers have good prospectives in the pharmaceutical industry for the production of antibiotics.[7]

The potential for the joint use of these inducers with the live whole microorganism, in a biocontrol package containing a growth stimulant combined with the fungal antagonist for field applications is very high. In some field situations, competition for space or specific infection sites, nutrients and other factors necessary for growth is the process used by antagonistic microbes to control plant pathogens (Chapter 8). Competition for space has not been clearly demonstrated as a major mechanism of antagonism for *Trichoderma*, but there is evidence that it plays a role in biocontrol when these antagonists establish their dominance in a specific environmental niche. For example, *T. harzianum* controls *Botrytis cinerea* on grapes by colonizing blossom tissue, excluding the pathogen from its infection site.[6] The most important mechanism involved in biocontrol of vascular disease caused by *F. oxysporum* f. sp. *melonis* is the carbon and nitrogen competition that occurs with *T. harzianum* during rhizosphere colonization.[2] A boost in spore germination or accelerated mycelial growth of the fungus by the addition of an inducer would aid in the establishment of the antagonist at target sites.

Low MW compounds, the various *Trichoderma* cell wall degrading enzymes or even the whole fungal organism could be used in the field as stimulators of induced systemic resistance or as plant growth promoters of crop plants. Soil incorporation of the *Trichoderma*-based substances would not only protect the germinating seeds from pathogen attack, as well as augment seed germinability and survival, and consequently improve development of the root system, plant growth and yield. Post-emergence spray treatments could aid plant and fruit development, and serve as inhibitors of disease expansion.

Finally, the further understanding of the mechanisms involved in the interaction between *Trichoderma* and the plant now provide the opportunity to genetically improve the ability of biocontrol strains to induce ISR. We transformed *Trichoderma* strain P1 with a glucose oxidase gene from *Aspergillus niger*.[34] In subsequent in vitro and in plant–soil experiments, the transformants constitutively producing GOX outperformed the wild type strain both as mycoparasites and ISR inducing agents.[19] Bean seeds coated with spores from the *Trichoderma* transformants produced plants that were more resistant to subsequent *B. cinerea* leaf infection in comparison to those treated with the wild type. This is probably because the high glucose oxidase activity expressed by the *Trichoderma* transformants catalyzed the production of hydrogen peroxide and reactive oxygen species that were able to systemically alert the plant related mechanism prior to pathogen attack.

References

1. S. R. Gliessman, Sustainable agriculture: An agroecological perspective, *Adv. Plant Pathol.* 11, 45–57 (1995).
2. I. Chet, *Innovative Approaches to Plant Disease Control* (Wiley, New York, 1987).
3. D. Hornby, *Biological Control of Soil-borne Plant Pathogens* (CAB International, Wallingford, UK, 1990), 479 p.
4. M. N. Burge, *Fungi in Biological Control Systems* (Manchester University Press, Manchester, UK, 1988), 269 p.
5. R. D. Lumsden and J. A. Lewis, Selection, production, formation, and commercial use of plant disease biocontrol fungi: problems and progress, in *Biotechnology of Fungi for Improving Plant Growth*, edited by J. M. Whipps and R.D. Lumsden (Cambridge University Press, Cambridge, UK, 1989), pp. 171–190.
6. G. E. Harman and C. P. Kubicek, *Trichoderma* and *Gliocladium*, *Vol. 2:. Enzymes, Biological Control and Commercial Applications* (Taylor and Francis, London, 1998), 278 p.
7. C. R. Howell, Mechanisms employed by *Trichoderma* species in the biological control of plant diseases: The history and evolution of current concepts, *Plant Dis.* 87, 4–10 (2003).
8. J. M. Whipps, Microbial interactions and biocontrol in the rhizosphere, *J. Exp. Bot.* 52, 487–511 (2001).
9. M. Lorito, S. Woo, M. Iaccarino, and F. Scala, *Microrganismi antagonisti*, in *Microrganismi benefici per le piante*, edited by M. Iaccarino (Idelson-Gnocchi, Napoli, Italy, 2006), pp. 146–175.
10. C. P. Kubicek and G. E. Harman, *Trichoderma* and *Gliocladium, Vol. 1: Basic Biology, Taxonomy and Genetics* (Taylor & Francis, London, 1998), 278 p.
11. G. E. Harman, Overview of mechanisms and uses of *Trichoderma* spp., *Phytopathology* 96, 190–194 (2006).
12. R. Weindling, *Trichoderma lignorum* as a parasite of other fungi, *Phytopathology* 22, 837–845 (1932).
13. G. C. Papavizas, *Trichoderma* and *Gliocladium*: Biology, ecology, and potential for biocontrol, *Annu. Rev. Phytopathol.* 23, 23–54 (1985).
14. G. E. Harman, M. Lorito, and J. M. Lynch, Uses of *Trichoderma* spp. to alleviate or remediate soil and water pollution, *Adv. Appl. Microbiol.* 56, 313–331 (2004).
15. G. E. Harman, Myths and dogmas of biocontrol: Changes in perceptions derived from research on *Trichoderma. harzianum* T-22, *Plant Dis.* 84, 377–393 (2000).
16. S. L. Woo, F. Scala, M. Ruocco, and M. Lorito, The molecular biology of the interactions between *Trichoderma*, phytopathogenic fungi and plants, *Phytopathology* 96, 181–185 (2006).
17. C. P. Kubicek, R. L. Mach, C. K. Peterbauer, and M. Lorito, *Trichoderma*: From genes to biocontrol, *J. Plant Pathol.* 83, 11–23 (2001).
18. G. E. Harman, C. R. Howell, A. Viterbo, I. Chet, and M. Lorito, *Trichoderma* species—Opportunistic, avirulent plant symbionts, *Nat. Rev. Microbiol.* 2, 43–56 (2004).

19. K. Brunner, S. Zeilinger, R.Ciliento, S. L. Woo, M. Lorito, C. P. Kubicek, and R. L. Mach, Genetic improvement of a fungal biocontrol agent to enhance both antagonism and induction of plant systemic disease resistance, *Appl. Environ. Microbiol.* 71(7), 3959–3965 (2005).
20. M. Lorito, S. L. Woo, I. Garcia Fernandez, G. Colucci, G. E. Harman, J. A. Pintor-Toro, E. Filippone, S. Muccifora, C. B. Lawrence, A. Zoina, S. Tuzun, and F. Scala, Genes from mycoparasitic fungi as a source for improving plant resistance to fungal pathogens, *Proc. Natl. Acad. Sci. USA* 95, 7860–7865 (1998).
21. M. Lorito, F. Scala, A. Zoina, and S. L. Woo, Enhancing biocontrol of fungal pests by exploiting the *Trichoderma* genome, in *Enhancing Biocontrol Agents and Handling Risks*, edited by M. Vurro and J. Gressel (IOS Press, Amsterdam, 2001), pp. 248–259.
22. E. Monte, Understanding *Trichoderma*: Between biotechnology and microbial ecology, *Int. Microbiol.* 4, 1–4 (2001).
23. S. Zeilinger, C. Galhaup, K. Payer, S. L. Woo, R. L. Mach, C. Fekete, M. Lorito, and C. P. Kubicek, Chitinase gene expression during mycoparasitic interaction of *Trichoderma harzianum* with its host, *Fung. Genet. Biol.* 26, 131–140 (1999).
24. M. Rey, A. Llobell, E. Monte, F. Scala, and M. Lorito, Genomics of *Trichoderma,* in *Appl. Mycol. Biotechnol.* 4, 225–248 (2004).
25. M. Lorito, Chitinolytic enzymes and their genes, in *Trichoderma and Gliocladium, Volume 2: Enzymes, Biological Control and Commercial Applications*, edited by G. E. Harman and C. P. Kubicek (Taylor and Francis, London, 1998), pp. 73–99.
26. S. L. Woo, B. Donzelli, F. Scala, R. Mach, G. E. Harman, C. P. Kubicek, G. Del Sorbo, and M. Lorito, Disruption of the *ech42* (endochitinase-encoding) gene affect biocontrol activity in *Trichoderma harzianum* P1, *Mol. Plant-Microbe Interact.* 12, 419–429 (1999).
27. C. Kullnig, R. L. Mach, M. Lorito, and C. P. Kubicek, Enzyme diffusion from *Trichoderma atroviride* (*T. harzianum* P1) to *Rhizoctonia solani* is a prerequisite for triggering of *Trichoderma ech42* gene expression before mycoparasitic contact, *Appl. Environ. Microbiol.* 66, 2232–2234 (2000).
28. S. Haran, H. Schickler, A. Oppenheim, and I. Chet, Differential expression of *Trichoderma harzianum* chitinases during mycoparasitism, *Phytopathology* 86, 980–985 (1996).
29. M. Lorito, G. E. Harman, C. K. Hayes, R. M. Broadway, A. Tronsmo, S. L. Woo, and A. Di Pietro, Chitinolytic enzymes produced by *Trichoderma harzianum*: Antifungal activity of purified endochitinase and chitobiosidase, *Phytopathology* 83, 302–307 (1993).
30. M. Lorito, C. K. Hayes, A. Di Pietro, S. L. Woo, and G. E. Harman, Purification, characterization and synergistic activity of a glucan 1,3-β-glucosidase and an *N*-acetyl-β-glucosaminidase from *Trichoderma harzianum*, *Phytopathology*, 84, 398–405 (1994).
31. A. Tronsmo, Biological and integrated controls of *Botrytis cinerea* on apple with *Trichoderma harzianum*, *Biol. Control* 1, 59–62 (1991).
32. E. Agosin and J. M. Aguilera, Industrial production of active propagules of *Trichoderma* for agricultural use, in *Trichoderma and Gliocladium, Volume 2: Enzymes, Biological Control and Commercial Applications*, edited by G. E. Harman and C. P. Kubicek (Taylor and Francis, London, 1998), pp. 205–227.
33. X. Jin, C. K. Hayes, and G. E. Harman, Principles in the development of biological control systems employing *Trichoderma* species against soil-borne plant pathogenic fungi, in *Frontiers in Industrial Mycology*, edited by G. F. Leatham (Chapman and Hall, New York, 1992), pp. 174–195.
34. R. L. Mach, C. K Peterbauer, K. Payer, S. Jaksits, S. L. Woo, S. Zeilinger, C. M. Kullnig, M. Lorito, and C. P. Kubicek, Expression of two major chitinase genes of *Trichoderma*

atroviride (T. harzianum P1) is triggered by different regulatory signals. *Appl. Environ. Microbiol.* 65, 1858–1863 (1999).
35. M. Lorito and F. Scala, Microbial genes expressed in transgenic plants to improve disease resistance. *J. Plant Pathol.*, Pisa, 81(2), 73–88 (1999).
36. M. Lorito, S. L. Woo, M. D'Ambrosio, G. E. Harman, C. K. Hayes, C. P. Kubicek, and F. Scala, Synergistic interaction between cell wall degrading enzymes and membrane affecting compounds, *Mol. Plant-Microbe Interact.* 9, 206–213 (1996).
37. M. Lorito, C. Peterbauer, C. K. Hayes, and G. E. Harman, Synergistic interaction between fungal cell wall degrading enzymes and different antifungal compounds enhances inhibition of spore germination, *Microbiology UK* (London, UK) 140, 623–629 (1994).
38. J. M. Baek, C. R. Howell, and C. M. Kenerley, The role of an extracellular chitinase from *Trichoderma virens* Gv29-8 in the biocontrol of *Rhizoctonia solani, Curr. Genet.* 35, 41–50 (1999).
39. C. Carsolio, N. Benhamou, S. Haran, C. Cortes, A. Gutierrez, I. Chet, and A. Herrera-Estrella, Role of the *Trichoderma harzianum* endochitinase gene, *ech42*, in mycoparasitism, *Appl. Environ. Microbiol.* 65, 929–935 (1999).
40. V. Fogliano, A. Ballio, M. Gallo, S. L. Woo, F. Scala, and M. Lorito, *Pseudomonas lipodepsipeptides* and fungal cell wall-degrading enzymes act synergistically in biological control, *Mol. Plant-Microbe Interact.* 15, 323–333 (2002).
41. M. Lorito, A. Di Pietro, C. K. Hayes, S. L. Woo, and G. E. Harman, Antifungal, synergistic interaction between chitinolytic enzymes from *Trichoderma harzianum* and *Enterobacter cloacae, Phytopathology* 83, 721–728 (1993).
42. F. Vinale, R. Marra, F. Scala, E. L. Ghisalberti, M. Lorito, and K. Sivasithamparam, Major secondary metabolites produced by two commercial *Trichoderma* strains active against different phytopathogens, *Lett. Appl. Microbiol.* 43(2), 143–148 (2006).
43. C. R. Howell and R. D. Stipanovic, Gliovirin, a new antibiotic from *Gliocladium virens*, and its role in the biological control of *Pythium ultimum*, *Can. J. Microbiol.* 29, 321–324 (1983).
44. M. Schirmböck, M. Lorito, Y. L. Wang, C. K. Hayes, I. Arisan-Atac, F. Scala, G. E. Harman, and C. P. Kubicek, Parallel formation and synergism of hydrolytic enzymes and peptaibol antibiotics, molecular mechanisms involved in the antagonistic action of *Trichoderma harzianum* against phytopathogenic fungi, *Appl. Environ. Microbiol.* 60, 4364–4370 (1994).
45. A. Viterbo, M. Harel, and I. Chet, Isolation of two aspartyl proteases from *Trichoderma asperellum* expressed during colonization of cucumber roots, *FEMS Microbiol. Lett.* 238, 151–158 (2004).
46. C. Altomare, W. A. Norvell, T. Bjorkman, and G. E. Harman, Solubilization of phosphates and micronutrients by the plant-growth-promoting and biocontrol fungus *Trichoderma harzianum* Rifai 1295-22, *Appl Environ Microbiol.* 65, 2926–2933 (1999).
47. I. Yedidia, A. K. Srivastva, Y. Kapulnik, and I. Chet, Effect of *Trichoderma harzianum* on microelement concentrations and increased growth of cucumber plants. *Plant Soil* 235, 235–242 (2001).
48. S. L. Woo, M. Ruocco, R. Ciliento, S. Lanzuise, F. Vinale, E. Formisano, V. Scala, D. Turrà, F. Scala, A. Zoina, K. Abadi, and M. Lorito, Molecular factors involved in the interaction between plants, pathogens and biocontrol fungi, in *11th International Congress on Molecular Plant–Microbe Interactions*, St. Petersburg, Russia, July 18–26, 2003, Volume of Abstracts, p. 368.
49. S. L. Woo, E. Formisano, V. Fogliano, C. Cosenza, A. Mauro, D. Turrà, I. Soriente, S. Ferraioli, F. Scala, and M. Lorito, Factors that contribute to the mycoparasitism stimulus in *Trichoderma atroviride* strain P1, *J. Plant Pathol.* 86(4), 337 (2004).

50. I. Yedidia, N. Benhamou, and I. Chet, Induction of defense responses in cucumber plants (*Cucumis sativus* L.) by the biocontrol agent *Trichoderma harzianum*, *Appl. Environ. Microbiol.* 65, 1061–1070 (1999).
51. I. Yedidia, N. Benhamou, Y. Kapulnik, and I. Chet, Induction and accumulation of PR proteins activity during early stages of root colonization by the mycoparasite *Trichoderma harzianum* strain T-203, *Plant Physiol. Biochem.* 38, 863–873 (2000).
52. G. De Meyer, J. Bigirimana, Y. Elad, and M. Hofte, Induced systemic resistance in *Trichoderma harzianum* T39 biocontrol of *Botrytis cinerea, Eur. J. Plant Pathol.* 104, 279–286 (1998).
53. L. E. Hanson and C. R. Howell, Elicitors of plant defense responses from biocontrol strains of *Trichoderma virens, Phytopathology* 94, 171–176 (2004).
54. M. T. Windham, Y. Elad, and R. Baker, A mechanism for increased plant growth induced by *Trichoderma* spp., *Phytopathology* 76, 518–521 (1986).
55. S. C. van Wees, M. Luijendijk, I. Smoorenburg, L. C. van Loon, and C. M. J. Pieterse, Rhizobacteria-mediated induced systemic resistance (ISR) in*Arabidopsis* requires sensitivity to jasmonate and ethylene but is not accompanied by an increase in their production, *Physiol. Mol. Plant Pathol.* 57, 123–134 (2000).
56. I. Yedidia, M. Shoresh, Z. Kerem, N. Benhamou, Y. Kapulnik, and I. Chet, Concomitant induction of systemic resistance to *Pseudomonas syringae* pv. *lachrymans* in cucumber by *Trichoderma asperellum* (T-203) and accumulation of phytoalexins, *Appl. Environ. Microbiol.* 69, 7343–7353 (2003).
57. A. Bolwerk, B. J. J. Lugtenberg, M. Lorito, and G. V. Bloemberg, Biocontrol of tomato foot and root rot by *Trichoderma, Mol. Plant-Microbe Interact.* in press.
58. Z. Lu, R. Tombolini, S. L. Woo, S. Zeilinger, M. Lorito, and J. K. Jansson, In vivo study of *Trichoderma*–pathogen–plant interactions with constitutive and inducible GFP reporter systems, *Appl. Environ. Microbiol.* 70, 3073–3081 (2004).
59. L. Navazio, B. Baldan, R. Moscatiello, A. Zuppino, S. L. Woo, P. Mariania, and M. Lorito, Early events in the molecular interaction between the biocontrol agent *Trichoderma harzianum* and soybean cells, in *XVII International Botanical Congress*, Vienna, Austria, 2005, p. 267.
60. J. Enkerli, G. Felix, and T. Boller, The enzymatic activity of fungal xylanase is not necessary for its elicitor activity, *Plant Physiol.* 121, 391–397 (1999).
61. A. Sharon, Y. Fuchs, and J. D. Anderson, The elicitation of ethylene biosynthesis by a *Trichoderma* xylanase is not related to the cell wall degradation activity of the enzyme, *Plant Physiol.* 102, 1325–1329 (1993).
62. P. J. G. M. De Wit, Molecular characterization of gene-for-gene systems in plant–fungus interactions and the application of avirulence genes in control of plant pathogens, *Annu. Rev. Phytopathol.* 30, 391–418 (1992).
63. W. Ning, F. Chen, B. Mao, Q. Li, Z. Liu, Z. Guo, and Z. He, *N*-Acetylchito-oligosaccharides elicit rice defence responses including hypersensitive response-like cell death, oxidative burst and defence gene expression, *Physiol. Mol. Plant Pathol.* 64, 263–271 (2004).
64. M. Ron and A. Avni, The receptor for the fungal elicitor ethylene-inducing xylanase is a member of a resistance-like gene family in tomato, *Plant Cell* 16, 1604–1615 (2004).
65. S. Orrenius, B. Zhivotovsky, and P. Nicotera, Regulation of cell death, the calcium-apoptosis link, *Nat. Rev. Mol .Cell Biol.* 4, 552–565 (2003).
66. A. V. Tiedemann, Evidence for a primary role of active oxygen species in induction of host cell death during infection of bean leaves with *Botrytis cinerea*, *Physiol. Mol. Plant Pathol.* 50, 151–166 (1997).

67. A. Zuppini, B. Baldan, R. Millioni, F. Favaron, L. Navazio, and P. Mariani, Chitosan induces Ca^{2+}-mediated programmed cell death in soybean cells, *New Phytol.* 161, 557–568 (2004).
68. J. Müller, C. Staehelin, Z. P. Xie, G. Neuhaus-Url, and T. Boller, Nod factors and chitooligomers elicit an increase in cytosolic calcium in aequorin-expressing soybean cells, *Plant Physiol.* 124, 733–739 (2000).
69. A. Mithöfer, J. Ebel, A. A. Bhagwat, T. Boller, and G. Neuhaus-Url, Transgenic aequorin monitors cytosolic calcium transients in soybean cells challenged with β-glucan or chitin elicitors, *Planta* 207, 566–574 (1999).
70. R. Marra, P. Ambrosino, V. Carbone, F. Vinale, S. L. Woo, M. Ruocco, R. Ciliento, S. Lanzuise, S. Ferraioli, I. Soriente, S. Gigante, D. Turrà, V. Fogliano, F. Scala, and M. Lorito, Study of the three-way interaction between *Trichoderma atroviride*, plant and fungal pathogens by using a proteomic approach, *Curr. Genet.* 50, 307–321 (2006).

7. THE MECHANISMS AND APPLICATIONS OF SYMBIOTIC OPPORTUNISTIC PLANT SYMBIONTS

Gary E. Harman* and Michal Shoresh
Department of Horticultural Sciences, Cornell University, Geneva, NY 14456, USA

Abstract. A number of fungi have evolved a symbiotic life style with plants, including some organisms that include similar strains or species that are plant pathogens. Some are obligate symbionts such as ecto- or endomycorrhizal fungi, while others are endophytes that have free-living capabilities. Still others are highly competitive in soil and proliferate there. These are the opportunistic plant symbionts. Fungi in the genus *Trichoderma* have long been considered as biocontrol agents, but they are highly successful plant symbionts as well. The critical step for establishment of the symbiotic life style begins with root colonization and infection of outer cortical layers. A zone of chemical interaction is established; some of the *Trichoderma* signaling molecules are known. As a result of this interaction, the fungus is walled off; in rare cases where components of this communication are lacking, *Trichoderma* can become a pathogen. The results of this interaction include induced systemic resistance, increased growth responses and yields, and increased nutrient uptake and fertilizer use efficiency. The interaction induces substantial changes in plant physiology. In the maize-*T. harzianum* strain T22 interaction, more than 300 proteins have altered expression, with a number of them being up-regulated. Included in this group are, most notably, enzymes of carbohydrate metabolism and proteins associated with pathogen resistance and stress. Multiple forms of several proteins are upregulated, including numerous examples of chitinases, β-glucosidases, proteins with nucleotide binding sites and leucine rich repeats associated with resistance to disease, sucrose synthase, and methionine synthase. The substantial increases in several of these are highly suggestive of changes in metabolic pathways or regulation.

Keywords: *Trichoderma*, mechanisms, resistance, increased plant growth, proteomics

*To whom correspondence should be addressed, e-mail: geh3@cornell.edu

M. Vurro and J. Gressel (eds.), Novel Biotechnologies for Biocontrol Agent Enhancement and Management, 131–155.

7.1. Introduction

A number of fungi have evolved to provide a symbiotic relationship with plants as opposed to a pathogenic one. These include representatives of fungal genera that are usually considered to be serious plant pathogens. Some fungi, such as endophytes and mycorrhizae, have a totally obligate symbiotic life style. Ecto- and/or endomycorrhizal fungi colonize roots of most plants. They provide advantages to their hosts through increased uptake of nutrients, enhanced resistance to diseases and increases in plant growth.[1,2] Other fungi are also free-living but exist also as plant endophytes. For example, strains of *Muscodor* are plant endophytes that produce remarkably potent volatile antibiotics that can be used to sterilize various media and to act as soil fumigants.[3–5]

Other fungi proliferate and readily survive in soil, as opposed to the obligate or near-obligate plant symbionts just mentioned. Some are plant pathogens, such as fungi in the genera *Fusarium* and *Rhizoctonia*, and cause serious diseases, especially of plant roots. However, there exist strains or species of these fungi that have adapted a nonpathogenic life style and that have become plant symbionts. For example, binucleate *Rhizoctonia* are biological control strains by virtue, at least in part, by their abilities to induce plant resistance.[6,7] Nonpathogenic *Fusarium* species cause soils to become suppressive to pathogenic *Fusarium* spp. when both are present, at least in part due to the abilities of the nonpathogenic strains to induce systemic resistance.[8–10] Several fungi, including strains of *Phoma*, *Penicillium* and a sterile fungus, induce systemic resistance.[11] Thus, fungi that are closely related may have both plant beneficial and plant pathogenic capabilities, which suggest that the difference between pathogen and symbiont is not great, but the differences in most cases are not yet known.

Frequently, *Trichoderma* spp. are among the most prevalent culturable fungi in soils, based upon the frequency of isolation on suitable media. They have long been investigated for their biocontrol abilities relative to plants,[12,13] and to induce increased plant growth and yield.[14–16] They are seldom pathogenic to plants, but there are exceptions, which in at least one case occurred because a single 18 kDa fungal protein that induces resistance was not expressed.[17] *T. virens* "P" strains are pathogenic to very disease susceptible cotton seedlings (but not to commercial ones) while "Q" strains are not. P and Q strains are generally quite similar but, in addition to the expression or nonexpression of the 18 kDa protein, they differ in the spectrum of antibiotics that they produce. However, even with *T. virens*, antibiotic production is not a major contributor to biocontrol.[18]

Given this apparent balance between plant pathogenic and plant beneficial life styles, an important question is why the nonpathogenic life style occurred. In the case of *Trichoderma*, this nonpathogenic life style is of considerable advantage. Recent evidence based upon experiences with green fluorescent protein (gfp) and electron microscopy indicate that the fungi penetrate the root cortex, and that in this regard they are similar to endomycorrhizal fungi[19] (the data in the cited paper is for *T. asperellum* and cucumber, but this has been extended to a number of other species and plants, for example, *T. atroviride* or *T. harzianum* on tomato or maize roots). None of these incite disease even though they have enzyme systems fully capable of macerating plant tissue.[20] When they have colonized plant root cortical cells, they have access to plant nutrients, which allows them to proliferate. Moreover, they significantly enhance plant root growth in many cases[15] and this increase in root mass provides more sites for growth of the fungi than in their absence. Thus, the ability of *Trichoderma* strains to induce greater root growth and enhance plant health provides more niches for growth of the organism. Clearly, this is a successful strategy for the fungus, and probably accounts for at least a portion of the abundance of these fungi in soils around the world. The plant benefits from this relationship through increased root and shoot growth, increased macro- and micronutrient uptake and protection from disease.[15,19–22] Therefore, this interaction clearly is mutually beneficial and is a true symbiosis.[20] Since *Trichoderma* spp. and other fungi also are capable of living freely in soil, they must be considered as opportunistic plant symbionts.

It is the purpose of this paper to describe the interactions of *Trichoderma* spp. with plants and microorganisms.

7.2. Mechanisms of Biocontrol by *Trichoderma*—Direct Interactions with Other Fungi

Most of the early work on biocontrol of plant diseases by *Trichoderma* revolved around the direct ability of these fungi to interact with, and control, other fungi. Some of the specific mechanisms are briefly described as follows.

7.2.1. MYCOPARASITISM

Mycoparasitism is the ability of *Trichoderma* spp., or other fungi, to directly parasitize other filamentous pathogens, such as other fungi or Oomycetes. This response was observed by Weindling more than 70 years ago[23] and has been an area of interest ever since. First, *Trichoderma* strains detect other fungi and grow tropically toward them.[24] This tropic response is initiated by remote sensing of the target fungi by *Trichoderma* spp. and is at least partially due to sequential expression of fungal cell wall degrading enzymes (CWDEs). These include various classes of chitinases, various classes of glucanases/glucosidases, proteinases and other enzymes.[25] Different strains may follow different patterns of induction but the fungi apparently always produce low levels of an extracellular exochitinase. Diffusion of this enzyme catalyzes release of cell wall fragments from target fungi. These cell wall fragments presumably react with receptors on the *Trichoderma* cell wall or membrane and this, in turn, induces expression of fungitoxic CWDEs[26] that also diffuse and begin the attack on the target fungi before contact is actually made.[27,28] Once the fungi come into contact, *Trichoderma* spp. attach and may coil about and form appressoria on the surface of the host.

Attachment is mediated by binding of *Trichoderma* cell wall carbohydrates to lectins on the target fungus.[29] The *Trichoderma* produces fungitoxic CWDEs and probably also peptaibol antibiotics.[30] The combined activity of these materials is necessary for parasitism of the target fungus and dissolution of the cell walls. These products damage the fungal host and render its nutrients available to the attacking fungus. There are at least 20–30 known genes, proteins or metabolites directly involved in this process.[31,32] This is typical of the complex systems employed by these fungi in their interactions with other organisms. Thus, a wide range of different genes are activated in this process, as is typical of cascade-type responses. This process does occur and no doubt has some effect upon biocontrol. However, while mycoparasitism does occur in vivo,[33,34] for many years the hypothesis that mycoparasitism is a major factor in biocontrol caused by *Trichoderma* has been questioned in at least protection of planted seeds by seed treatments. Mycoparasitism was infrequently observed on seedcoats of planted seeds. In addition the seed-rotting pathogen, *Pythium*, infects seeds before *Trichoderma* spores germinate.[35] Further, it is difficult to conduct precise gene deletion studies to investigate the role of mycoparasitism since there are many gene products that are involved in synergistic interactions. Indeed, single gene knock-out or overexpression studies have given mixed results.[36–38] However, in some cases, mycoparasitism may be of fundamental importance, for example, in the parasitism and destruction of sclerotia.

7.2.2. ANTIBIOSIS

Some *Trichoderma* spp. produce antibiotics; the principal ones have been summarized,[39] and for many years, the production of these metabolites was considered to be a primary mechanism of biocontrol. Indeed, the literature is replete with papers discussing selection of strains for biocontrol testing based on in vitro paired plate tests of *Trichoderma* strains versus other pathogens. Studies strongly suggesting a role for antibiosis in biocontrol were especially prevalent for *T. virens*.[39,40] However, a series of mutants of *T. virens* were prepared that were deficient in mycoparasitic ability and/or in the ability to produce antibiotics.[41] Deletion of these capabilities had no effect upon the biocontrol abilities of these strains. However, there was a very strong correlation between biocontrol and the capability of the strains to induce plant resistance as indicated by the levels of phytoalexins (terpenes) in cotton seedlings.[20,41] It may be that antibiotic production is not strongly linked to the abilities of strains to accomplish biocontrol, although it would seem logical that antibiosis might be very useful to the fungi in competing within soil microcommunities.

7.2.3. OTHER MECHANISMS

There are a number of other mechanisms whereby *Trichoderma* spp. or other microbes directly affect plant pathogens. For example, one of the classical mechanisms proposed for biocontrol is competition for nutrients or space. Induction of soil suppressiveness to pathogens has been induced by proliferation of bacteria that produce chelators that bind iron very tightly. This chelation in soils with limited iron can make this essential nutrient so scarce as to prevent growth of pathogenic fungi.[42] Moreover, in some cases, this same effect can strongly inhibit the growth of biocontrol agents; strain T95 of *T. atroviride* was very capable of controlling seed rots caused by *Pythium* spp. as a seed treatment in some soils but not in others. This difference was explained by the fact that planted seeds are very rapidly colonized by bacteria, with high numbers developing within the first 24 h after planting. In the soils where T95 was effective, iron was available in the form of the carbonate but in soils where it was ineffective, iron was limiting. If iron was added in a chelated form to the "*Trichoderma* suppressive" soil, then T95 regained its efficacy. Moreover, when a strain that produced it own chelators was isolated from that soil, then biocontrol efficacy was obtained.[33] *T. harzianum* strain T22 was produced by protoplast fusion between T12 and T95[43] and one of its useful attributes is the ability to produce iron chelating compounds.

In another specialized form of competition, *Trichoderma* spp. have been demonstrated to produce proteases on leaf surfaces even when spores do not fully germinate. Fungi such as *Botrytis cinerea* require plant cell wall degrading enzymes in order to penetrate the leaves; the proteases degrade these enzymes and prevent cell wall penetration.[44,45]

There are other cases where *Trichoderma* spp. can metabolize nutrients from seeds that stimulate and enhance plant diseases. Seeds of cotton that are particularly susceptible to seed rots can be protected by *Trichoderma* spp. more effectively than with chemical fungicides, primarily because the bicontrol agent metabolizes pathogen-stimulating seed exudates and removes them from the system,[46] thereby preventing seed decay. Also, cells in dry seeds contain poorly organized mitochondia that must reorganize and regain function upon imbibition before aerobic respiration and the tricarboxylic acid cycle can begin. Reorganization and functionality of mitochondria from seeds of poor quality proceeds slowly[47] and so these seeds produce relatively high levels of acetaldehyde and ethanol for a longer time period than do good quality seeds. These materials are (a) very effective stimulants of microbial propagule germination and growth and (b) are themselves toxic to seeds. *Trichoderma* spp. efficiently scavenge and metabolize these materials, which probably reduces susceptibility of seeds to attack by pathogens and also prevents them from accumulating at toxic levels in germinating seeds.[48,49] This capability might explain the apparent ability of *T. harzianum* seed treatments to dramatically improve maize seed germination, even that of seemingly dead seeds.[50]

Several of these last examples clearly involve the entire system, which includes the pathogen, the plant and the biocontrol agent. The next section will describe even more intimate connections between biocontrol fungi, the plant, and indirectly, the pathogen. It seems likely to the authors that the reactions that will be described below may be more important than direct effects on pathogens such as mycoparasitism and antibiosis.

7.3. Colonization of Plant Roots by *Trichoderma* spp. and Establishment of an Area of Chemical Interaction between the Plant and the Biocontrol Fungus

Until recently, it was unclear whether *Trichoderma* spp. colonized roots only on their exterior or whether internal colonization also occurred. However, with electron microscopy and production of mutant strains that produce green fluorescent protein, it has been possible to examine this interaction in detail. It is now clear, as first demonstrated with *T. asperellum* on cucumber roots, that the interaction with plant roots has many features in common with mycoparasitism. The fungi produce appressoria-like structures on root surfaces that are similar to those observed in mycoparasitism[24] and they coil about root hairs. They also directly infect the root.[19]

Once the fungi infect the cortical cells, they then induce the plant to form thickened cell walls and to produce phenolic depositions that limit the *Trichoderma* strains to the area of infection. Thus, the *Trichoderma* strains have evolved to induce plants to form localized resistance to *Trichoderma* and that prevents them from becoming pathogenic. At least in cotton, *T. virens* can become a pathogen if localized resistance does not occur.[18]

7.3.1. THE ZONE OF CHEMICAL INTERACTION

Trichoderma spp., once they establish infection, then interact with the plant by means of chemical signals. A range of these are chemically quite distinct, as described below and as summarized elsewhere.[20]

7.3.1.1. *Proteins with Enzymatic or Other Functions*

The first chemical communicator produced by *Trichoderma* that was discovered was a 22 kDa xylanase that is secreted by several species.[51–53] This protein induces localized resistance in plants. *T. virens* produces a series of proteins and peptides that induce terpenoid phytoalexins involved in resistance in cotton. The peptides or proteins that were effective had masses of 6.5, 18, 20, 32 and 42 kDa. The 20 kDa band was cross reactive to the 22 kDa xylanase just mentioned, along with another smaller band, indicating that this elictor of plant response is produced by many strains. Many of the materials retained their activity as denatured proteins, which suggests that, at least for some, particular amino acid sequences are the important factor in their activity rather than enzymatic function.[54] In addition, a *T. longibrachiatum* cellulase also induced resistance responses when infiltrated into melon cotyledons; the nature of the response depended upon whether the cellulase was active or denatured. The nonactive protein stimulated the ethylene/jasmonic acid pathway while the active cellulase appeared to induce both that pathway and the salicylic acid pathway.[55] At least one other protein with a size of 42 kDa induces resistance. This active protein is the principal endochitinase of several strains.[32] This enzyme when inserted transgenically into plants induces resistance,[56,57] which is an effect that has been attributed to the antifungal activity of this enzyme. However, in our studies, this protein also induced expression of plant proteins that are part of resistance mechanisms.

All of the proteins noted above are secreted from the fungus. This means that they are released into the plant cortical cells that they infect and so, in this regard, the *Trichoderma* thallus can be envisioned as a delivery vehicle for highly bioactive chemicals. Thus, it can be modified to secrete other bioactive molecules that can potentially increase the biocontrol capability of organisms. In a first step in this direction, *T. atroviride* was engineered to secrete the *Aspergillus* glucose oxidase protein. The reaction of glucose oxidase with glucose, which of course is plentiful in cells, results in the production of hydrogen peroxide, which activates the oxidative burst that leads to resistance. This modified strain was much more effective in biocontrol than was the parental strain.[58] Thus, the concept of *Trichoderma* as an intracellular delivery vehicle for plants was validated.

7.3.1.2. *Avr Homologues*

The Avr proteins are produced by a variety of fungal and bacterial plant pathogens. They usually function as race- or pathovar-specific elictors of hypersensitive and other defense-related responses in plant species that contain the corresponding resistance gene. *Trichoderma* species produce a range of Avr-like proteins including ones that are homologues of Avr4 and 9 from

Cladosporium fulvum.[59] These Avr-like proteins may play a role in inducing changes in plant physiology and gene expression. *Trichoderma* spp. were transformed with the gene encoding the *C. fulvum* Avr4 protein and then applied to tomatoes that contain or lack the corresponding *Cf4* gene; those with the gene were sensitive to the Avr4 protein. The *Cf4* plants were stunted with necrotic spots, while the plants without the gene were not affected.[59] These data demonstrate again the concept that *Trichoderma* spp. may be used as vehicles to deliver bioactive molecules within plants. This information will be more completely described elsewhere in this volume.[60]

7.3.1.3. *Oligosaccharides and Similar Low-molecular Weight Compounds*

Trichoderma spp. are rich sources of a variety of cell wall degrading enzymes that function both on fungal and plant cell walls. They release oligo- and monomeric carbohydrate molecules. Some of these, typically oligosaccharides with or without amino acid moieties, enhance the growth and development of *Trichoderma* spp., enhance their level of enzyme production and induce resistance in plants.[59]

Thus, there is a range of materials produced directly or indirectly by *Trichoderma* spp. that communicate with plants, and no doubt, more will be discovered. Thus, the zone of chemical interaction has a rich vocabulary and is critically important in the effects of *Trichoderma* spp. on plants.

7.4. The Effects of *Trichoderma* spp. on Plants

The abilities of *Trichoderma* spp. to improve plant perfomance begins, we believe, as a series of reactions that are initiated in the chemical interaction zone in root cortical cells, and results in several general changes in the plants, including the following:

- *Induced systemic or localized resistance*. There are at least ten reports, on an assortment of *Trichoderma* spp. and plants that range from monocots to dicots, where systemic resistance has been demonstrated. The pathogens controlled range from fungi to Oomycetes to bacteria and even one virus.[20] In one case, *T. virens* on cotton, localized resistance rather than systemic resistance was induced.[18] Essentially all of the data regarding induced resistance has dealt with disease control, but there is a good prospect that these systems may also increase resistance to, or enhance predation of, insect pests, especially since the ethylene/jasmonate pathway is involved in plant resistance to insects.[61,62] Similar pathways, such as the jasmonate/ethylene pathway of induced resistance, are induced by insect herbivory, so if this effect was enhanced by the presence of *Trichoderma*, then greater insect control would probably result. In addition, as described later, *Trichoderma* spp. have abilities to limit nematode damage.

- *Increased growth of plants including roots.* This topic has been described[14,16] and the older literature has been thoroughly reviewed[15,20] and will not be discussed in detail here. In general, growth of the entire plant is increased, with substantial increases in root growth.[15,50] In a later section, however, we will describe newer findings with wheat and maize.
- *Increased nutrient uptake.* There is a significant amount of data indicating that *Trichoderma* spp. have significant abilities to solubilize a range of plant nutrients that may be present in insoluble, and therefore unavailable, forms in soils. This includes phosphorus and minerals including iron, copper, zinc and manganese.[63] In addition, even when these nutrients are fully soluble and available to plants, the presence of *T. asperellum* on cucumber roots grown hydroponically increased uptake of a similar range of plant nutrients.[22] Thus, these root symbionts may be able both to solubilize insoluble plant nutrients and also to induce plants to take up more of already soluble nutrients, which implies at least two general mechanisms by which availability to the plant may be enhanced. Further, nitrogen fertilizer is used more efficiently in maize when roots are colonized by *T. harzianum* T22.[15,21,64] Maize requires abundant nitrogen fertilizer and only a fraction is taken up by the plant. The plant in the field responds to increases in nitrogen fertility with increased plant growth and yield up to a point called the yield plateau where additional nitrogen has no effect. *T. harzianum* strain T22 added as a seed treatment has season-long effects due to rhizosphere competence and decreases the amount of nitrogen fertilizer required to reach the yield plateau by as much as 50%. This is significant since (a) it substantially reduces farmer's costs and (b) nitrate pollution of waterways from unused N is a signficant pollution problem.[15,21,64] No doubt the increased root development that is frequently associated with colonization of plant roots by *Trichoderma* spp. contributes to this and other benefits to plants.
- *Increased photosynthetic rate.* Plants colonized by *Trichoderma* freqeuently are not only larger but also greener. However, this frequently is not reported in papers because of the difficulty of assessing this (although SPAD meters manufactured by Minolta do this efficiently). In our experiments with maize, if there is an increased growth response, there is almost always a corresponding increase in leaf greenness, although we have only one published report.[15] If leaves are greener, then it is reasonable to assume that there is an increased photosynthetic rate, but, insofar as we know, this has not been reported, either positively or negatively. If it is the case, however, then this probably contributes to overall vigor and productivity of plants and needs to be assessed.

7.4.1. EFFECTS OF *TRICHODERMA* SPP. ON NEMATODES

Root colonization by *Trichoderma* spp. can control nematodes, although this literature is much more scarce than reports of control of diseases caused by microorganisms. *T. harzianum* strain T12, which was one of the parental strains used to prepare T22, enhanced shoot and root growth of a nematode-susceptible maize hybrid when *Meloidogyne arenaria* was present.[65] The presence of the biocontrol agent also suppressed reproduction of the nematode by at least 50%. In the absence of the nematodes, the biocontrol agent increased shoot and root growth of the hybrid, which is frequently observed.

More recently, an extensive study on the abilities of *T. asperellum* T-203 and other strains was reported.[66] These strains provided a high level of control of galling caused by *M. javanica* on tomato roots and inhibited nematode penetration of roots, but not development within the roots. Extracts from soils containing one strain inhibited hatching of nematode eggs. Some strains were able to directly colonize immature nematode eggs, and all strains attacked J2 (stage 2 juvenile nematodes), but not older, larvae. Based on these results, they concluded that these strains must be present at high rates in soil and that induction of resistance was not the major mechanism of control. However, this report dealt with very high levels of inoculum that may be impractical in commercial agriculture.

We have investigated the abilities of *T. harzianum* T22 to reduce nematode damage in beans (Table I). T22 was applied to planting medium (25 mg of a dry formulation at 1×10^9 cfu/g mixed with 1.4 l of planting medium per individual pot) or as a seed treatment that provided about 10^5 cfu/g seeds in a starch-based adhesive. Plants were harvested and growth and yield

TABLE I. Plant growth and yield characteristics

T22 treatment	Nemat treatment*	Plant height (cm)	Leaf greenness (SPAD)†	Flowers and pods	Roots (gFW)	Total root L (m/plant)‡	Root surface (cm²/plant)‡
None	0	43b	32b	10ab	5.3c	51b	1252c
T22 seed	0	44b	31ab	11bc	3.9ab	53b	1102c
T22 soil	0	43b	30ab	11bc	5.3bc	68c	1365c
None	7	38a	28a	8a	3.0a	33a	672a
T22 seed	7	41ab	33b	11bc	3.3a	40ab	689ab
T22 soil	7	43b	30ab	12c	4.4abc	47b	1043bc

*Nematode eggs per cc soil were added at the start of the experiment at the levels indicated.
†Leaf greenness in the distal leaf of the bean plants was measured at the end of the experiment with a Minolta SPAD 502 meter. This meter primarily measures chlorophyll content.
‡Total root length, together with diameter of roots, was measured with an image analysis system MacRhizo Ver. 3.8, Régent Instruments, Quebec City, Quebec (www.regent.qc.ca). Root surface area was calculated from the length of roots and their diameters.

TABLE II. Nematode control

Treatment	Galling index*	Number of nematode eggs†	Eggs/m roots
None	6.4 a	4415 a	149 a
T22 seed treatment	4.9 b	3688 a	93 a
T22 soil treatment	5.0 b	2982 a	63 a

*Nematode gall ratings were on a 1 to 9 scale.[68]
†Eggs were recovered from roots of beans at the end of the experiment using NaOCl treatment.[69]

characteristics were measured when the experiment was terminated at early pod fill (approx. 7 weeks after seeding). In all cases, numbers followed with the same letter are not significantly different (Fisher's Protected LSD). The Table I presents the results of this study.

Treatment with T22 increased root growth; this was especially evident when root surface area was considered. Nematode infestation substantially reduced root surface area but when T22 was added, this reduction was largely reversed. Further, there was a tendency for leaf greenness to be increased by T22 and reduced by nematodes. Leaf greenness as measured by the SPAD meter primarily measures chlorophyll and is frequently directly related to nitrogen uptake.[67]

These effects occurred even though T22 did not have strong abilities to control nematodes, as indicated in Table II. These data are suggestive that, in this case, the reduction of nematode damage was due, at least in part, to an enhanced ability of the bean plants to grow and proliferate, and thereby minimize the effects of the nematodes on final plant performance. It is probable that the lower galling index with *T. harzianum* did not reflect a lowered numbers of galls, but instead a greater area and length of roots over which the galls were distributed. Thus, there may be multiple mechanisms of nematode control, including direct parasitism and an increased tolerance of plants to attack due to enhanced growth and vigor induced by *Trichoderma* spp.

7.4.2. SYNERGY WITH OTHER ROOT COLONIZING SYMBIONTS

Field data usually indicate that colonization of roots with mycorrhizal fungi and *Trichoderma* together is more beneficial to the plant than colonization by either one alone, even though *Trichoderma* spp. may parasitize hyphae of mycorrhizae in petri dish assays.[70] For example, control of Fusarium crown and root rot of tomatoes[71] in the field was greater with the combination of both *T. harzianum* strain T22 and *Glomus intraradices* than with either organism alone. More studies of this sort would be beneficial.

Trichoderma spp. also may be synergistic with *Rhizobium* or *Bradyrhizobium* spp. In the bean-nematode trial just described, a surprising result was noted. The trial was conducted in field soil and so rhizobia capable of colonizing and nodulating bean roots were naturally present. When the roots were examined for size and nematode damage, differing levels of rhizobial nodules were observed. With plants grown in the presence of nematodes alone, no nodules were observed. With plants grown without *Trichoderma* or nematodes, or on plants grown with *Trichoderma* plus nematodes, a moderate level of nodule numbers was observed. However, with plants with *Trichoderma* alone, the roots were very heavily nodulated.

The effects of T22 with or without *B. japonicum* were assessed in field trials in a factorial randomized block design. At the end of the trial, plants were dug with a backhoe and roots were separated from the sandy soil. In the soil with adequate levels of fertility, the roots that grew from plants whose seeds were treated with both organisms were substantially larger than those with any other treatment (Figure 1). Clearly, in this circumstance, root growth was synergistically enhanced by the presence of both organisms; this result was statistically validated by image analysis of replicated trials. Soybean yields were 1704 kg/ha without either organism, 2280 kg/ha with T22, 2640 kg/ha

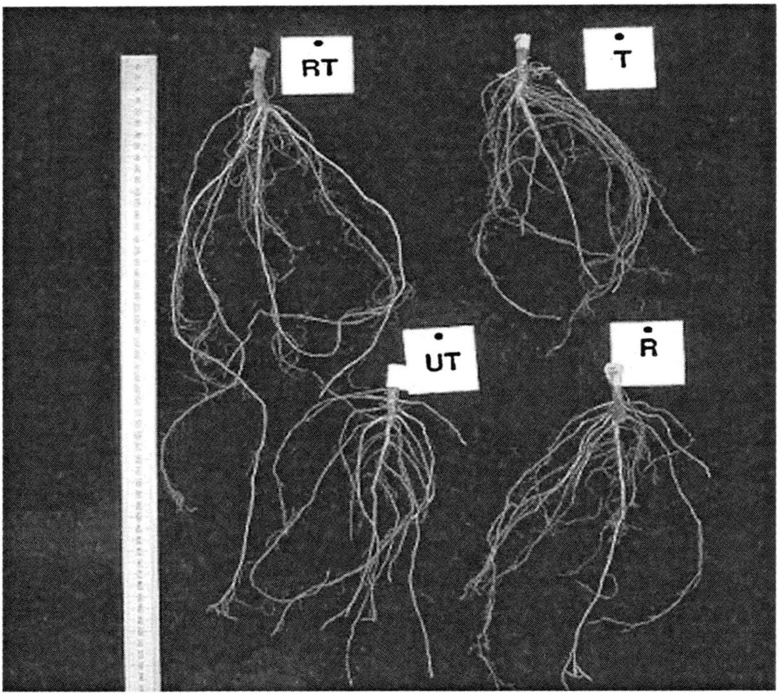

Figure 1. Root growth of mature soybeans grown from seeds treated with nothing (UT), with *T. harzianum* strain T22 (T), with *B. japonicum* (R), or with both organisms together (RT)

with the combination and 3000 kg/ha with *B. japonicum* alone. Yields were significantly enhanced over the untreated control ($P = 0.05$) by the combination treatment and by *B. japonicum* alone, but not with T22 alone. Thus, even though root growth was greatest with the combination, this did not result in a yield increase.

7.5. Effects of *Trichoderma* spp. on Plant Processes and Proteomes

Earlier sections have described the infection of plant cortical cells, establishment of the zone of chemical interaction between *Trichoderma* spp. and the plant, and genotypic plant effects, such as induced localized and systemic resistance, increased shoot and root growth including enhanced tolerance to root-attacking pathogens, and increased nutrient uptake and/or enhanced efficiency of fertilizer usage. These changes logically infer that there is a substantial alteration in gene regulation in affected plants, together with direct effects of the fungus on its environment, such as increased solubilization of plant nutrients in soil.

7.5.1. INDUCTION OF PLANT RESISTANCE

Much, but not all, of the biological evidence for induced resistance comes from the observation that *Trichoderma* spp. and other fungi colonize roots and resistance to a wide range of pathogens occurs on the above ground parts of the plants.[11,20,72–77] These changes frequently are associated with enhanced levels of pathogenesis-related proteins (PR proteins) and/or with accumulation of phytoalexin-type compounds.[72,75,77,78]

The physiological systems that are activated by *T. asperellum* in cucumber have been particularly well studied. *T. asperellum* on roots induces resistance to *Pseudomonas syringae* pv. *lachrymans* (*Psl*) on foliage. Addition of the biocontrol agent to roots led to a transient increase in the defense related protein phenylalanine ammonia lyase in both shoots and roots, but within two days, the level of the enzyme returned to background levels in both organs. However, if leaves were subsequently inoculated with *Psl*, the expression of several genes encoding several PR proteins, including hydroperoxide lyase, chitinase, β-1,3 glucanase, and peroxidase, was increased more than if just the pathogen or the *Trichoderma* strain was used singly.[20,77] This increase in enzyme level was also associated with an increase in phenolic glycoside levels; the aglycones of these materials are strongly antibiotic to a wide range of microorganisms and *Psl* cell numbers in leaves from plants with both organisms were dramatically decreased compared to plants without *Trichoderma* treatment.[20,77] Thus, it is very important to note that inoculation with the biocontrol agent resulted in only a short burst of detectable PR proteins; however, when leaves of these plants were subsequently inoculated with *Psl*, the proteins were highly up-regulated. Thus, the presence of *T. asperellum* potentiates the systemic resistance system but the entire pathway is not constantly turned on. This implies that certain up-stream regulatory genes are activated to provide a much more rapid response than would occur in the absence of *T. asperellum*. A similar mechanism occurs when roots are inoculated with plant growth enhancing bacteria, including species of *Pseudomonas*.[79,80] However, the direct effect of *T. harzianum* T22 in inducing systemic resistance is more long-lasting[75] than in the *T. asperellum*-cucumber system.

The pathways induced by *T. asperellum* are beginning to be fairly well understood. The presence of the biocontrol agent induces the jasmonate/ethylene pathway, but not the salicylate pathway, of induced resistance. This was shown by the following[81]:

- Treatment of plants with the jasmonate biosynthesis inhibitor diethyldithiocarbamic acid abolished the protective effect of *T. asperellum*. Treatment of different plants with silver thiosulfate, which inhibits ethylene action, also reduced the protective effect. However, neither treatment affected colonization of roots by the fungus;
- The presence of *T. asperellum* had no effect on salicylate concentrations in plants even though *Psl* infection did increase salicylate concentrations;
- In roots, real time RT-PCR indicated that *Lox1*, which encodes a lipoxygenase involved in jasmonate synthesis, was upregulated by inoculation with *T. asperellum*;
- Other genes expected to be up-regulated by the jasmonate pathway, including *Pal1* (encoding phenylalanine ammonia lyase), were expressed at higher levels in treated plants.
- The expression of two genes that negatively regulate ethylene, *ETR1* and *CTR1*, were also examined. The proteins encoded by these two products work together; ethylene binding to the receptor would down-regulate the activity of the complex and result in de-repression of the response pathway. For ETR1, there was a transitory increase in expression followed by a reduction to below control levels and the expression of CTR1 was almost abolished in plants inoculated with both organisms.[81]

Thus, there is strong evidence that when *T. asperellum* is applied to roots the jasmonate/ethylene pathway is potentiated, but only fully activated when the pathogen is inoculated on foliage. As noted earlier, *Trichoderma* spp. can be considered as vehicles for the delivery of bioactive molecules, and by changing the proteins that are secreted by the fungus in the zone of interaction, then the pathways that are activated can be changed or multiple pathways may be induced.[58]

The mechanisms by which these pathways are regulated are clearly controlled by the interaction of the signal molecules from *Trichoderma* with particular plant receptor molecules in the zone of chemical interaction. Very recently, one plant regulatory protein involved has been identified in the *T. asperellum*-cucumber system. A kinase that is essential to signal transduction leading to expression of this pathway has been discovered. This protein has been named *Trichoderma*-Induced MAPK (TIPK). The gene is homologous to *WIPK*, *MPK3* and *MPK3a*. TIPK is also induced by wounding. Unique attenuated virus vectors based on zucchini yellow mosaic virus (ZYMV-AGII) was used to overexpress TIPK protein and antisense RNA. Plants overexpressing *TIPK* were more resistant to pathogenic bacterial attack than control plants, even in the absence of *T. asperellum* pre-inoculation. Conversely, plants expressing *TIPK*-antisense revealed increased sensitivity to pathogen attack. Moreover, *Trichoderma* pre-inoculation could not protect these antisense plants against subsequent pathogen attack. These results demonstrate that *T. asperellum* exerts its protective effect on plants through activation of the *TIPK* gene, a MAPK that is involved in signal-transduction pathways of defense responses. It appears to function downstream from jasmonate expression.[82] This particular kinase acts fairly far downstream from the initial signal response, but by using similar systems, it will be possible to discover regulatory proteins that act earlier in the system. The ideal situation is to discover the regulatory links between elicitors produced by *Trichoderma* in the zone of chemical interaction and the receptors in plants.

7.6. Elucidation of Other Plant Responses Including Plant Growth Promotion

There has been substantial recent progress in elucidating the mechanisms of plant resistance induced by *Trichoderma* spp. With induced resistance, we have an advantage since the resistance pathways in plants are at least generally known. However, the mechanisms whereby increased growth responses, enhanced nutrient uptake and the like are not yet defined. Therefore, we decided that the most efficient methods of examining these effects are through whole plant examination of genes and gene products.

We chose a proteomics approach to elucidate the changes that occur and to develop hypotheses regarding the mechanisms by which these complex alterations in plant performance occur.

Very well-defined systems that we could analyze were needed; we used the maize-*T. harzianum* T22 system. Originally, we had observed effects of T22 on maize in the field. In order to do molecular analyses we needed rapid, high throughput systems with seedlings that would be predictive of the field situation. We screened a number of inbreds and settled on Mo17, which gave very reliable growth promotion at the seedling phase that continued on into the maturing plant. Moreover, systemic resistance occurred against the serious maize foliar pathogen, *Colletotrichum graminicola*.[75]

7.6.1. PROTEOME ANALYSES OF THE MAIZE-T22 SYSTEM

Six-day-old seedlings grown from T22-treated seeds had elevated levels of proteins, and increased activity levels of chitinase and β-1,3 glucanase in both shoots and roots. The effects of the seedling pathogen *Pythium ultimum* was also examined; its effects were in large part opposite to that of T22, especially in roots.[75]

We isolated proteins from six-day-old seedlings in the presence or absence of T22 and subjected them to two-dimensional gel electrophoresis. To limit the complexity of spots, we used narrow range isoelectric focusing strips of pH 5.3–6.5 and 6.3–7.5. The range of pI between 5.3 and 7.5 contains most of the maize proteins based upon earlier experiments.

Proteins were identified by peptide mass fingerprinting (PMF) using MALDI-TOF MS (matrix assisted laser desorption/ionization-time of flight mass spectroscopy), and which also used PSD (post source decay) acquisition to identify sequences of selected peptides. We also performed peptide sequencing using nanospray ion-trap tandem mass spectrometry (nESI-IT MS/MS) to identify proteins that proved problematic with MALDI-based techniques. Protein identification by PMF or nanospray sequencing was carried out using the PMF-GPS Explorer, ESI—Analyst (Applied Biosystems) software. Non-redundant NCBI (National Center for Biotechnology Information, W) and SwissProt (European Bioinformatics Institute, Heidelberg, Germany) databases were used for the search. Searches were performed in the full range of *M*r and p*I*. When an identity search had no matches, the homology mode was used. The maximum number of missed cleavages was set at two. Variable modifications selected for searching were carbamidomethyl-cys and oxidation of methionine. Only candidates that appeared at the top of the list and had protein C. I. % over 99.5 were considered positive identifications. 57.8% of the spots were identified using more than 10 peptides and 27.4% with 4–10 peptides; 7.4% were subjected to LC/MS/MS for identification. In nearly all cases, identity of specific peptides was confirmed by PSD fragmentation ion scores. Only 7.4% of the proteins were not identified.

Categorization of proteins was done using DAVID 2.1.[83] Gene ontology (http://www.geneontology.org/GO.tools.shtml) and KEGG terms (http://www. genome.ad.jp/kegg/). For gene families study we used Data mining tools from NCBI, EBI, ExPASy, and Softberry. These processes and systems provided unambiguous protein identifications.

Using these systems, we have thus far been able to obtain solid identifications of more than 200 differentially expressed proteins, with identities coming most frequently from the maize databases, but with frequent identities also from the rice and *Arabidopsis* databases. Very large changes in the maize seedling proteome were induced by inoculation with T22. In total, over both roots and shoots there were over 300 differentially expressed spots, with more differences in the shoots than in the roots, even though T22 is present only on roots. Since we were able to identify the proteins, it was possible to provide identities of specific functional proteins and also to develop hypotheses regarding up- or down-regulated pathways or systems.

The most commonly affected enzymes were those involved in carbohydrate metabolism, especially those in the glycolytic, TCA or similar pathways, and, of these, 37 of 42 were up-regulated in the shoot. These data suggest strongly that, in the presence of T22, the shoots are metabolizing substrates at an enhanced rate, which would be reasonable given the overall increase in growth rate. These seedlings are green at the time of seedling harvest and increased greenness frequently occurs as a consequence of seed treatment with T22. An increase in photosynthetic rate is consistent with the observed increases in overall metabolic rate and the field observations of more rapid growing, greener plants. About 20% of the differentially expressed proteins were those whose functions are related to defense or stress responses. The other large group of proteins that were differentially regulated were enzymes associated with genetic information processing.

Multiple sizes and forms of some proteins with the same function were discovered. For example, in proteins involved with carbohydrate metabolism, fifteen separate spots were glyceraldehyde 3-phosphate dehydrogenases, six were sucrose synthases, five were β-glucosidases, and at least two spots each of malate dehydrogenase, fructokinase and fructose bisphosphate aldolase. These all were up-regulated and are all involved in carbohydrate metabolism.

Proteins with multiple forms with defense/stress related functions include eight spots that are methionine synthases and five that are proteins with nucleotide binding sites and leucine rich repeats. Other proteins that were up-regulated included a peroxidase, a heat shock protein and an oxalate oxidase. Again, all were up-regulated.

Proteins with multiple spots may occur because they are products of different genes or because of post-translational modifications to single gene products. We still are analyzing the data but both phenomena occurred in data analyzed to date.

One of the up-regulated functional groups (eight protein spots), methionine synthase, is strongly suggestive of enhanced production of the growth hormone ethylene. Other amino acid synthetic enzymes were not up-regulated frequently, and methionine synthase functions in the pathway to ethylene synthesis. This provides an important clue that ethylene regulated systems are important in the plant interaction with *Trichoderma*, as was already suggested by recent studies[81] cited above.

In addition, some proteins are down-regulated. One with intriguing properties is a regulatory protein that, in *Arabidopsis*, results in smaller plants. This regulatory protein is readily detectable in plants grown in the absence of T22, but was not detectable in its presence. This is but one example of an interesting regulatory protein that, in addition to TIPK, can begin to provide the crucial direct link between elicitors from *Trichoderma* and regulatory proteins of plants.

One other group of proteins of interest not seen in analyses noted thus far is exo- and endochitinases. As noted earlier, chitinase activity is significantly higher in T22 treated than untreated maize.[75] However, most plant chitinases have either acidic or basic isoelectric points, and so they would not have been seen in the gels run as described above. The diversity of plant chitinases is remarkable; in *Arabidopsis* there are more than 20 separate chitinase genes, and in rice, more than 40. A search of the maize databases indicated only about four endochitinases and a similar number of exochitinases. Since this seemed unreasonable, we used published rice and *Arabidopsis* sequences to BLAST search the abundant EST databases of maize, then used contig and domain analyses to "build," in silico, entire genes. In this way we increased the number of endochitinases to 22, along with four exochitinases. We can readily detect chitinase activity on gels. We therefore ran one dimensional gels with chitinases and detected activity bands, which were invariably more intense from T22 treated than from untreated plants. We then conducted LC/MS/MS on these and have detected differential expression of six endochitinases in different classes and three exochitinases. It is likely that these different classes of enzymes have separate or, more intriguingly, synergistic activity with other chitinases or other defense related enzymes.

In the studies just described, proteomics were used to investigate the maize-*T. harzianum* interaction that was conducted in soil. In a different approach, the interaction between *T. atroviride*, pathogens and plants was analyzed also using proteomic approaches.[84] In this case, the organisms were separated by cellophane barriers, and so only the diffusable chemical interactions were tested, and the *Trichoderma* strain did not penetrate into the root. Nonetheless, complex interactions were revealed, with proteins from each organism strongly affected by the presence of the other organisms. No doubt, for the pathogen, defense reactions are initiated by the *Trichoderma* strain, and that diffusible pathogen and *Trichoderma* elicitors affects the plant response.[84] This research is described more fully elsewhere in this book.[60]

7.6.2. RELATIONSHIP TO BIOCONTROL PROBLEMS AND ISSUES

One factor that limits the use of biocontrol and plant growth promotion is variability in performance. For example, there have been more than 800 field trials conducted in maize with an average yield increase of about 5%. However, it has become apparent that there are significant varietal differences, with some maize lines giving neutral or even negative growth responses.[85] When we consider the complex changes that occur, as evidenced by the data just described for changes in the proteome, this is not surprising—large changes in the proteome or attendant gene expression are likely to result in different types of responses since it will be the effects, on balance, that provide the final result.

Since we have observed the substantial changes in methionine synthase, we have become more interested in changes in volatiles that are produced by the plant including ethylene and methyl jasmonate, as well as volatiles involved in insect resistance.[86] We are screening hundreds of maize genotypes to determine their growth responses. These have been conducted in growth chambers with high levels of air movement/exchange. Given the consideration of volatiles, we have developed air baffles surrounding the growing plants and have determined that the growth response is strongly influenced by the level of air movement around the plants. We will measure the volatiles that are released and factor this into the overall responses. Thus, the basic information developed above is already providing practical biocontrol results.

Interestingly, while the T22-wheat system has been less studied, it appears to be extremely robust in the field. There have been 52 field trials in the USA and in 49 of these positive yield responses were obtained, in association with increased tillering (the probability that the overall yield is increased is <0.0001). The trials were conducted over a wide range of geographical regions ranging from the Dakotas to New York State, and over a range of wheat types. The basic responses of wheat and maize are likely to be similar but clearly there is less variability in the wheat than in the maize response.

Very reliable and consistent results are essential if biocontrol is ever to become more widely used. It is the expectation that once the specific control mechanisms of the *Trichoderma*-plant interaction are known, then very specific genetic lines that have favorable outcomes can be readily identified and used. Moreover, knowledge of specific critical gene products that are associated with favorable outcomes will permit rapid assays of the expression of critical proteins or genes even on a field scale. This will provide a major management tool that will afford a reliable assessment of the interaction.

Finally, it is apparent that some proteins, and their attendant genes, are up-regulated in favorable outcomes while others are down-regulated. Since increased growth and yield occurs, this may provide a useful tool for improvement of plant growth and yield even in the absence of the biocontrol organism through breeding and genetic engineering techniques.

Acknowledgments

This research was supported in part by the US–Israel Binational Agricultural Research and Development Fund (BARD) grant US-3704R. We thank Kristen Ondik for editorial advice.

References

1. P. Chakravarty and T. Unestam, Differential influence of ectomycorrhizae on plant growth and disease resistance in *Pinus sylvestris* seedlings, *J. Phytopathology* 120, 104–120 (1987).
2. R. G. Linderman, Mycorrhizal interactions with the rhizosphere microflora, the mycorrhizosphere effect, *Phytopathology*, 78, 366–371 (1988).
3. D. Ezra, W. M. Hess, and G. A. Strobel, New endophytic isolates of *Muscodor albus*, a volatile-antibiotic-producing fungus, *Microbiology* 150, 4023–4031 (2004).
4. J. Mercier and D. C. Manker, Biocontrol of soil-borne diseases and plant growth enhancement in greenhouse soilless mix by the volatile-producing fungus *Muscodor albus*, *Crop Protect.* 24, 355–362 (2005).
5. J. Mercier and J. L. Smilanick, Control of green mold and sour rot of stored lemon by biofumigation with *Muscodor albus*, *Biol. Contr.* 32, 401–407 (2005).
6. L. J. Herr, Biocontrol of rhizoctonia crown and root rot of sugar beet by binucleate *Rhizoctonia* spp. and *Laetisaria arvalis*, *Ann. App. Biol.* 113, 107–118 (1988).
7. J. Hwang and D. M. Benson, Expression of induced systemic resistance in poinsettia cuttings against rhizoctonia stem rot by treatment of stock plants with binucleate *Rhizoctonia*, *Biol. Contr.* 27, 73–80 (2003).
8. N. Benhamou, C. Garand, and A. Goulet, Ability of nonpathogenic *Fusarium oxysporum* strain Fo47 to induce resistance against *Pythium ultimum* infection in cucumber, *Appl. Environ. Microbiol.* 68, 4044–4060 (2002).
9. J. G. Fuchs, Y. Moenne-Loccoz, and G. DeFago, Nonpathogenic *Fusarium oxysporum* strain Fo47 induces resistance to Fusarium wilt in tomato, *Plant Dis.* 81, 492–496 (1997).
10. B. J. Duijff, D. Pouhair, C. Olivain, C. Alabouvette, and P. Lemanceau, Implication of systemic induced resistance in the suppression of Fusarium wilt of tomato by *Pseudomonas fluorescens* WCS417r and by nonpathogenic *Fusarium oxysporum* Fo47, *Eur. J. Plant Pathol.* 104, 903–910 (1998).
11. N. Koike, M. Hyakumachi, K. Kageyama, S. Tsuyumu, and N. Doke, Induction of systemic resistance in cucumber against several diseases by plant growth-promoting fungi: Lignification and superoxide generation, *Eur. J. Plant Pathol.* 107, 523–533 (2001).
12. I. Chet, in *Innovative Approaches to Plant Disease Control*, edited by I. Chet (Wiley, New York, 1987), pp. 137–160.
13. R. Weindling, Studies on a lethal principle effective in the parasitic action of *Trichoderma lignorum* on *Rhizoctonia solani* and other soil fungi, *Phytopathology* 24, 1153–1179 (1934).
14. Y.-C. Chang, Y.-C. Chang, R. Baker, O. Kleifeld, and I. Chet, Increased growth of plants in the presence of the biological control agent *Trichoderma harzianum*, *Plant Dis.* 70, 145–148 (1986).

15. G. E. Harman, Myths and dogmas of biocontrol. Changes in perceptions derived from research on *Trichoderma harzianum* T-22, *Plant Dis.* 84, 377–393 (2000).
16. D. L. Lindsey and R. Baker, Effect of certain fungi on dwarf tomatoes grown under gnotobiotic conditions, *Phytopathology* 57, 1262–1263 (1967).
17. C. R. Howell, Mechanisms employed by *Trichoderma* species in the biological control of plant diseases: The history and evolution of current concepts, *Plant Dis.* 87, 4–10 (2003).
18. C. R. Howell, Understanding the mechanisms employed by *Trichoderma virens* to effect biological control of cotton diseases, *Phytopathology* 96, 178–180 (2006).
19. I. Yedidia, N. Benhamou, and I. Chet, Induction of defense responses in cucumber plants (*Cucumis sativus* L.) by the biocontrol agent *Trichoderma harzianum*, *Appl. Environ. Microbiol.* 65, 1061–1070 (1999).
20. G. E. Harman, C. R. Howell, A. Viterbo, I. Chet, and M. Lorito, *Trichoderma* species—Opportunistic, avirulent plant symbionts, *Nature Rev. Microbiol.* 2, 43–56 (2004).
21. G. E. Harman and B. G. G. Donzelli, in *Enhancing Biocontrol Agents and Handling Risks*, edited by M. Vurro, J. Gressel, T. Butt, G. Harman, R. St. Leger, D. Nuss, and A. Pilgeram (IOS Press, Amsterdam, 2001), pp. 114–125.
22. I. Yedidia, A. K. Srivastva, Y. Kapulnik, and I. Chet, Effect of *Trichoderma harzianum* on microelement concentrations and increased growth of cucumber plants, *Plant Soil* 235, 235–242 (2001).
23. R. Weindling, *Trichoderma lignorum* as a parasite of other soil fungi, *Phytopathology* 22, 837–845 (1932).
24. I. Chet, G. E. Harman, and R. Baker, *Trichoderma hamatum*: Its hyphal interactions with *Rhizoctonia solani* and *Pythium* spp., *Microb. Ecol.* 7, 29–38 (1981).
25. I. Chet, N. Benhamou, and S. Haran, in *Trichoderma and Gliocladium*, edited by G. E. Harman and C. P. Kubicek (Taylor and Francis, London, 1998), vol. 2, pp. 153–172.
26. K. Brunner, C. K. Peterbauer, R. L. Mach, M. Lorito, S. Zeilinger, and C. P. Kubicek, The Nag1 *N*-acetylglucosaminidase of *Trichoderma atroviride* is essential for chitinase induction by chitin and of major relevance to biocontrol, *Curr. Genet.* 43, 289–295 (2003).
27. A. Viterbo, M. Montero, O. Ramot, D. Friesem, E. Monte, A. Llobell, and I. Chet, Expression regulation of the endochitinase chit36 from *Trichoderma asperellum* (*T. harzianum* T-203), *Curr. Genet.* 42, 114–122 (2002).
28. S. Zeilinger, C. Galhaup, K. Payer, S. L. Woo, R. L. Mach, C. Fekete, M. Lorito, and C. P. Kubicek, Chitinase gene expression during mycoparasitic interaction of *Trichoderma harzianum* with its host, *Fung. Genet. Biol.* 26, 131–140 (1999).
29. J. Inbar, A. Menendez, and I. Chet, Hyphal interaction between *Trichoderma harzianum* and *Sclerotinia sclerotiorum* and its role in biological control, *Soil Biol. Biochem.* 28, 757–763 (1996).
30. M. Schirmböck, M. Lorito, Y. L. Wang, C. K. Hayes, I. Arisan-Atac, F. Scala, G. E. Harman, and C. P. Kubicek, Parallel formation and synergism of hydrolytic enzymes and peptaibol antibiotics, molecular mechanisms involved in the antagonistic action of *Trichoderma harzianum* against phytopathogenic fungi, *Appl. Environ. Microbiol.* 60, 4364–4370 (1994).
31. T. Benitez, C. Limon, J. Delgado-Jarana, and M. Rey, in *Trichoderma and Gliocladium*, edited by G. E. Harman and C. P. Kubicek (Taylor and Francis, London, 1998), vol. 2, pp. 101–127.
32. M. Lorito, in *Trichoderma and Gliocladium*, edited by G. E. Harman and C. P. Kubicek (Taylor and Francis, London, 1998), vol. 2., pp. 73–99.
33. J. P. Hubbard, G. E. Harman, and Y. Hadar, Effect of soilborne *Pseudomonas* spp. on the biological control agent, *Trichoderma hamatum*, on pea seeds, *Phytopathology* 73, 655–659 (1983).

34. Z. Lu, R. Tombolini, S. Woo, S. Zeilinger, M. Lorito, and J. K. Jansson, In vivo study of *Trichoderma*-pathogen-plant interactions, using constitutive and inducible green fluorescent protein reporter systems, *Appl. Environ. Microbiol.* 70, 3073–3081 (2004).
35. R. Lifshitz, M. T. Windham, and R. Baker, Mechanism of biological control of preemergence damping-off of pea by seed treatment with *Trichoderma* spp., *Phytopathology* 76, 720–725 (1986).
36. C. Carsolio, N. Benhamou, S. Haran, C. Cortes, A. Gutierrez, I. Chet, and A. Herrera-Estrella, Role of the *Trichoderma harzianum* endochitinase gene, *ech42*, in mycoparasitism, *Appl. Environ. Microbiol.* 65, 929–935 (1999).
37. M. C. Limon, J. M. Lora, I. Garcia, J. De La Cruz, A. Llobell, T. Benitez, and J. A. Pintor-Toro, Primary structure and expression pattern of the 33-kDa chitinase gene from the mycoparasitic fungus *Trichoderma harzianum*, *Curr. Genet.* 28, 478–483 (1995).
38. S. L. Woo, B. Donzelli, F. Scala, R. Mach, G. E. Harman, C. P. Kubicek, G. Del Sorbo, and M. Lorito, Disruption of the *ech42* (endochitinase-encoding) gene affects biocontrol activity in *Trichoderma harzianum* P1, *Molec. Plant-Microbe Interact.* 12, 419–429 (1999).
39. C. R. Howell, in *Trichoderma and Gliocladium*, edited by G. E. Harman and C. P. Kubicek (Taylor and Francis, London, 1998), vol. 2, 173–184.
40. C. R. Howell and R. D. Stipanovic, Gliovirin, a new antibiotic from *Gliocladium virens*, and its role in the biological control of *Pythium ultimum*, *Can. J. Microbiol.* 29, 321–324 (1983).
41. C. R. Howell, L. E. Hanson, R. D. Stipanovic, and L. S. Puckhaber, Induction of terpenoid synthesis in cotton roots and control of *Rhizoctonia solani* by seed treatment with *Trichoderma virens*, *Phytopathology* 90, 248–252 (2000).
42. F. M. Scher, and R. Baker, Effect of *Pseudomonas putida* and a synthetic iron chelator on induction of soil suppressiveness to Fusarium wilt pathogens, *Phytopathology* 72, 1567–1574 (1982).
43. T. E. Stasz, G. E. Harman, and N. F. Weeden, Protoplast preparation and fusion in two biocontrol strains of *Trichoderma harzianum*, *Mycologia* 80, 141–150 (1988).
44. Y. Elad, D. Rav David, T. Levi, A. Kapat, B. Kirshner, E. Gorin, and A. Levine, in *Modern Fungicides and Antifungal Compounds II,* edited by H. Lyr (Intercept Ltd., Andover, Hampshire, UK, 1999), pp. 459–467.
45. Y. Elad and A. Kapat, The role of *Trichoderma harzianum* protease in the biocontrol of *Botrytis cinerea*, *Eur. J. Plant Pathol.* 105, 177–189 (1999).
46. C. R. Howell, Cotton seedling preemergence damping-off incited by *Rhizopus oryzae* and *Pythium* spp. and its biological control with *Trichoderma* spp., *Phytopathology* 92, 177–180 (2002).
47. G. E. Harman and A. L. Granett, Deterioration of stored pea seed: Changes in germination, membrane permeability and ultrastructure resulting from infection by *Aspergillus ruber* and from aging, *Physiol. Plant Pathol.* 2, 271–278 (1972).
48. R. J. Gorecki, and G. E. Harman, Effects of antioxidants on viability and vigour of ageing pea seeds, *Seed Sci. Technol.* 15, 109–118 (1987).
49. J. M. Norton and G. E. Harman, Responses of soil microorganisms to volatile exudates from germinating pea seeds, *Can. J. Bot.* 63, 1040–1045 (1985).
50. T. Bjorkman, L. M. Blanchard, and G. E. Harman, Growth enhancement of shrunken-2 sweet corn with *Trichoderma harzianum* 1295-22: Effect of environmental stress, *J. Am. Soc. Hort. Sci.* 123, 35–40 (1998).
51. J. D. Anderson, B. A. Bailey, R. Taylor, A. Sharon, A. Avni, A. K. Mattoo, and Y. Fuchs, in *Cellular and Molecular Aspects of the Plant Hormone Ethylene*, edited by J.C. Pech, A. Latche', and C. Balague (Kluwer, Dordecht, 1993), pp. 197–204.

52. T. Lotan and R. Fluhr, Xylanase, a novel elicitor of pathogenesis-related proteins in tobacco, uses a non-ethylene pathway for induction, *Plant Physiol.* 93, 811–817 (1990).
53. Y. Fuchs, A. Saxena, H. R. Gamble, and J. D. Anderson, Ethylene biosynthesis-inducing protein from cellulysin is an endoxylanase, *Plant Physiol.* 89, 138–143 (1989).
54. L. E. Hanson and C. R. Howell, Elicitors of plant defense responses from biological control strains of *Trichoderma virens*, *Phytopathology* 94, 171–176 (2004).
55. C. Martinez, F. Blanc, E. Le Claire, O. Besnard, M. Nicole, and J. C. Baccou, Salicylic acid and ethylene pathways are differentially activated in melon cotyledons by active or heat-denatured cellulase from *Trichoderma longibrachiatum*, *Plant Phsiol.* 127, 334–344 (2001).
56. J. P. Bolar, J. L. Norelli, K. W. Wong, C. K. Hayes, G. E. Harman, and H. S. Aldwinckle, Expression of endochitinase from *Trichoderma harzianum* in transgenic apple increases resistance to apple scab and reduces vigor, *Phytopathology* 90, 72–77 (2000).
57. M. Lorito, S. L. Woo, I. G. Fernandez, G. Colucci, G. E. Harman, J. A. Pintor-Toro, E. Filippone, S. Muccifora, C. B. Lawrence, A. Zoina, S. Tuzun, and F. Scala, Genes from mycoparasitic fungi as a source for improving plant resistance to fungal pathogens, *Proc. Natl. Acad. Sci. USA* 95, 7860–7865 (1998).
58. K. Brunner, S. Zeilinger, R. Ciliento, S. L. Woo, M. Lorito, C. P. Kubicek, and R. L. Mach, Improvement of the fungal biocontrol agent *Trichoderma atroviride* to enhance both antagonism and induction of plant systemic resistance, *Appl. Environ. Microbiol.* 71, 3959–3965 (2005).
59. S. L. Woo, F. Scala, M. Ruocco, and M. Lorito, The molecular biology of the interactions between *Trichoderma* spp., phytopathogenic fungi, and plants, *Phytopathology* 96, 181–185 (2006).
60. S. Woo and M. Lorito, in *Novel Biotechnologies for Biocontrol Agent Enhancement and Management*, edited by M. Vurro and J. Gressel (Springer, Amsterdam, 2007).
61. E. A. Schmelz, H. T. Alborn, and J. H. Tumlinson, The influence of intact-plant and excised-leaf bioassay designs on volicitin- and jasmonic acid-induced sesquiterpene volatile release in *Zea mays*, *Planta-(Berlin)* 214, 171–179 (2001).
62. T. C. J. Turlings, J. H. Tumlinson, and W. J. Lewis, Exploitation of herbivore-induced plant odors by host-seeking parasitic wasps, *Science* 250, 1251–1253 (1990).
63. C. Altomare, W. A. Norvell, T. Björkman, and G. E. Harman, Solubilization of phosphates and micronutrients by the plant-growth-promoting and biocontrol fungus *Trichoderma harzianum* Rifai 1295-22, *Appl. Environ. Microbiol.* 65, 2926–2933 (1999).
64. G. E. Harman, in *Proceedings of International Symposium on Biological Control of Plant Diseases for the New Century—Mode of Action and Application Technology*, edited by D.D.-S. Tzeng and J.W. Huang (National Chung Hsing University, Taichung City, Taiwan, 2001), pp. 71–84.
65. G. L. Windham, M. T. Windham, and W. P. Williams, Effects of *Trichoderma* spp. on maize growth and *Meloidogyne arenaria* reproduction, *Plant Dis.* 73, 493–495 (1989).
66. E. Sharon, M. Bar-Eyal, I. Chet, A. Herrera-Estrella, O. Kleifeld, and Y. Spiegel, Biological control of the root-knot nematode *Meloidogyne javanica* by *Trichoderma harzianum*. *Phytopathology* 91, 687–693 (2001).
67. W. P. Piekielek and R. H. Fox, Use of a chlorophyll meter to predict sidedress nitrogen requirements for maize, *Agron. J.* 84, 59–65 (1992).
68. N. M. Viaene and G. S. Abawi, Damage threshold of *Meloidogyne hapla* to lettuce in organic soil, *J. Nematol.* 28, 537–545 (1996).
69. R. S. Hussey and K. R. Barker, A comparison of methods of collecting inocula of *Medoidogyne* spp., including a new technique, *Plant Dis. Repr.* 57, 1025–1028 (1973).

70. A. Rousseau, N. Benhamou, I. Chet, and Y. Piche, Mycoparasitism of the extramatrical phase of *Glomus intraradices* by *Trichoderma harzianum*, *Phytopathology* 86, 434–443 (1996).
71. L. E. Datnoff, S. Nemec, and K. Pernezny, Biological control of Fusarium crown and root rot of tomato in Florida using *Trichoderma harzianum* and *Glomus intraradices*, *Biol. Control* 5, 427–431 (1995).
72. S. A. Ahmed, C. P. Sanchez, and M. E. Candela, Evaluation of induction of systemic resistance in pepper plants (*Capsicum annuum*) to *Phytophthora capsici* using *Trichoderma harzianum* and its relation with capsidiol accumulation, *Eur. J. Plant Pathol.* 106, 817–824 (2000).
73. J. Bigirimana, G. De Meyer, J. Poppe, Y. Elad, and M. Hofte, Induction of systemic resistance on bean (*Phaseolus vulgaris*) by *Trichoderma harzianum*, *Med. Fac. Landbouww. Univ. Gent* 62, 1001–1007 (1997).
74. G. De Meyer, J. Bigirimana, Y. Elad, and M. Hofte, Induced systemic resistance in *Trichoderma harzianum* T39 biocontrol of *Botrytis cinerea, Eur. J. Plant Pathol.* 104, 279–286 (1998).
75. G. E. Harman, R. Petzoldt, A. Comis, and J. Chen, Interactions between *Trichoderma harzianum* strain T22 and maize inbred line Mo17 and effects of this interaction on diseases caused by *Pythium ultimum* and *Colletotrichum graminicola*, *Phytopathology* 94, 147–153 (2004).
76. C.-T. Lo, T. F. Liao, and T. C. Deng, Induction of systemic resistance of cucumber to cucumber green mosaic virus by the root-colonizing *Trichoderma* spp., *Phytopathology* 90, S47 (2000).
77. I. Yedidia, M. Shoresh, Z. Kerem, N. Benhamou, Y. Kapulnik, and I. Chet, Concomitant induction of systemic resistance to *Pseudomonas syringae* pv. *lachrymans* in cucumber by *Trichoderma asperellum* (T-203) and accumulation of phytoalexins, *Appl. Environ. Microbiol.* 69, 7343–7353 (2003).
78. I. Yedidia, N. Benhamou, Y. Kapulnik, and I. Chet, Induction and accumulation of PR proteins activity during early stages of root colonization by the mycoparasite *Trichoderma harzianum* strain T-203, *Plant Physiol. Biochem.* 38, 863–873 (2000).
79. C. M. J. Pieterse, J. A. Van Pelt, J. Ton, S. Parchmann, M. J. Mueller, A. J. Buchala, J. P. Metraux, and L. C. Van Loon, Rhizobacteria-mediated induced systemic resistance (ISR) in *Arabidopsis* requires sensitivity to jasmonate and ethylene but is not accompanied by an increase in their production, *Physiol. Molec. Plant Pathol.* 57, 123–134 (2000).
80. C. M. J. Pieterse, J. A. Van Pelt, S. C. M. Van Wees, J. Ton, K. M. Leon-Kloosterziel, J. J. B. Keurentjes, B. W. M. Verhagen, M. Knoester, I. Van der Sluis, P. A. H. M. Bakker, and L. C. Van Loon, Rhizobacteria-mediated induced systemic resistance: triggering, signalling and expression, *Eur. J. Plant Pathol.* 107, 51–61 (2001).
81. M. Shoresh, I. Yedidia, and I. Chet, Involvement of the jasmonic acid/ethylene signaling pathway in the systemic resistance induced in cucumber by *Trichoderma asperellum* T203, *Phytopathology* 95, 76–84 (2005).
82. M. Shoresh, A. Gal-on, and I. Chet, Characterization of a MAPK gene from cucumber required for *Trichoderma*-conferred plant resistance, *Plant Physiol.* 142, 1169–1179 (2006).
83. G. Dennis Jr., B. T. Sherman, D. A. Hosack, J. Yang, W. Gao, H. C. Lane, and R. A. Lempicki, DAVID: Database for annotation, visualization, and integrated discovery, *Gen. Biol.* 4, P3 (2003).
84. R. Marra, P. Ambrosino, V. Carbone, F. Vinale, S. L. Woo, M. Ruocco, R. Ciliento, S. Lanzuise, I. Ferraioli, S. Gigante, D. Turra, V. Fogliano, F. Scala, and M. Lorito, Study of

the three-way interaction between *Trichoderma atroviride*, plant and fungal pathogens by using a proteomic approach, *Curr. Genet.* 50, 307–321 (2006).
85. G. E. Harman, Overview of mechanisms and uses of *Trichoderma* spp., *Phytopathology* 96, 190–194 (2006).
86. E. A. Schmelz, H. T. Alborn, and J. H. Tumlinson, Synergistic interactions between volicitin, jasmonic acid and ethylene mediate insect-induced volatile emission in *Zea mays*, *Physiol. Plant.* 117, 403–412 (2003).

8. USING STRAINS OF *FUSARIUM OXYSPORUM* TO CONTROL FUSARIUM WILTS: DREAM OR REALITY?

Claude Alabouvette,* Chantal Olivain, Floriane L'Haridon, Sébastien Aimé, and Christian Steinberg
INRA Umr Microbiologie Géochimie des Sols, F 21065 Dijon Cedex, France

Abstract. Soil-borne strains of *F. oxysporum* are involved in the mechanisms of soil suppressiveness to Fusarium wilts, and many attempts have been made to use strains of *Fusarium oxysporum* to control Fusarium diseases. The modes of action of the protective strains are diverse; they include direct antagonism, competition for nutrients, and indirect antagonism through induced resistance of the plant. The use of newer tools has enabled a reconsideration of these modes of action; e.g., competition for infection sites whose importance has been minimized, and to make progress in the understanding of the interactions between the plant and either pathogenic or protective strains of *F. oxysporum*. Even though the mechanisms of biocontrol of *F. oxysporum* are far from being understood, several processes of mass production have been developed to enable field application of the biocontrol strains. These strains possess a strong ecological fitness and establish in soil of different physico-chemical properties. Their introduction into the soil does not durably modify the structure of the soil-borne communities of fungi and bacteria, indicating that their use does not present any risk to the environment.

Keywords: competition, induced systemic resistance, suppressive soils

8.1. Introduction

Soil-borne diseases are among the most difficult to control, as it is not possible to directly apply fungicides to the roots or to the soil. Until recently, growers could only eliminate the plant pathogenic organisms by biocidal treatments such as methyl bromide fumigation. This practice, which destroys both pathogenic and beneficial soil organisms, has been banned because it was dangerous to man and the environment. This led to a renewed interest in biological control. In its broad sense, biological control includes the choice of

*To whom correspondence should be addressed, e-mail: ala@dijon.inra.fr

M. Vurro and J. Gressel (eds.), Novel Biotechnologies for Biocontrol Agent Enhancement and Management, 157–177.

tillage practices, crop rotation sequences, organic amendments, and application of biological control agents to decrease disease incidence or disease severity.[1,2] In our case we will mainly focus on microbiological control of fusarium wilts.

Diseases induced by pathogenic formae speciales of *Fusarium oxysporum* have always been difficult to control. The best solution is to use resistant cultivars or to graft the susceptible cultivars on resistant root stocks. For many crops, these solutions are not available or not economically feasible and other biological practices have to be developed. Among them microbiological control, based on the use of the so-called "nonpathogenic" strains of *F. oxysporum*, represents an alternative coming from the study of soil suppressive to Fusarium wilts.

The existence of soils that naturally limit the incidence of Fusarium wilts has been recognized for more than a century. During the sixties, the role of abiotic factors, mainly clays, was studied in relation to the reduction of disease incidence in the so-called "long-life" soils in banana plantations in Central America.[3,4] More attention was later given to the role of the biological factors, especially to the saprophytic microflora including the soil-borne population of *Fusarium* spp.[5] Soil suppressive to Fusarium wilts supported large populations of nonpathogenic species of *Fusarium* spp.[6] Similarly, the suppressive soil from Châteaurenard harbored large populations of *F. oxysporum* and *F. solani*.[7] A form of Koch's postulates confirmed the involvement of the soil-borne populations of *Fusarium* spp. as part of the mechanism of soil suppressiveness. Indeed, suppressiveness was destroyed by heat treatment at a temperature above 55° that eliminated the thermo-sensitive microflora including the *Fusarium* spp., and suppressiveness was restored by artificial introduction of strains of *F. oxysporum* or *F. solani* in the heat-treated soil.[8] Numerous studies then clearly identify a role for nonpathogenic *Fusarium* spp. in suppressive soils from different parts of the world.[9–14] Strains of *F. oxysporum* were much more efficient in establishing suppressiveness in soil than other species of *Fusarium*. The strains of *F. oxysporum*, isolated from soil and not able to induce disease on the plant species of interest, are called nonpathogenic. But it must be underlined that these strains might be pathogenic to an unknown plant species. Indeed, the pathogenic strains of *F. oxysporum* inducing wilts show a narrow host-specificity: strains pathogenic to a given plant species are typically nonpathogenic to other plant species. This led to the concept of *forma specialis* equivalent to pathovar and including all the strains able to infect the same plant species. In this paper "nonpathogenic" strains of *F. oxysporum* refer to this restricted definition: i.e., strains nonpathogenic to the plant species to which they are applied. Indeed, not only the soil-borne isolates of *F. oxysporum* possess a biocontrol capacity; most of the pathogenic strains applied to a non host plant are able to protect this plant against contamination by its specific *forma specialis*.[15] This phenomenon initially described as cross-protection is an expression of induced systemic resistance, a general plant defense response to microbial infection.[16] Beside its scientific interest, this resistance induced by *F. oxysporum* might be used as a mechanism of biological control.

Nonpathogenic strains of *F. oxysporum* have also been isolated from the stems of healthy plants. These strains usually possess biological control activity against the pathogenic *formae speciales* of *F. oxysporum*.[17,18] Therefore, most of the research dealing with biological control of Fusarium diseases has focused on nonpathogenic strains of *F. oxysporum*. However, as these strains are either soil-borne or pathogenic strains applied to a non host plant, it would be more appropriate to call them "protective" or "biocontrol" strains of *F. oxysporum*.

8.2. Modes of Action of the Protective Strains *of F. oxysporum*

The mechanisms of action associated with nonpathogenic *F. oxysporum* can be divided in two broad categories: direct antagonism of the nonpathogenic strains to the pathogen and indirect antagonism mediated through the host plant.

8.2.1. DIRECT ANTAGONISM

Generally speaking, mechanisms of direct microbial antagonism include parasitism, antibiosis and competition. Thus far, there is no evidence of either parasitism or antibiosis among strains of *F. oxysporum*, but data from many studies support the role of competition. Competition can be divided into saprophytic competition for nutrients in the soil and rhizosphere, and competition for infection sites on the root and inside the vessels.

8.2.1.1. *Competitive Interactions in the Soil and Rhizosphere*

In the absence of any evidence of antibiosis between nonpathogenic and pathogenic strains of *F. oxysporum*, a hypothesis of trophic interactions was proposed to explain the role of nonpathogenic *F. oxysporum* in the mechanisms of soil suppressiveness. More precisely, the hypothesis of competition for carbon sources was proposed based on the fact that a single addition of glucose to a suppressive soil was sufficient to make the soil conducive.[7] The validity of the hypothesis of competition for carbon between strains of *F. oxysporum* was demonstrated by comparing the growth kinetics of a small collection of strains of *F. oxysporum* introduced into a sterilized soil amended with various amounts of glucose.[20] Modeling of the growth curve[21] enabled calculation of the growth rate and the yield coefficient (i.e., the number of propagules formed per unit of glucose consumed) for each strain. Results showed a great diversity among the seven strains compared. The yield coefficient varied from one to eight million propagules formed per milligram of glucose consumed. Six of these strains were then confronted with a 7th strain, the pathogenic strain *F. oxysporum.* f. sp. *lini* (Foln3) resistant to benomyl. Each strain was introduced into sterilized soil in combination with the pathogenic strain Foln3 at five different inoculum ratios. By following the kinetics of growth of each strain in mixture it was possible to calculate a "competitiveness index" for each strain. These indices ranged from 1.3 to 3.5, indicating a large diversity in the ability of these six strains to compete in soil with the pathogenic strain *F.o.* f. sp. *lini.* It was confirmed, *in vitro*, that carbon was the major nutrient that a pathogenic strain of *F.o.*f. sp. *dianthi* was competing for in soil-less culture with the biocontrol agent Fo47.[22] These results were further confirmed by demonstrating that isolate Fo47 significantly inhibited chlamydospore germination of the pathogen in soil at 0.2 or more mg g^{-1} soil of glucose.[19] Germ tube growth was also significantly reduced in soil containing Fo47 compared with untreated soil. Competition for nutrients has also been shown to be involved in the mode of action of other isolates of nonpathogenic *F. oxysporum* such as strain 618.12[18] and strains C5 and C14.[23] In contrast, the biocontrol isolate *F. oxysporum* CS-20 had no effect on germination or germ tube development of the pathogen.

The competitive ability of a biocontrol strain partly determines its capacity to establish in soil and in the plant rhizosphere, and is probably involved in its capability to colonize the root surface. Different biocontrol strains have different capacities to colonize a heat-treated soil; as well as different capacities to colonize the plant root growing in the pre-colonized soil.[24] There was no correlation between the population density of the biocontrol strains in soil and their capacity to effectively colonize the roots. Thus if competition for nutrients, especially carbon, is one mode of action of the protective strains of *F. oxysporum*, it is not the only mechanism of action.

8.2.1.2. *Competitive Interactions on the Root Surface and in the Plant Tissues*

Competition for infection sites was considered as an important mechanism as it was postulated[23] that the root surface had a finite number of infection sites that could be protected by increasing the inoculum density of the nonpathogenic strain. Many studies have been conducted that supported this hypothesis of competition for infection sites.

An indirect method demonstrated that pathogenic and nonpathogenic strains were competing for root colonization[25]: A *gus* transformed strain of a pathogenic *F. oxysporum* (*F.o.* f. sp. *lini*) was confronted to the wild type biocontrol strain Fo47 in the presence of the plant root. The total biomass of fungi having colonized the plant root was estimated by ELISA and the metabolic activity of the pathogen was measured by the glucuronidase (*gus*) activity in root tissues. The results demonstrated that the co-inoculation of the pathogen with the biocontrol strain resulted in the same total fungal biomass as when the pathogenic or the biocontrol strains were inoculated alone, but the biocontrol agent induced a significant decrease of the pathogen as measured by *gus* activity. The same experiment was conducted with a small collection of five protective strains of *F. oxysporum*. The results showed that different strains have different abilities to compete with the same pathogen at the surface and in the root tissues of the host plant.

Both a pathogenic and a biocontrol strain were clearly able to actively colonize the surface of the tomato root, as shown with *gus*-transformed strains.[26] Both the pathogenic and the protective fungi penetrated the epidermal cells and colonized the upper layers of cortical cells. A serial dilution technique was used to quantify the colonization of the root by the fungi[23,27]; there was a reduction of colonization intensity of the root tissues by the pathogen in the presence of the biocontrol strain, but no competition at the root surface for infection sites. The pathogen was inside the vessels and the nonpathogenic strain at the root surface and in the upper layers of cortical cells.[27] It was concluded that the strains can exclude each other from the same ecological niche. These observations demonstrated that the presence of the nonpathogenic strain in the upper layers of root tissues did not prevent the development of the pathogen in the vessels, but they did not prove the existence of a limited number of infection sites and competition for these infection sites.

The use of transformed strains expressing either the *GFP* or the *DsRed2* genes and confocal laser microscopy recently enabled observation of the behavior of a pathogenic and a biocontrol strains of *F. oxysporum* and their growth on the root surface.[28] The conidia of both the pathogenic and the biocontrol strain introduced into the soil before transplanting tomato seedlings, germinated 18 h after seedling transplantation and germ tubes were observed reaching the root surface. Both fungi intensively and evenly colonized the root surface. There was no evidence of specific sites of infection. Moreover, in contrast to what has been reported earlier[29,30] the apical root zone was never colonized by these fungi. The previous studies had been conducted in hydroponics or in sand culture, and the fungi were applied to the seeds or into the nutrient solution. In such conditions, an intense colonization of the root apices was observed, which was explained by an intense production of exudates at the root apex. When the conidia were introduced into the soil before transplanting the seedlings, the conidia adhering to soil particles were induced to germinate only when the root exudates produced at the root apex reached them. As the root is growing faster than the germ tube of the conidia, the germ tubes always reached the root surface behind the apex.

When both fungi were applied together, images showed the two fungi growing together exactly at the same spot on the root surface. When the protective strain was introduced at a concentration 100 times greater than that of the pathogen, it was dominant on the root surface. However it never completely excluded the pathogen, which was observed reaching the root surface at spots already heavily colonized by the nonpathogenic strain. It appeared that the surface of the root available for colonization by the fungi was not a limiting factor. Thus competition for colonization of the root surface does not take place or plays a little role in the interaction between pathogenic and nonpathogenic *F. oxysporum*. As the probability of a successful infection depends on the ratio between the population densities of the pathogen and the biocontrol strain, one must admit that competition plays a role, but our results minimized its importance, and competition relates to nutrients rather than to space on the root surface.

The hypothesis of direct competition between two strains of *F. oxysporum* in the vessels of the host plant was considered by comparing the growth in the stele of carnation of a pathogenic strain of *F. oxysporum* f. sp. *dianthi* and of several nonpathogenic strains after artificial inoculation of these strains, alone or in combination into the vessels of the plant.[31] Some nonpathogenic strains were able to reduce the stem colonization by the pathogen resulting in a decrease of disease severity. Locally induced resistance or direct competition between strains within the vessels could cause this disease suppressive effect after mixed inoculation into the stem. These observations are in agreement with others who selected a nonpathogenic strain of *F. oxysporum* able to control Fusarium wilt of sweet potato when introduced into the stem of the plant.[17] These results were obtained by direct inoculation of the protective strain into the xylem vessels or by dipping cuttings into a conidia suspension of the protective strain. One might doubt of the role of competition in the vessels of a plant as there is no evidence that the protective strain inoculated to the roots reaches the vessels in more normal growing conditions.

Colonization of the root surface and root tissues probably depends not only on the fungal strain but also on the plant species and plant cultivar. The compatibility between biocontrol strains of *F. oxysporum* and the plant species or plant cultivar has not been thoroughly investigated. Still, the watermelon cultivar "Crimson Sweet" created its own suppressive soil via its root exudates, which increased populations of beneficial *F. oxysporum* while other watermelon cultivars did not.[32]

In summary, competition, either for nutrients or for root surface colonization does occur when the resource for which the two strains are competing is limited. The carbon source can be a limiting factor for fungal germination and growth, especially in suppressive soils, but there was no experimental evidence that infection sites on the root surface are a limiting factor. Increasing the population density ratio of the protective strain over the pathogen always resulted in an increased competition between the two strains, but it does not demonstrate that the protective strain is more competitive than the pathogen.

8.2.2. INDIRECT ANTAGONISM: INDUCTION OF SYSTEMIC RESISTANCE

It is well established that pre-inoculation of a plant with an incompatible strain of *F. oxysporum* results in the mitigation of disease symptoms when the plant is later inoculated with a compatible pathogen.[15] This phenomenon, which is now considered as an expression of induced systemic resistance has been extensively studied, as it could explain the disease control provided by nonpathogenic strains of *F. oxysporum*. Induced systemic resistance for the control of Fusarium wilt of watermelon was first achieved by several strains of nonpathogenic *F. oxysporum*.[32] Many investigators have used a split root method to study induced resistance in *Fusarium*,[19,23,33–36] where a nonpathogenic strain applied to some roots of a host plant can delay symptom expression induced by the pathogen separately applied to other roots or directly into the stem of the plant. As there is no direct interaction between the two microorganisms, the observed disease reduction is attributed to increased plant defense reactions in response to root colonization by the nonpathogenic strain.

When competition is the main mode of action, typically the population of the biocontrol fungus must be larger than that of the pathogen population to achieve control; whereas, when induced resistance is the main mode of action, control can be achieved when the pathogen population is much greater than that of the biocontrol fungus. Wilt incidence in tomato was reduced when the pathogen population was up to 1000 times greater than that of the protective strain CS-20.[19] In contrast, strain Fo47, which functions mainly through competition, is only effective when it is introduced at concentrations 10 to 100 times higher than the pathogen concentration.[37]

Induced systemic resistance[16] is correlated with enzymatic changes in the plant and with the formation of the physical barriers discussed below. Results presented in the literature are quite confusing about changes in the expression of defense related proteins. An increased activity of several plant enzymes related to plant defense reactions (laminarinase, chitinase, other glycosidases, peroxidase and polyphenoloxidase) was reported in tomato plants transplanted in sterilized soil infested with a biocontrol strain of *F. oxysporum*.[38] The biocontrol activity of the protective strain Fo47 was attributed to induced resistance in tomato,[34] and correlated with an increased activity of chitinase, ß 1–3 glucanase and ß 1–4 glucosidase. Although the nonpathogenic strain Fo47 was not very effective in inducing systemic resistance in tomato, it induced an increase of PR proteins.[39] An overall increased activity of constitutive glycosidase isoforms in response to infection by *F.o.* f. sp. *lycopersici* was found that did not occur in roots colonized with a biocontrol strain.[40] These contradictory results were all obtained with the biocontrol strain Fo47 applied to tomato, and demonstrate that the biochemical response of the plant is not clearly understood. The biochemical events induced by inoculation of biocontrol strains of *F. oxysporum* to the plant must be accurately understood, before this system can be compared to other plant pathogen models where the cascade of biochemical events is better known.

8.3. Interactions between Pathogenic or Protective Strains of *F. oxysporum* and the Plant

As presented above, induced systemic resistance is partly responsible of the biocontrol activity of *F. oxysporum*. It was thus necessary to reconsider the interactions between pathogenic or protective strains of *F. oxysporum* and the plant. This is the aim of a project initiated by our group in Dijon. In a first step, the interactions of the plant with either a pathogenic or a biocontrol strain were compared using several approaches: (i) microscopy to describe the process of root colonization and the plant defense reactions observed at the tissue and cell levels; (ii) biochemistry to characterize the early physiological events of plant cells in response to inoculation by *F. oxysporum*; (iii) molecular biology to compare gene expression during infection by either a pathogenic or a biocontrol strain.

8.3.1. PROCESS OF ROOT COLONIZATION BY PATHOGENIC AND PROTECTIVE STRAINS OF *F. OXYSPORUM*

The protective strains actively colonized the root surface, and hyphae penetrated into epidermis and induced intense defense reactions in cells of the hypodermis (the cell layer just below the epidermis) and sometimes in the first layer of cortical cells. These responses led to the formation of a barrier made of necrotic tissues, preventing the entrance of the fungus towards the inner cortex. In these necrotic areas, cells appeared flattened and wall coiling entrapped hyphae. Wall appositions were frequently observed and electron dense material was formed in cells as well as in the intercellular spaces. Papillae were formed in reaction to penetration pegs between cells. In some cases as early as 24 h after inoculation, cells infected by the protective strains appeared dead among non infected healthy cells. Comparing these observations with others,[41] the protective strains are similar to endophytic fungi as they are able to establish in the root cortex. However, as previously reported[26,29] these endophytes have a reduced capacity to colonize the roots.

Roots inoculated with the pathogenic strains also showed an intense surface colonization by the fungus. Hyphae penetrated through epidermis cells and colonized the hypodermis and the first layer of cortical cells, which did not appear heavily disturbed. However plant defense reactions could be observed in some areas. They appeared similar but were both less abundant and less intense than those described above for the nonpathogenic strains. Colonized cells showed either little or no reaction to fungal invasion. The main difference between plants inoculated by the pathogenic or the biocontrol strains was that the barrier made of defense reactions stopped the ingress of the protective strains, although the hyphae of the pathogens reached the stele where they penetrated into the vessels of the xylem.

8.3.2. EARLY PHYSIOLOGICAL EVENTS INDUCED IN PLANT CELLS INOCULATED WITH EITHER A PATHOGENIC OR A PROTECTIVE STRAIN OF *F. OXYSPORUM*

As both pathogenic and nonpathogenic strains penetrated into the roots and induced defense reactions, it was interesting to analyze the early physiological events induced in plant cells by inoculation of both types of *F. oxysporum*. Indeed, based on the results obtained with other plant–pathogen models (*Pseudomonas*/*Phytophthora*/ tobacco,[42,43] *Pseudomonas*/soybean[44]) it has been well established that plant physiological events expressed by plant cells after infection by a bio-aggressor enable differentiating the compatible from the incompatible reactions. These physiological early events: ion fluxes, production of reactive oxygen species, reactive nitrogen species production, and cell death, have been correlated with defense reactions to the hypersensitive response (HR). An experimental model was developed in which cell suspensions were inoculated with germinated conidia of *F. oxysporum*.[45] The experiments were conducted with the three model plants: flax, tomato and melon and several strains of *F. oxysporum* either pathogenic or protective to these plant species (see Table I).

Production of ROS was determined by measuring the production of H_2O_2 by chemiluminescence. As in most of the models studied,[46,47] the protective strains always induced a biphasic production of H_2O_2. The second burst reached its maximum between 150 and 300 min post-inoculation, and the quantity of H_2O_2 produced was much greater than that produced during the first burst. The pathogenic strains induced either a single burst (Foln3/flax), or two bursts (Fol8/tomato and Fom24/melon). However, the quantity of H_2O_2 produced during the second burst was always much lower than the quantity produced in response to inoculation of the plant cells by the protective strains. Both protective and pathogenic strains induced a Ca^{2+} influx in plant cells,

TABLE I. Protective or pathogenic capacity of *Fusarium oxysporum* strains cited in this chapter on three plant species

	Soil-borne, nonpathogenic isolates		Pathogenic isolates			
	Fo47	CS20*	Foln3 f.sp.*lini*	Fol8 f. sp. *lycopersici*	Fom24 f. sp. *melonis*	Rev 157mutant of Fom24
Flax	Protective	Protective	Pathogenic	Protective	Protective	Non-protective
Tomato	Protective	Protective	Protective	Pathogenic	Protective	Non-protective
Melon	Protecttive	Protective	Protective	Protective.	Pathogenic	Non-pathogenic Non-protective

*Provided by D. Fravel.[19]

which was always more intense in response to inoculation by the protective strains. This difference between the two treatments increased with time starting 120 min after inoculation. Cells responded to inoculation by germinated microconidia of both protective and pathogenic strains of *F. oxysporum* by alkalinization of the extracellular medium. This alkalinization was greater with the protective than with the pathogenic strains. Finally, inoculation of plant cell suspensions with germinated microconidia of both the protective and the pathogenic strains induced cell death at the same rate until 14 h after inoculation. Later on, higher percentages of dead cells were observed in cell suspensions inoculated by the protective than the pathogenic strains, indicating that the protective strain induced programmed cell death related to induced resistance (see Table II).

All these results suggest that plant cells recognize the protective versus the pathogenic capacity of the fungus. Indeed, after a first phase during which the cells reacted similarly, the early physiological events are more intense in response to the protective than the pathogenic strains. This specific recognition should induce a cascade of molecular signals leading to either the disease or the protection of the plant.

TABLE II. Percentage of dead cells 23 h after inoculation with germinated conidia of protective or pathogenic strains of *F. oxysporum*

	Fo47	Fom24	Fol8	Foln3
Flax	86	77		52
Tomato	88	94	56	
Muskmelon	83	51		

8.3.3. ACCUMULATION OF mRNAS ENCODING DEFENSE PROTEINS IN CELL CULTURES INOCULATED WITH EITHER A PATHOGENIC OR A PROTECTIVE STRAIN OF *F. OXYSPORUM*

As stated above, studies dealing with resistance induced by biocontrol strains of *F. oxysporum* led to contradictory results regarding the accumulation of defense related proteins. The accumulation of pathogenesis related proteins (PR proteins) is an indicator of Systemic Acquired Resistance (SAR).[16] Thus, new experiments were conducted to assess the accumulation of several plant defense related proteins in plant cells inoculated with either a pathogenic or a protective strain of *F. oxysporum*. The expression of mRNAs encoding: phenylalanine-ammonialyase (PAL), PR-1, extracellular acidic β-1,3-glucanase, intracellular basic β-1,3-glucanase, extracellular acidic chitinase (CHI3), and intracellular basic chitinase (CHI9) was followed from 1 to 12 h after inoculation of the cells with germinated microconidia.

mRNA encoding PAL and chitinase 9 accumulated in a similar manner in cells inoculated with either the pathogen or the protective strain. The other defense proteins studied began to accumulate 4 h after inoculation. This occurred to a greater extent in the cells inoculated by the pathogen than in cells inoculated by the biocontrol strain. These results are in agreement with other studies[40] showing an accumulation of chitinase and β-1,3-glucanase in the tissues of the diseased plant but not in the tissues of the plant protected by Fo47. These results contradict the attribution of the biocontrol activity of Fo47 to its ability to induce accumulation of some PR proteins.[34] Considering all the contradictory results, one must agree with conclusions from studies of the expression of defense genes in chick pea infected by *F. oxysporum* f. sp. *ciceri*: "Fusarium wilt resistance in chickpea did not require induction of the defense-related genes after Fusarium infection. The systemic regulation of the defense-related genes at transcription level associated with disease resistance in other model plant species such as *Arabidopsis*, might not confer resistance in chickpea against *F.o.* f. sp. *ciceri*, and further studies focused on constitutive or unknown related systems independent of salicylic acid and jasmonic acid mediated systemic resistance mechanisms are required to understand fungal resistance mechanisms in chickpea."[48] Our results are leading to the same conclusion.

8.4. Differential Gene Expression During Interaction between Tomato Cells and a Protective or a Non Protective Strain of *F. oxysporum*

In the studies reported above, the plants were inoculated by either a strain pathogenic to this plant species or by a different strain able to protect it.

The two fungal strains had many different traits; therefore it was difficult to associate the differences observed with the biocontrol capacity of the strain. To address the question of the determinism of the biocontrol capacity of the strain, it was decided to follow another approach and to produce mutants of the biocontrol strains having lost the capacity to protect the plant. The strategy chosen was to utilize transposon mutagenesis to induce mutations in the strain and to screen the revertants obtained for their capacity to protect the plant. This strategy has been used successfully with Fo47, a model strain with biocontrol capacity,[49] and with a strain of *F.o.* f. sp. *melonis* (Fom24) pathogenic on muskmelon, but having a protective effect on other plant species such as flax and tomato. Two mutants of Fom24 that lost their virulence on muskmelon[50] also lost the capacity to protect flax and tomato against their specific pathogen (Olivain et al., unpublished data). Transposon mutagenesis has been proposed as a tool to tag genes.[51] Unfortunately in the case of the mutants of Fo47 and Fom24, it has not yet been possible to identify the gene in which the transposon has been reinserted.

The mutants that have lost their biocontrol capacity may still be useful for identifying genes involved in the biocontrol capacity, but another strategy has to be used. For this purpose we chose to study differential gene expression during the interactions between the plant and either a protective strain and its mutant having lost the biocontrol capacity. The model chosen consists of tomato cell suspensions inoculated with germinated conidia of either Fom24 or its mutant Rev 157. The method chosen to compare gene expression is the Rapid Amplified Subtractive Hybridization technique (RaSH).[52,53] Cells interacting with germinated conidia of the fungi were sampled from 0.5 to 12 h after inoculation. Total RNAs were extracted from the cell cultures inoculated with the fungi and the RNAs from different samples of the same treatment were pooled. From these pooled samples the mRNAs were purified and cDNAs were synthesized using a commercial kit. RaSH cDNA libraries were prepared from double-stranded cDNAs that were enzymatically digested into small fragments, ligated to adapters, and PCR amplified followed by incubation of tester and driver PCR fragments.[52,53] Based on the biphasic production of H_2O_2 by tomato cells inoculated with germinated conidia of Fom24 and Rev 157, two experiments targeted the early stages (30–90 min) or the later stages (0.5–12 h) of the interaction. Two probes were prepared by pooling all the RNAs extracted from samples taken from 30 to 90 min or from 0.5 to 12 h.

The preliminary results are interesting; several plant and fungal genes were differentially expressed. The most promising putative plant genes include: *Rin4* a gene involved in the interactions between *Arabidopsis* and *Pseudomonas*; a gene encoding an endochitinase involved in plant defense reactions, a gene encoding a porin implicated in cell death and a ferredoxin-NADP oxydoreductase encoding gene. From the fungal side the interesting sequences corresponded to ESTs of *Fusarium* and of sequences encoding putative proteins. Further analyses are needed to verify the involvement of these sequences in the interactions between the plant and the protective strains of *F. oxysporum*.

8.5. Production and Formulation of Protective Strains

The state of knowledge presented above shows clearly that we are far from a complete understanding of the mechanisms responsible for the biocontrol efficacy of any strain of *F. oxysporum*. However, we need not wait on a full understanding of the modes of action of the strains before evaluating their potential as biocontrol agents. To evaluate the efficacy of a biocontrol agent in large field experiments it is necessary to produce the inoculum in quantity sufficient for these experiments. Fo47 has to be introduced at a concentration from 10 to 100 times that of the pathogen to be effective under well controlled conditions in greenhouses. The inoculum density of the pathogen in field soil is not known, thus the first target was to protect vegetable and horticultural crops grown in soil-less culture on artificial substrates or on peat based growing substrates. The objective was to reach a concentration of 1×10^4 CFU ml^{-1} of substrate, therefore it was necessary to find a production process enabling the production of high quantity of inoculum. Two different processes have been used to produce Fo47: both submerged and solid state fermentation.

Submerged fermentation is achieved in bio-reactors using an appropriate liquid medium. At the end of the growth phase, the medium is removed by filtration. The microchlamydospores produced are mixed with talcum powder, an inert carrier. This talc inoculum is dried at 20°C by forced air and then stored at 4°C. The propagules of *Fusarium* remain viable for at least 18 months when kept at 4°C and for several months when kept at room temperature. Therefore, this talc inoculum could be commercialized without any specific problems. The inoculum of Fo47 can be mixed directly with soil as a powder or after having been suspended in water.[54]

The second process can be very useful for soil application as Fo47 is directly produced in γ irradiated peat inoculated with a conidia suspension. The fungus grew rapidly and at 1×10^3 or 1×10^6 CFU g^{-1} initial concentration, after 28 days, the protective agent has reached the carrying capacity of the substrate, more than 10^7 CFU g^{-1} peat. Thereafter, it maintained a stable density close to 6.9×10^7 CFU g^{-1} of peat for 11 months.This stored inoculum was as effective in controlling the disease as a freshly produced suspension of conidia.[55]

8.6. Fate in the Environment and Effects on the Soil Microflora

The biocontrol agent must establish in the soil or the substrate in which it is introduced to be effective. Thus, it must be adapted to the soil type and to the environmental conditions.

In order to study the fate of a biological control agent in the environment one must be able to characterize it among the soil-borne, indigenous populations of the same species. Two strategies are available: use of a mutant resistant to a fungicide or characterization of a genomic sequence specific of that strain (SCAR) that will enable its specific detection. It was quite easy to obtain mutants of *F. oxysporum* resistant to benomyl. Mutant Fo47b10 was chosen as it was as effective and as competitive as the parental strain. This biocontrol strain established itself in soil at densities close to the concentrations at which it has been introduced followed by a very slow progressive decline. The physico-chemical properties of soils (pH 7.8 or pH 4.5) did not affect its survival nor did the temperature ranging from 5°C to 25°C or the water potential ranging from pF 0.1 to pF 15.

A better approach to study the fate of a biological control strain would be to develop a SCAR marker as it has recently been done for a strain of *Trichoderma atroviride*.[56] In that case, the marker not only enabled the specific recognition of the strain among other strains, but the use of real time PCR after direct extraction of total soil DNA permitted following the population dynamics in soil without having to isolate the fungus. This type of technology should be preferable as it does not require any modification of the biocontrol agent and permits studying its fate in any type of environment.

In Europe, to be put on the market, a biocontrol fungus has to satisfy the requirements of directive 91/414. Among them, the effects on non targets organisms have to be considered, and in the case of a biological control agents applied to soil, effects on the soil microorganisms have to be studied. An experimental approach, based on extraction of total soil DNA has been developed[57] to determine whether or not a biocontrol strain affects the structure of the communities of the indigenous microorganisms. Bacterial and fungal community structures were analyzed by T-RFLP of 16S and 18S rRNA genes, respectively. In one experiment, two soils of different physico-chemical properties were inoculated with the biocontrol strain Fo47. The structures of both the bacterial and fungal communities of these two soils were analyzed 24 h after introduction of Fo47 into the soils, and 30 days later. Results analyzed by principal component analysis (PCA) are presented in Figure 1. The structures of the bacterial communities were different in the two control (non-infested) soils, which appeared clearly separated along the axis1. Immediately after introduction of the biocontrol agents, the structure of the bacterial communities appeared modified in both soils. But one month later, the differences

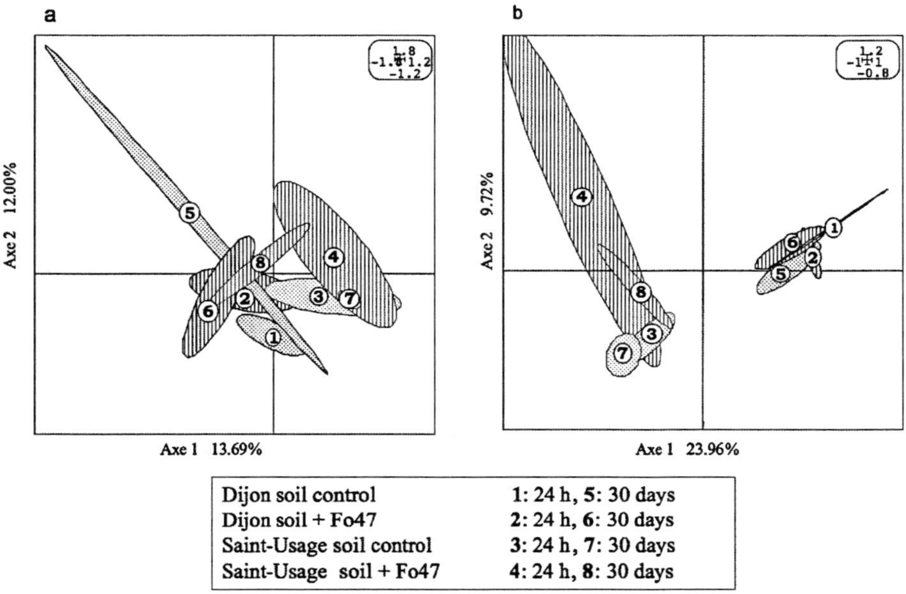

Figure 1. The changing structure of the fungal (a) and bacterial (b) communities of two soils infested or not with the protective strain Fo47. The analyses were performed by tRFLP 24 h and 30 days after soil inoculation

were not as clear, indicating that the structures of the bacterial communities after inoculation of biocontrol agents tended to come back to the initial state.

The structure of the fungal communities in the two soils appeared more similar than the structure of the bacterial communities. Introduction of the biocontrol agent induced changes in the structure of the fungal communities in both soils. These changes were persistent, as 1 month later the non infested controls were separated from the infested soils. As the method used is based on extraction of total DNA from soil, it seems logical that the biocontrol agent introduced at a concentration of 1×10^6 CFU g^{-1} soil will be detected 24 h after inoculation. The persistence of the effect on the structure of the fungal communities is related to the establishment of this biocontrol strain in the soil as demonstrated above. These preliminary results have to be confirmed, but they show that this type of method will be useful to follow the effects of biological control agents on non-target organisms, especially on the soil microbial communities.

8.7. Conclusions

Studies conducted on the modes of action of protective strains of *F. oxysporum* clearly show that we are far from understanding the mechanisms by which some strains of *F. oxysporum* can protect a plant against further infection by a pathogenic strain of *F. oxysporum*. It also underlines the originality of this model, as the protective and the pathogenic strains belonging to the same species have many traits in common; both are able to colonize the soil and the plant root and to induce defense reaction in the plant. The protective strains of *F. oxysporum* have many traits in common with the antagonistic strains of *Trichoderma* spp. that have been qualified as "opportunistic plant symbionts." Both fungi have evolved towards an endophytic life style, which is beneficial to the plant (Chapter 7). The study of the interactions between the protective strains and the plant should help understand what differentiates between a pathogenic and a nonpathogenic strain, i.e., will contribute to the understanding of the mechanisms of pathogenicity in *F. oxysporum*.

The idea to use "nonpathogenic" strains of *F. oxysporum* to control Fusarium diseases came from the demonstration that autochthonous populations of soil-borne *F. oxysporum* were involved in the mechanisms of soil suppressiveness. The modes of action of the "nonpathogenic" strains of *F. oxysporum* isolated from suppressive soils focused, three decades ago,[7] on the ecological fitness of the protective strains, which were thought to be more competitive than the pathogenic ones. The addition of glucose to a suppressive soil was sufficient to provoke the disease, and the hypothesis of competition for nutrients seemed very realistic. The soil is an oligotrophic milieu, and the protective and the pathogenic strains share the same ecological niche and have to compete for the limited nutrients. Indeed, the comparison of the saprophytic ability of the model strain Fo47 with that of a few pathogenic strains showed that the protective strain has a greater competitive saprophytic ability. Results of many experimental studies have confirmed that competition for nutrients in soil and the rhizosphere is one of the modes of action of the protective strains of *F. oxysporum*. Still, under field conditions, the biocontrol strain always has to be applied at a concentration several times that of the pathogen, to increase its competitive advantage.

The second hypothesis proposed to explain the efficacy of the biocontrol strains was their competitive ability at the root surface. The use of transformed strains expressing fluorescent reporter genes demonstrated that competition for infection sites does not play the role proposed by several authors.[19,23,26] Indeed, it was demonstrated that there are no specific sites of infection at the root surface.[28] On the contrary, the entire root surface can be colonized by strains of *F. oxysporum*, and under experimental conditions even when the soil is infested with high inoculum concentration of *F. oxysporum* the root surface did not appear to be a limited resource. Although appealing, the image of a root fully protected by a shield made of hyphae of the biocontrol strains is not realistic.

The third hypothesis considered was the capacity of protective strains of *F. oxysporum* to induce systemic resistance in the plant. Indirect methods such as a split root system, demonstrated that both the soil-borne nonpathogenic strains and the pathogenic strains applied to a non host plant induced resistance in the plant. In contrast to the numerous studies dealing with induced resistance against aerial pathogens, basic studies on resistance induced by protective strains of *F. oxysporum* are limited. The hypersensitive response induced by these organisms has been neglected due to the difficulties inherent in experimentation with soil-borne pathogens infecting the roots. The protective strains of *F. oxysporum* penetrating into the root cortex induced plant defense reactions characterized by the formation of necrotic areas, similar to the hypersensitive responses on leaves, even if they are more difficult to detect. This comparison between the root response to infection by protective strains of *F. oxysporum* and the hypersensitive response invite us to study the early physiological events associated with the hypersensitive response. The use of cell cultures inoculated with germinated conidia of protective or pathogenic strains of *F. oxysporum* allowed the characterization of these early events. The kinetics of accumulation of mRNAs encoding defense related proteins were different between the compatible and the incompatible reactions, but the results suggested that these molecules might not be involved in plant resistance induced by protective strains of *F. oxysporum*. Mutants having lost their biocontrol capacity have recently been selected and are being used to determine the protective capacity of *F. oxysporum* strains. Differential gene expression was followed during interaction between tomato cells and germinated conidia of either a biocontrol strain or a mutant having lost its biocontrol capacity. The preliminary results obtained are in agreement with the accumulation of mRNAs encoding defense related proteins, i.e., differences in the kinetics of expression of defense related genes. Therefore, one can speculate that the protective strains induce a priming effect in the plant that will express an increased resistance when confronted to a pathogenic strain. Obviously there is a long way to go before being able to demonstrate this hypothesis.

Progress has been made in the practical aspects of production and application of protective strains of *F. oxysporum*. Mass production can be achieved in submerged or solid state fermentation. The propagules produced have a shelf life of over 1 year, sufficient to be compatible with commercial distribution. *F. oxysporum* is a soil-borne fungus that is well adapted to survival in soil, and experiments demonstrated that it established well in soils of different physicochemical properties and under a wide range of pHs, temperatures, and water potentials. One must wonder why there is not any preparation based on protective *F. oxysporum* on the market. The very expensive and time consuming registration process was the reason given by some small European companies.

One must wonder if the registration authorities will allow a protective strain belonging to the same species as the pathogen, as horizontal transfer of virulence can not be totally excluded. Finally one must admit that biological control lacks consistency. The only solution would be to better understand the conditions that are required for success of biological control, requiring more experiments under field conditions. Thus, today, despite progress made in the understanding of the modes of action of the protective strains of *F. oxysporum* and in the production and application methods, the use of strains of *F. oxysporum* to control Fusarium diseases remains a dream. But one must remember that natural soil suppressiveness to Fusarium wilts is not only based on the activity of the autochthonous population of *F. oxysporum*. It has been well established that other antagonistic microorganisms such as the fluorescent pseudomonads play a role, alone or in association with the nonpathogenic fusaria.[55] Therefore, using a single antagonistic microorganism might never give as good control of the disease as an association of strains. The attempts to use selected strains of *Pseudomonas fluorescens* to control Fusarium wilt were not more successful than the attempts to use protective strains of *F. oxysporum* alone. On the contrary, the uses of mixtures of protective strains of *F. oxysporum* with fluorescent pseudomonads always improve the control provided by a single organism.[22,55] Interactions between the antagonists and the pathogens are controlled by the environmental factors. Specific conditions might be required for full expression of the beneficial effects of the biological control agents. More experimentation under field conditions is needed to determine the conditions needed for successful control of the disease.

References

1. R. Cook and K. F. Baker, *The Nature and Practice of Biological Control of Plant Pathogens* (American Phytopathological Society, St Paul, MN 1983).
2. C. Alabouvette, C. Olivain, and C. Steinberg, Biological control of plant diseases: The European situation, *Eur. J. Plant. Pathol.* 114, 329–341 (2006).
3. R. H. Stover, Fusarial wilt (Panama disease) of bananas and other *Musa* species. CMI, *Phytopathol. Papers* 4 (1962)
4. G. Stotzky and R. T. Martin, Soil mineralogy in relation to the spread of Fusarium wilt of banana in Central America, *Plant Soil* 18, 317–337 (1963).
5. S. N. Smith and W. C. Snyder, Relationship of inoculum density and soil types to severity of fusarium wilt of sweet potato, *Phytopathology* 61, 1049–1051 (1971).
6. T. A. Toussoun, Fusarium-suppressive soils, in *Biology and Control of soil-borne Plant Pathogens*, edited by G. W. Bruehl (The American Physiological Society, St Paul, MN, 1975), pp. 145–151.
7. J. Louvet, F. Rouxel, and C. Alabouvette, Recherches sur la résistance des sols aux maladies, I: Mise en évidence de la nature microbiologique de la résistance d'un sol au développement de la fusariose vasculaire du melon, *Ann. Phytopathol.* 8, 425–436 (1976).

8. F. Rouxel, C. Alabouvette C, and J. Louvet, Recherches sur la résistance des sols aux maladies, II : Incidence de traitements thermiques sur la résistance microbiologique d'un sol à la fusariose vasculaire du melon, *Ann. Phytopathol.* 9, 183–192 (1977).
9. R. P. Larkin, D. L. Hopkins, and F. N. Martin, Effect of successive watermelon plantings on *Fusarium oxysporum* and other microorganisms in soils suppressive and conducive to Fusarium wilt of watermelon. *Phytopathology* 83, 1097–1105 (1993).
10. R. P. Larkin, D. L. Hopkins, and F. N. Martin, Suppression of fusarium wilt of watermelon by nonpathogenic *Fusarium oxysporum* and other microorganisms recovered from a disease suppressive soil, *Phytopathology* 86, 812–819 (1996).
11. T. C. Paulitz, C. S. Park, and R. Baker, Biological control of Fusarium wilt of cucumber with nonpathogenic isolates of *Fusarium oxysporum*, *Can. J. Microbiol.* 33, 349–353 (1987).
12. R. W. Schneider, Effects of nonpathogenic strains of *Fusarium oxysporum* on celery root infection by *Fusarium oxysporum* f. sp. *apii* and a novel use of the Lineweaver-Burk double reciprocal plot technique, *Phytopathology* 74, 646–653 (1984).
13. G. Tamietti and C. Alabouvette, Résistance des sols aux maladies, XIII : Rôle des *Fusarium oxysporum* non pathogènes dans les mécanismes de résistance d'un sol de Noirmoutiers aux fusarioses vasculaires, *Agronomie* 6, 541–548 (1986).
14. G. Tamietti and R. Pramotton, La réceptivité des sols aux fusarioses vasculaires: Rapport entre résistance et microflore autochtone avec référence particulière aux *Fusarium* non pathogènes, *Agronomie* 10, 69–76 (1990).
15. A. Matta, Induced resistance to fusarium wilt diseases, in *Vascular Wilt Diseases of Plants-Basic Studies and Control*, edited by E. C. Tjamos and C. H. Beckman (NATO ASI Series, Springer-Verlag, Berlin, Germany, 1989), pp. 175–196.
16. L. C. Van Loon, Systemic induced resistance, in *Mechanims of Resistance to Plant Diseases*, edited by A. J. Slusarenko, R. S. S. Fraser, and L. C. van Loon (Kluwer Academic Publishers, Dordrecht, The Netherlands, 2000, pp 521–574.
17. K. Ogawa and H. Komada, Biological control of Fusarium wilt of sweet potato by non-pathogenic *Fusarium oxysporum*, *Ann. Phytopathol. Soc. Japan* 50, 1–9 (1984).
18. J. Postma and H. Rattink, Biological control of fusarium wilt of carnation with a non-pathogenic isolate of *Fusarium oxysporum*, *Can. J. Bot.* 70, 1199–1205 (1992).
19. R. P. Larkin and D. R. Fravel, Mechanism of action and dose-response relationships governing biological control of Fusarium wilt of tomato by nonpathogenic *Fusarium* spp., *Phytopathology* 89, 1152–1161 (1999).
20. Y. Couteaudier and C. Alabouvette, Quantitative comparison of *Fusarum oxysporum* competitiveness in relation with carbon utilization, *FEMS Microbiol. Ecol.* 74, 261–268 (1990).
21. Y. Couteaudier and C. Steinberg, Biological and mathematical description of growth pattern of *Fusarium oxysporum* in sterilized soil, *FEMS Microbiol. Ecol.* 74, 253–260 (1990).
22. P. Lemanceau, P. A. H. M. Bakker, W. J. De Kogel, C. Alabouvette, and B. Schippers, Antagonistic effect on nonpathogenic *Fusarium oxysporum* Fo47and and pseudobactin 358 upon pathogenic *Fusarium oxysporum* f. sp dianthi, *Appl. Environ. Microbiol.* 59, 74–82 (1993).
23. Q. Mandeel and R. Baker, Mechanisms involved in biological control of Fusarium wilt of cucumber with strains of nonpathogenic *Fusarium oxysporum*, *Phytopathology* 81, 462–469 (1991).
24. H. Nagao, Y. Couteaudier, and C. Alabouvette, Colonization of sterilized soil and flax roots by strains of *Fusarium oxysporum* and *Fusarium solani*, *Symbiosis* 9, 343–354 (1990).
25. A. Eparvier and C. Alabouvette, Use of ELISA and *GUS*-transformed strains to study competition between pathogenic and nonpathogenic *Fusarium oxysporum* for root colonization, *Biocontrol Sci. Technol.* 4, 35–47 (1994).

26. C. Olivain and C. Alabouvette, Process of tomato root colonization by a pathogenic strain of *Fusarium oxysporum* f. sp. *lycopersici* in comparison with a nonpathogenic strain, *New Phytol.* 141, 497–510 (1999).
27. J. R. Bao and G. Lazarovitz, Differential colonization of tomato roots by nonpathogenic and pathogenic *Fusarium oxysporum* strains may influence Fusarium wilt control, *Phytopathology* 91, 449–456 (2001).
28. C. Olivain, C. Humbert, J. Nahalkova, J. Fatehi, F. L'Haridon, and C. Alabouvette, Colonization of tomato roots by pathogenic and nonpathogenic *Fusarium oxysporum* together and separately in the soil, *Appl. Environ. Microbiol.* 72, 1523–1531 (2006).
29. C. Olivain and C. Alabouvette, Colonization of tomato root by a nonpathogenic strain of *Fusarium oxysporum*, *New Phytol.* 137, 481–494 (1997).
30. A. L. Lagopodi, A. F. L. Ram, G. E. M. Lamers, P. J. Punt, C. A. M. J. J. Van den Hondel, B. J. J. Lugtenberg, and G. V. Bloemberg, Novel aspects of tomato root colonization and infection by *Fusarium oxysporum* f. sp. *radicis-lycopersici* revealed by confocal laser scanning microscopic analysis using the green fluorescent protein as a marker, *Mol. Plant-Microbe Interact.* 15, 172–179 (2002).
31. J. Postma and A. J. G. Luttikholt, Colonization of carnation stems by a nonpathogenic isolate of *Fusarium oxysporum* and its effect on *Fusarium oxysporum* f. sp *dianthi*, *Can. J. Bot.* 74, 1841–1851 (1996).
32. R. P. Larkin, D. L. Hopkins, and F. N. Martin, Ecology of*Fusarium oxysporum* f. sp. *niveum* in soils suppressive and conductive to Fusarium wilt of watermelon, *Phytopathology* 83, 1105–1116 (1993).
33. C. J. Biles and R. D. Martyn, Local and systemic induced watermelons by formae speciales of *Fusarium oxysporum*, *Phytopathology* 79, 856–860 (1989).
34. J. G. Fuchs, Y. Moënne-Loccoz, and G. Défago, Nonpathogenic *Fusarium oxysporum* strain Fo47 induces resistance to Fusarium wilt in tomato, *Plant Dis.* 81, 492–496 (1997).
35. B. A. M. Kroon, R. J. Scheffer, and D. M. Elgersma, Induced resistance in tomato plants against Fusarium wilt invoked by *Fusarium oxysporum* f. sp. *dianthi*, *Neth. J. Plant Pathol.* 97, 401–408 (1991).
36. C. Olivain, C. Steinberg, and C. Alabouvette, Evidence of induced resistance in tomato inoculated by nonpathogenic strains of *Fusarium oxysporum*, in *Environmental Biotic Factors in Integrated Plant Disease Control*, edited by M Manka (The Polish Phytopathological Society, Poznan, Poland, 1995), pp. 427–430.
37. A. Bolwerk, A. L. Lagopodi, B. J. J. Lugtenberg, and G. V. Bloemberg, Visualization of interactions between a pathogenic and a beneficial *Fusarium* strain during biocontrol of tomato foot and root rot, *Mol. Plant-Microbe Interact.* 78, 710–721 (2005).
38. G. Tamietti, L. Ferraris, A. Matta, and I. Abbattista Gentile, Physiological responses of tomato plants grown in *Fusarium* suppressive soil, *J. Phytopathol.* 138, 66–76 (1993).
39. B. J. Duijff, D. Pouhair, C. Olivain, C. Alabouvette, and P. Lemanceau, Implication of systemic induced resistance in the suppression of fusarium wilt of tomato by *Pseudomonas fluorescens* WCS417r and by nonpathogenic *Fusarium oxysporum* Fo47, *Eur. J. Plant Pathol.* 104, 903–910 (1998).
40. G. Recorbet, G. Bestel-Corre, E. Dumas-Gaudot, S. Gianinazzi, and C. Alabouvette, Differential accumulation of β-1,3-glucanase and chitinase isoforms in tomato roots in response to colonization by either pathogenic or nonpathogenic strains of *Fusarium oxysporum*, *Microbiol. Res.* 153, 257–263 (1998).
41. N. Benhamou and C. Garrand, Cytological analysis of defense-related mechanisms induced in pea root tissues in response to colonization by nonpathogenic *Fusarium oxysporum* Fo47, *Phytopathology* 91, 730–740 (2001).

42. L. D. Keppler, C. J. Baker, and M. M. Atkinson, Active oxygen production during a bacteria-induced hypersensitive reaction in tobacco suspension cells, *Phytopathology* 79, 974–978 (1989).
43. A. J. Able, D. I. Guest, and M. W. Sutherland, Use of a new tetrazolium-based assay to study the production of superoxide radicals by tobacco cell cultures challenged with avirulent zoospores of *Phytophtora parasitica* var *nicotianae*, *Plant Physiol.* 117, 491–499 (1998).
44. C. J. Baker, N. Mock, J. Glazener, and E. Orlandi, Recognition responses in pathogen/non-host and race/cultivar interactions involving soybean (*Glycine max*) and *Pseudomonas syringae* pathovars, *Physiol. Mol. Plant Pathol.* 43, 81–94 (1993).
45. C. Olivain, S. Trouvelot, M. N. Binet, C. Cordier, A. Pugin, and C. Alabouvette, Colonization of flax roots and early physiological responses of flax cells inoculated with pathogenic and nonpathogenic strains of *Fusarium oxysporum*, *Appl. Environ. Microbiol.* 69, 5453–5462 (2003).
46. L. De Gara, M. C. De Pinto, and F. Tommasi, The antioxidant systems vis-à-vis reactive oxygen species during plant-pathogen interaction, *Plant Physiol. Biochem.* 41, 863–870 (2003).
47. C. Lamb and R. Dixon, The oxidative burst in plant disease resistance, *Annu. Rev. Plant Physiol. Plant Mol. Biol.* 48, 251–275 (1997).
48. S. Cho and F. J. Muehlbauer, Genetic effect of differentially regulated fungal response genes on resistance to necrotrophic fungal pathogens in chickpea (*Cicer arietinum* L.), *Physiol. Mol. Plant Pathol.* 64, 57–66 (2004).
49. S. Trouvelot, C. Olivain, G. Recorbet, Q. Migheli, and C. Alabouvette, Recovery of *Fusarium ioxysporum* Fo47 mutants affected in their biocontrol activity after transposition of the *Fot1* element, *Phytopathology* 92, 936–945 (2002).
50. Q. Migehli, C. Steinberg, P. M. Davière, C. Olivain, C. Gerlinger, N. Gautheron, C. Alabouvette, and M. J. Daboussi, Recovery of mutants impaired in pathogenicity after transposition of *impala* in *Fusarium oxysporum* f. sp. *melonis*, *Phytopathology* 90, 1279–1284 (2000).
51. F. Villalba, M. H. Lebrun , A. Hua-Van, M. J. Daboussi, and M. C. Grosjean-Cournoyer, Transposon *impala*, a novel tool for gene tagging in the rice-blast fungus *Magnaporthe grisea*, *Mol. Plant-Microbe Interact.* 14, 308–315 (2001)
52. H. Jiang, D. Kang, D. Alexandreand, and P. B. Fisher, RASH, a rapid subtraction hybridization approach for identifying and cloning differentially expressed genes, *Proc. Natl .Acad. Sci. USA* 97, 12684–12689 (2000).
53. M. Schmoll, S. Zeilinger, R. L. Mach, and C. P. Kubicek, Cloning of genes expressed early during cellulose induction in *Hypocrea jecorina* by a rapid subtraction hybridization approach, *Fungal Genet. Biol.* 41, 877–887 (2004).
54. C. Alabouvette, D. De la Broise, P. Lemanceau, Y. Couteaudier, and J. Louvet, Utilisation de souches non pathogènes de *Fusarium* pour lutter contre les fusarioses: Situation actuelle dans la pratique, *Bul. OEPP-EPPO* 17, 665–774 (1987).
55. C. Olivain, C. Alabouvette, and C. Steinberg, Production of a mixed inoculum of *Fusarium oxysporum* Fo47 and *Pseudomonas fluorescens* C7 to control fusarium diseases, *Biocontrol Sci. Technol.* 14(3), 227–238 (2004).
56. C. Cordier, V. Edel-Hermann, F. Martin-Laurent, B. Bachar, C. Steinberg, and C. Alabouvette, SCAR-based real time PCR to identify a biocontrol strain (T1) of *Trichoderma atroviride* and study its population dynamics in soils, *J. Microbiol. Methods*, in press.
57. V. Edel-Hermann, C. Dreumont, A. Pérez-Piqueres, and C. Steinberg, Terminal restriction fragment length polymorphis manalysis of ribosomal RNA genes to assess changes in fungal community structure in soils, *FEMS Microbiol. Ecol.* 47, 397–404 (2004).

9. *METARHIZIUM ANISOPLIAE* AS A MODEL FOR STUDYING BIOINSECTICIDAL HOST PATHOGEN INTERACTIONS

Raymond J. St. Leger*
Department of Entomology, University of Maryland, College Park MD, USA

Abstract. Molecular biology methods have elucidated pathogenic processes in several biocontrol agents including one of the most commonly applied entomopathogenic fungi, *Metarhizium anisopliae*. In this article I will describe how a combination of EST and microarray approaches, gene disruption strategies, manipulation of gene expression and use of marker genes has: (1) identified and characterized genes involved in infection; (2) manipulated the genes of the pathogen to improve biocontrol performance; (3) allowed expression of a neurotoxin from the scorpion *Androctonus australis*; (4) allowed assessments of environmental risks posed by these modifications and (5) identified differences in genic constituents and gene expression that account for differences between strains.

Keywords: *Metarhizium anisopliae,* insect pathogen, microarrays, strain diversity

9.1. Introduction

Due to the well publicized environmental and pest-resistance problems associated with chemical pesticides, there is increasing interest in the exploitation of fungi for the control of invertebrate pests, weeds and plant diseases, as evidenced by the number of commercial products available and under development.[1]

Insect pathogenic fungi are key regulatory factors in insect pest populations. Unlike bacteria and viruses that have to be ingested to cause disease, fungi infect insects by direct penetration of the cuticle. They therefore provide the only practical means of microbial control of insects which feed by sucking plant or animal juices, as well as for the many coleopteran pests that have no known viral or bacterial diseases. They are best employed either as one component of an integrated pest management strategy or as inundative mycoinsecticides.[2,3] The aim of using inundative mycoinsecticides is to

*To whom correspondence should be addressed, e-mail: stleger@umd.edu

M. Vurro and J. Gressel (eds.), Novel Biotechnologies for Biocontrol Agent Enhancement and Management, 179–204.

maximize the kill from the initial application, in the same way as with a chemical pesticide.[2] This strategy developed with the realization that with proper formulation, infection from an initial application can occur independent of humidity, but high humidity is required for the production of spores and it is this constraint that limits the spread of disease.[4]

However, the slow speed of kill and inconsistent results of biologicals in general compared with chemicals has deterred development. For example, it usually takes 10 days for *M. anisopliae* sf. *acridum* ("green muscle") to kill locusts and this is constraining successful commercialization, even though it consistently provides >80% control.[5] Consequently any consideration of the suitability of a pathogen for commercial development inevitably leads to the possibility of improving its performance.[6] Ultimately, various traits of fungal pathogens, including host range, production capacity, stability and virulence, will be enhanced through genetic manipulations.[7]

This chapter outlines studies on the molecular and biochemical interactions between fungi and insects that have utilized *Metarhizium anisopliae* as a model system. *M. anisopliae* is the best studied entomopathogenic fungi in terms of biochemical/molecular data and its application to genetic engineering. However, it is an underlying assumption that work on *M. anisopliae* will enrich understanding of the *ca.* 1,000 other species of entomopathogens, and accelerate the genetic manipulation of pathogenicity in the nine species besides *M. anisopliae* currently being developed or utilized for insect control.[1]

9.2. *M. anisopliae* As a Model Pathogen

M. anisopliae is one of the most commonly isolated insect pathogenic fungi with over 200 insect-host species and cosmopolitan distribution.[9] *M. anisopliae* is commercially available for the control of pests on pasture turf and its proposed future applications in soil include white grubs, mole crickets, caterpillars, fire ants, ticks and the $1 billion p.a termite problem.[9–11] Many of these insects provide a particular challenge to pest control specialists as there are few microorganisms available for use against them.[3,8] *M. anisopliae* strain F52 (registered for use by the EPA in 2003) is being targeted against various ticks, beetles and flies in residential and institutional lawns, landscape perimeters and greenhouses. However, traditionally, soil based inoculums of entomopathogenic fungi have needed to be very high to achieve effective control of pests as compared to applications against aerial pests such as locusts (Section 9.2.3).

M. anisopliae is a tractable model system offering EST collections, microarray analyses,[12–17] promoters that allow expression of foreign genes[18,19] and gene disruption technology.[17] *M. anisopliae* also produces many different cell types for developmental studies including conidia, hyphae, appressoria (pre-penetration swellings produced by many plant and insect pathogens) and uni-cellular blastospores that closely resemble budding yeast. Identification of types of genes whose manipulation would have potential in mycoinsecticide development is easier in an organism such as *M. anisopliae*, for which there is extensive physiological and biochemical data.[9] However, genetic studies of *M. anisopliae* were traditionally hampered by low transformation frequencies. This has been remedied by adapting a method of *Agrobacterium* mediated transformation,[20] to generate insertional mutants of *M. anisopliae* using a vector pFBENGFP from the Bidochka lab. Inherent advantages of working with *M. anisopliae* also include significant ecological and genetic differences between strains to facilitate comparative studies on life strategies.[21,22]

M. anisopliae is in many respects a typical pathogenic fungus but with some strains being rhizosphere competent it has more lifestyle options than most. This may be because of its heritage as the basal lineages of clavicipitaceous ascomycete fungi are grass pathogens and *M. anisopliae* clusters with clavicipitaceous grass endophytes (*Epichloe*) in phylogenetic studies.[23] The phylogeny of the *Metarhizium* genus is well characterized.[24] *M. anisopliae* has a clonal population structure (strains persist over time and space); no sexual stage is known in N. America (but has been identified in Thailand) and heterokaryon incompatibility precludes parasexuality except between very closely related strains.[21,22] Thus, gene exchange is likely to be a rare event,[4] but this has not been properly investigated in field conditions. *M. anisopliae* contains strains with wide host ranges (e.g., *M. anisopliae* sf. *anisopliae* 2575), and strains that like sf. *acridum* strain 324 (used for locust control) show specificity for certain locusts, beetles, crickets, hemipterans, etc, and are unable to infect other insects. While some specialized lineages, such as sf. *acridum*, are phylogenetically distant from generalist strains implying evolutionarily conserved host use patterns, closely related strains can also differ greatly in host range.[21,25,26] Evidence that most specialists arose from generalists includes: (1) the vast majority of isolates found in nature belong to the genetically very diverse sf. *anisopliae* and typically demonstrate wide host ranges; (2) specialist strains are scattered among generalists in phylogenies and have independently adapted to different insects, and (3) specialization is associated with conditions that are assumed to be derived including reduced breadth of diet.[21,27] Being a generalist does not rule out their showing adaptations to nutrients on frequently met hosts. For example, nutrients on Hemiptera (i.e., aphids) are supplemented by insect secretions rich in sugars while beetles carry low levels of nitrogenous nutrients. Consistent with this, many lines isolated from Coleoptera require low levels of complex nitrogenous nutrients to induce appressoria, while hemipteran-derived lines also produce appressoria in glucose medium.[21,26] Closely related strains isolated from beetles or hemipterans show these differences indicating that there are genetic mechanisms allowing rapid adaptation.[26]

9.2.1. STRAIN SELECTION

Ascertaining which isolate(s) should be mass produced for a given pest situation is of key importance at the beginning of a pest control project.[1,28] To date, strains employed for pest control have been obtained by screening natural populations. This can be a daunting task because of the large number of isolates to choose from, and each step of the selection process can be time consuming.[9] If the pathogen is being applied as an inundative mycoinsecticide then environmental persistence is not required, and might be regarded as a drawback by a company seeking repeat sales. However, if the pathogen is to be employed for classical biocontrol and is expected to persist in the environment, then laboratory virulence tests may not be well correlated with field effectiveness. In addition to virulence, the isolate must be "in tune" with the habitat of the target insect and in fact, natural selection on a pathogen may be as much by environmental factors as by specific hosts.

9.2.2. ENVIRONMENT/HABITAT

Salient factors influencing the success of entomopathogens as pest control agents include a wide range of climatic (solar radiation, temperature, water availability, precipitation and wind), edaphic (soil types) and biotic (antagonists) conditions.[9,28,29] Genetically based resistance to these parameters would be a distinct advantage, both during infection and during product preparation and storage. Considerable variability exists among taxa and strains within species in their thermal characteristics, requirements for relative humidity and susceptibility to irradiation.[29–32] This provided evidence for strong selective pressures and the existence of a range of naturally available tools for developing tolerance to environmental constraints. The genetic mechanisms of resistance to environmental parameters are not well understood but are probably governed by polygenic factors that may therefore be too complex to be readily amenable to genetic manipulation. However, progress has been made in understanding susceptibility to damage by the UV-B (290–315 nm) portion of the solar spectrum; a major impediment to the successful commercialization of entomopathogens for field crops. Recent studies have shown that the degree of conidial pigmentation and levels of DNA repair enzymes contribute to tolerance and that there is a relationship between this tolerance and the geographical origin of the insect host.[33]

Inspite of the potential for genetic manipulation, immediate advances are likely to come from improved formulations, such as the use of sun screens, and by careful strain selection. Unprotected *B. bassiana* spores are almost completely inactivated by exposure to 60 min of direct sunlight. The most effective substrates tested were egg albumin and skimmed milk powder which extended persistence of *B. bassiana* threefold.

9.2.3. SOME RECENT EXAMPLES OF BIOLOGICAL CONTROL USING *M. ANISOPLIAE*

The best publicized product has exploited sf. *acridum* to control locust populations. Traditional locust control involves large quantities of chemicals being applied to vast areas of land. LUBILOSA (Lutte Biologique contre les Locustes et Sauteriaux: http://www.lubilosa.org/) was set up in response to environmental concerns over the heavy use of these chemicals. They focused on disease causing agents. Locusts were considered to mobile and to reproduce too quickly for classical biological control so they needed to develop an inundative insecticide. This required a pathogen that was reproducible in artificial cultures in large quantities. LUBILOSA were also looking for a specific pathogen that did not hurt non-targets including natural enemies of the pest. After extensive screening they identified an African strain of sf. *acridum* (Green muscle) that fulfilled these criteria. It is also important when looking for a biological control agent amongst natural strains that consideration be given to how the pathogen fits into the environment of the pest. Green muscle is adapted to desert conditions by producing spores within the cadaver to avoid UV. In addition, its spores are comparatively resistant to UV. During the course of the program it became clear that key technical challenges in the development of a mycoinsecticide were mass production of spores and development of a delivery system, which were linked by a critical process: separation of the spores from the growth media. Large mechanical mycoharvesters were developed that allowed high quality spore separation after mass production from solid substrates (e.g., rice) in a form that is easy to desiccate, formulate and package. Fungi are traditionally seen as needing humid conditions to work well. A critical discovery by Chris Prior at LUBILOSA changed this. He observed that spores of these fungi were more infectious when formulated in oil with their action more independent of environmental conditions.[2,4]

The first field trial targeted a 2000-hectare area in Niger. An important part of the trial was to evaluate the attitude of local farmers as they are ultimately the consumers who will decide the fate of the product. The slower kill by the fungus compared to the chemicals was considered a problem, although farmers appreciated that the fungus is much more persistent compared with a standard acridicide. Its non-toxicity to farmers and livestock was also seen as a big advantage.[34] Unfortunately, two field trials conducted in 2004 on 400-hectare plots in Mauritania and Niger had inconclusive results. This was due to several logistical problems including the products thick formulation that made spraying difficult. Trials with biocontrol agents in general have been plagued by quality control issues in part because as living things they usually require more knowledge to use effectively than competing insecticides.

However, sf. *acridum* can be used for successful locust control as spraying an Australian strain (Green guard) achieved 65–97% reductions within 8–11 days in populations of the oriental migratory locust in Tianjin and Henan provinces, China.[35] Compared to application in Africa, a higher concentration of spores (50 to 125 g spores/1125 ml oil/ha) was required due to thick vegetation protecting most locusts from direct impact during spraying.

Given that UV degrades most microbial insecticides, there has been recent emphasis in applying the pathogens in a UV protected site frequented by the pest. An example is the use of black cloth treated with *M. anisopliae* inside Tanzanian houses (black is attractive to mosquitoes). This reduces the number of bites fourfold.[36] The effect on malaria may be more pronounced than this sounds as lab studies suggest that *Plasmodium* infected mosquitoes are much less likely to survive.[37]

Another example is the use of *M. anisopliae* to attack *Varroa* mites. These infest honey bee colonies across most of N. America and can destroy a colony in a few months which is of considerable import as bees add $10 billion per year to N. American agriculture through pollination, not including honey, beeswax etc. The mites have developed resistance to the only approved chemicals—fluvalinate and coumaphos—now used for control. After screening various disease agents USDA scientists identified a strain of *M. anisopliae* that is very potent against mites but has no effect on individual bees, colony development or population size. In field trials the fungus was coated onto plastic strips that were placed into hives. Bees attack anything that gets into the hive and their attempts to chew up the strips spread the fungus throughout the colony. Most of the mites on them died within 3–5 days. The fungus was as effective as fluvalinate even 42 days after application.[38]

9.2.4. SOIL ADAPTATION

M. anisopliae is recoverable from soil world-wide[39] but is most abundant (10^6 propagules per gram) in undisturbed pastures, 2–6 cm deep.[3] These fungi could genuinely flourish in soil or survive there in a dormant state awaiting a susceptible host as it is not clear whether what is being recovered are conidia, mycelia surviving on insect remains, or mycelia living on non-insect substrates.[4,29] Aside from a report that many soil isolates are non-pathogenic to scarab beetles,[3] there is little information available on the relative virulence of isolates from soil and from insects. There may be two diverse sets of selection pressures on *Metarhizium* spp., one for optimum characteristics for soil survival and another for virulence to insects.[4] If so, it is unlikely that the same characteristics will be optimum for both insects and soil. Thus, genetic groups of *M. anisopliae* are linked to habitat type rather than insect host, suggesting that selection for survival in the soil is more important in shaping the population genetics of *M. anisopliae* than is selection for pathogenicity.[40]

Presumably a large population of insect hosts could contribute to *Metarhizium* soil populations. However, populations as large as those characteristic of *M. anisopliae* are normally the result of organic substrates in rhizospheres of the upper layers of the soil. Given that rooting density is particularly high in grasses and cereal crops, i.e., <3 mm spaces between roots,[41] the *Metarhizium* community must be living in overlapping rhizospheres ("rhizosphere" is defined as the zone of soil immediately adjacent to plant roots in which the kinds, numbers, or activities of microorganisms differ from that of the bulk soil, and "rhizosphere competence" is the ability of an organism to colonize the rhizosphere).

Clearly, interactions between organisms have an important role in shaping organismal diversity. Yet except for some limited aspects of host–pathogen and predator–prey interactions, the nature of evolutionary forces acting during these processes are particularly poorly understood.[42] Thus, even for mycoparasitic *Trichoderma* spp where rhizosphere competence is known to be strongly related to biocontrol,[43] the genetic and physiological factors controlling rhizosphere competence are little understood compared to those controlling pathogenicity.[44]

9.3. Field Testing a Transgenic Strain of *M. anisopliae*

We conducted a field trial on a patch of cabbage with an engineered hypervirulent strain carrying extra protease genes plus the gene for EGFP1 (a variant of the green fluorescent protein).[45] The *gfp* gene is driven by a constitutive promoter and the cytoplasmically located protein strongly labels the whole fungus, with no detectable effects on fungal growth and pathogenicity. Use of GFP to monitor survival and distribution was essential because: (a) there were no precedents for the release of such fungal products, and (b) there is an inherent paucity of knowledge concerning the fate of fungal genotypes at the population and ecosystem level. This ignorance has helped stir controversy concerning the risks and benefits of releasing transgenic (or foreign) fungi for disease control, insect, and plant pest management or bioremediation, and provides a powerful motivation for studies on their ecology.[40,46] The field test confirmed that GFP is a very convenient way to monitor pathogen strains in field populations and demonstrated short term effects of insect transmission (non-target insects). The constitutively expressed subtilisin provided an additional marker during this trial. We are currently field testing transgenic strains of 2575 expressing the *gus* gene (*Escherichia coli* β-D glucoronidase gene) described before[47] as well as GFP. We used CHEF's technology (a form of pulsed field gel electrophoresis) to identify transformants with marker genes on different chromosomes.[45,47] The idea behind multiple markers is that while integrative transformants are very stable when grown for long periods in the absence of selection in pure culture under lab conditions,[47] stability may be different in a complex environment. In such a case it is unlikely that both unlinked markers (GUS and GFP) would be lost at once. The frequency of loss of each phenotype relative to the other could be determined, and there should usually be at least one marker remaining to positively distinguish a transformant from a native organism.

The most interesting result of our original field trial was that it documented rhizosphere competence of an entomopathogenic fungus. This emphasizes that for many economically important pathogens the most understudied aspect of their biology involves the extended periods they survive in soil in the absence of a suitable host.[48] Such knowledge is clearly of crucial importance for being able to predict and control outbreaks of plant or animal disease. The generality of rhizosphere competence in other entomopathogenic fungi commonly regarded as insect pathogens is still being investigated but the study places sharp focus on the soil/root interphase as a site where plants, insects, and pathogens will interact to determine fungal efficacy, cycling and survival.

In retrospect, we realized there was evidence in the literature before our study that *M. anisopliae* was rhizosphere competent. Thus, general surveys have shown that while *M. anisopliae* is ubiquitous, it is most abundant in grass root soils.[3] This abundance would have been very suggestive of rhizosphere competence to a soil microbiologist. The failure to appreciate the relationship between *M. anisopliae* and plants seems to be an example of scientists that belong to different scientific disciplines not being familiar with each other's literature.

9.4. The Relevance of Rhizophere Competence for Biological Control

Rhizosphere competence is particularly important when considering the potential commercial use of biocontrol agents toward soilborne plant pathogens,[49] and presumably the same could apply to pathogens of root insects. The fact that many genotypes of *M. anisopliae* appear specialized to different soils, e.g., grassland soils versus forest soils[31] suggests that the impact of rhizosphere competence by *M. anisopliae* on plant ecology in general could be considerable with implicit co-evolutionary implications. It may need to be considered as a feature for selecting fungal strains for biocontrol and this also raises the possibility of managing the rhizosphere microflora to achieve insect control. This would dovetail with attempts in IPM to manipulate the environment of the plant and insect to enhance insect biocontrol.[50] If a good root colonizer is chosen, that is capable of being transported by the root through the soil profile, then seed treatment would be an attractive method for introducing it into the soil–plant environment where it may have the opportunity to be the first colonizer of roots. The seed has already proved an important delivery vehicle for a variety of beneficial microbes for plant growth enhancement and biological disease control.[51,52] Species of clavicipitaceous (i.e., related to *M. anisopliae*) endophytes are used commercially in turf grass seeds in this manner. The development of appropriate combinations that included insect pathogens would obviously provide a higher level of plant protection and constitute a very promising research area.

However, there are many environmental and economic reasons why researchers and industry would not seek to permanently establish an engineered microbial agent in the environment.[46,51] In particular, the public is wary of biological control efforts due to potential unforeseen environmental impacts, and rhizosphere competence might increase the difficulty of eliminating the pathogen following unanticipated and deleterious environmental effects. Many crop plants are grasses where rhizosphere competence might be expected and, in any event it appears to be non-specific as rhizosphere competence was established with cabbages.[45] It is also likely that an entomopathogen applied to fields could drift to neighboring pastures and woodlands. Nevertheless, a key advantage of classical biocontrol over the use of synthetic insecticides is the ability of pathogens to replicate and persist in the environment providing long-term control. Ideally, therefore, we would want a strain to persist in the environment long enough to kill pest insects and short enough not to survive more than one season.

Unfortunately, the current predictive data base for risk assessment issues regarding future releases of genetically engineered fungi remains small and very little is known concerning the survival of individual genotypes in the field. We still need to identify the lifestyle (saprotrophy or pathogenicity) responsible for maintaining the large populations of insect pathogens in soil. We also need to provide the knowledge required to predict and improve fungal responses to various environmental stimuli. In particular, to determine side-effects of genetic alterations on the survival of transgenics in soil, their interactions with other soil organisms, transmission to insects and genetic stability. Such knowledge might facilitate genetically based containment by reducing the ability of the organism to spread through a lack of saprophytic competence.

9.5. Functional Genomics of *M. anisopliae*

Our earlier, pre-functional genomics work uncovering the genes and core signaling pathways regulating infection processes in *M. anisopliae* is reviewed.[46] Classical genetics and conventional gene analysis have been powerful tools for dissecting host pathogen interactions that are affected by the gain or loss of function of single proteins. Some of these genes encode enzymes and toxins with demonstrated targets in the insect. Other genes have been identified as virulence determinants because of their role in signal transduction during the production of infection structures.[46]

Such strategies have been less fruitful for understanding disease processes that are controlled by many genes. In addition side effects occurring in constructed strains are hard to predict and access and the full range of engineering possibilities cannot be exploited, due to lack of knowledge about the interrelated regulatory and metabolic processes going on in cells. So, the analysis of differential gene expression-known as functional genomics-has become one of the most widely used strategies for discovering and understanding the molecular circuitry underlying disease processes. Several of the ingenious techniques available[53] have been applied to insect pathogens.

We have assembled a *M. anisopliae* strain 2575 dataset containing about 11,000 ESTs (i.e., partial sequencing of randomly selected cDNA clones) from which we defined 3,563 EST unigenes (ca. 30% of 2575's total genes).[13,14] These include root exudate induced transcripts[15] to assess differences, overlaps and networking in secreted products (enzymes/toxins etc) and physiological parameters (protein phosphorylation events, transcriptional regulatory factors and physiological cues, etc.) that define the life of strain 2575 as a pathogen and as a saprophyte.

Focusing on EST approaches we compared gene expression patterns between strains 2575 and 324.[13] These are two of the most distantly related strains and essentially span the range of variation within *M. anisopliae.*[21,24] About 60% of the ESTs expressed by 2575 during growth on insect cuticle putatively encode secreted enzymes and toxins. We speculated that the large number and diversity of these effectors may be the key to the ability of strain 2575 to infect a wide variety of insects. In contrast, strain 324 expresses fewer putative hydrolytic enzymes and very few toxins. Those missing include some previously demonstrated to be required for the virulence of 2575 in various hosts.[13]

9.6. Microarray Studies

A long-term goal is to identify and determine the role of all the genes involved in host pathogen interactions. This daunting task is only feasible if the total number of experiments is limited by using a hierarchical approach to group genes of related function. We have used *Metarhizium* microarrays to putatively identify the large number of genes involved in colonization of hosts and then constructed smaller and smaller sub groups (e.g., fungal genes modulated by the chemistry of host cuticle, physical stimuli, etc.) to achieve a closer and closer approximation of the function of each gene. Having arrived at a manageable number of putative virulence genes we are using techniques for disrupting or overexpressing individual pathogen genes to confirm the roles suggested by their expression profiles. Four microarray studies have been published showing how sets of functionally related genes are coordinately induced or repressed by *M. anisopliae* in response to host related stimuli.[14–17] To date, we have identified more than 700 up-regulated genes in 2575 during adaptation to host cuticle or hemolymph. Some provide great insight into the very intricate mechanisms by which *M. anisopliae* has adapted to survive in these environments. Various aspects of hyphal growth in cuticle and hemolymph are associated with up regulation of different genes encoding components of signal transduction. Genes involved in membrane biogenesis, synthesis of cell wall components, storage or mobilization of nutrient reserves and protein folding are also highly expressed, indicative of manufacture and "remodeling" of cell structures. Other features highlighted by this work include the production of antimicrobial molecules and the very early cuticle-induced production of a variety of transporters and permeases that allow the fungus to "sample" the cuticle and then respond with secretion of a plethora of proteins. Multiple mechanisms involved in adaptation to hemolymph include dramatic remodeling of cell walls and lipid composition, the accumulation of solutes that increase internal osmotic pressure and up-regulation of non-oxidative respiratory pathways. A diverse range of genes encode virulence factors that help defend against possible host defenses such as oxidative and nitrosative (e.g., production of nitric oxide) stress and phenolics. These are up-regulated on cuticle and/or hemolymph along with a plethora of genes for extracellular enzymes and toxins that contribute to host damage.

The adaptive significance of many of the up-regulated genes involved in detoxification is clear (e.g., phenol hydroxylase) but others are surprising as they suggest, for example, that insects may employ cyanogenic compounds and propionate as defensive compounds.[15] If this turned out to be the case it would demonstrate that pathogen counter-responses can be used to predict host defenses.

We used 2575 arrays to probe the causes of sectorization (non-sporulating cultures) in two strains of sf. *anisopliae*. We demonstrated that sectorization was associated with mutations that produced oxidative stress and altered regulation of downstream aging-related genes.[16] Sectorization is a major problem for long term culturing and manufacture of many fungi, including entomopathogens. Having identified probable causes we wish to see if we can prevent or cure sterile cultures.

Using specific expression patterns for developing hypotheses on gene function has worked very well for us. For example, two of the most highly expressed genes during growth in hemolymph encode cell wall proteins; a collagen and an adhesin.[15] Construction of deletion strains showed these to be involved in evading host immunity and adhering to host surfaces, respectively.[17] These results illustrate the power of expression profiling for revealing previously unsuspected stratagems of infection.

9.6.1. STRAIN-SPECIFIC DIFFERENCES IDENTIFIED BY MICROARRAYS

We have verified that an array of ESTs from 2575 can be used for heterologous hybridization with DNA or RNA from diverse strains of *M. anisopliae.*[16,17] There are more examples in specialists than generalists where only select members of gene families respond to a component of cuticle or hemolymph.[16] The divergent transcriptomes of strains correlated with important biological differences and offered explanations for these. Unlike 2575, when 324 is grown in submerged cultures, it up-regulates transcripts involved in sporulation. This relates to the unusual ability of 324 to produce spores inside host cadavers as an adaptation to desert living. Demonstrations of the role that regulatory variation can play in providing the raw material for adaptive evolution of a pathogen is especially intriguing and timely with the new realization of the extent to which gene expression is a major vehicle used by evolution to produce new phenotypes of metazoans (including our own species).[54,55] Yeast provides the current model for such processes in fungi as the heritability of transcription,[56] changes in gene expression levels in response to selection,[57] and regulatory variation in four natural isolates, has been demonstrated.[58] However, this variation has not been related to adaptation to different environments. The host-adapted subtypes of *M. anisopliae* provide a model where genetic variation can be related to adaptation to particular hosts.

Patterns of gene duplication, divergence and deletion in several generalist and specialist strains were specifically determined by heterologous hybridization of total genomic DNA. DNA from each strain was competitively hybridized to an array of strain 2575 genes (Leclerque and St. Leger, in preparation). For most genes for major life processes, differences in genomic hybridization averaged less than 5%. One group of genes in 2575 that seem to lack counterparts in the other strains is mainly composed of putative mobile genetic elements. Exceptionally, there was an expansion in the number of insertion elements in the specialist strain 443 suggesting that evolution could occur in leaps. This has implications for strain stability, including the possibility of alterations in virulence and host range, that could impact commercial development. Other poorly conserved genes in specialist strains include some that putatively function in transporting and catabolizing sugars, non-ribosomal peptide synthases, a P450 cytochrome, a polyketide synthase and several secreted enzymes including a chymotrypsin. This implies that specialists are losing genes primarily required to live in alternative hosts or as saprophytes.

However, gene loss has also been proposed as an important force driving the evolution of recently evolved novel lineages.[59] The trypsin pseudogene in 324 provides an example of how this "less is more" hypothesis could have applied to *M. anisopliae*. Trypsins are the major transcripts expressed by strain 2575 on cuticle, and one of the transcripts is also expressed in hemolymph.[13,14] Trypsins presumably confer considerable selectable functions for 2575, but either provide no benefits to 324 or are detrimental. Injecting 2575 trypsin into grasshoppers (but not caterpillars) activates the host defense prophenoloxidase system (unpublished data). An active trypsin may therefore have placed a specific grasshopper pathogen at a selective disadvantage that could drive inactivation of the gene. Corollaries of this are that loss of function mutations will be deleterious to 324 if it returned to its ancestral habitat, and could also constrain opportunistic host switching.

9.7. Horizontal versus Vertical Transfer of Genes

During EST analysis of strain 2575 we identified transcripts putatively encoding at least 15 enzymes and toxins that were most similar to proteins produced by various streptomycetes (bacteria). Some of these genes were limited to *M. anisopliae* among eukaryotes, while others had also been found in some related plant and insect pathogens. One family, the trypsins, had homologs in streptomycetes, four other pathogenic ascomycetes as well as animals. We related the presence or absence of these genes to the phylogenetic relationship of 35 representative fungi to determine if: (1) components of the genetic apparatus of *M. anisopliae* were derived from an ancestor of the proto-streptomyces via horizontal gene transfer; or (2) gene diversity derived from duplication, divergence and gene loss in different fungal lineages. Our results support the second hypothesis-if horizontal gene transfer was involved these genes originated from a common ancestor of fungi and animals and the direction of gene transfer was to streptomycetes.[60]

A theme emerged from this work of niche-specific traits, i.e., traits shared by fungi that occupy the same niche irrespective of their phylogenetic position.This was apparent with respect to several activities, demonstrating the dynamism of fungal genomes. The trypsin genes, for example, are lacking in most saprophytes, but are present in a basidiomycete insect symbiont, most zygomycetes and many ascomycete plant and insect pathogens. The phylogenetic distribution of the trypsins was congruent with fungal phylogeny, indicating that these proteins have diverged in parallel with the organisms in which they are expressed.[60] Overall, our comparative studies suggest that individual genes, such as the trypsins have been lost many times independently in different lineages, and that the flux of genes is an ongoing process. There are multiple deletions in the 324 trypsin sequence; the rate of DNA loss as compared to its 2575 ortholog was 11% pseudogene DNA in approximately 11 MY (as *cf.* 6% of mammalian DNA deleted over 22MY).[61]

9.8. The Evolution of Gene Families

The variability and redundancy found in *Metarhizium* genomes presents major challenges to understanding pathogen ecology strictly by considerations of homology and function. It is clear from EST studies that many of the molecules involved in pathogenicity are members of large gene families.

For example, strain 2575 produces 13 subtilisins. *M. anisopliae* subtilisins are its best known examples of pathogenicity related genes and are the principal agents involved in solubilizing the proteinaceous insect cuticle. They presumably would be under evolutionary pressure to respond to hosts that themselves may undergo relatively rapid changes in levels and types of protease inhibitor that can provide a barrier to infection.[62] As there are very limited data on gene duplication and divergence in fungi, we used these subtilisins as a convenient model system to tackle the controversial issue of whether gene diversity occurs by selective pressure or fixation of neutral mutations. PCR was used to obtain their orthologs from *M. a* sf. *anisopliae* strain 820 (generalist strain) and sf. *acridum* 324 (locust pathogen). Sequence data, including the intron/exon structures of the subtilisins were used in their reconstruction.[63] Major findings include: (1) diversification by tandem gene duplication is an ongoing process in the generalist strains but not in strain 324; (2) most amino acid substitutions were neutral, and (3) the subtilisins differ in their interactions with protease inhibitors, secondary substrate specificities, adsorption properties and alkaline stability. This allows them to act synergistically for more efficient hydrolysis of cuticle and to provide backup systems in the presence of the numerous proteolytic inhibitors in insect hosts.[63]

We performed a phylogenomic study to put *M. anisopliae* in context of fungi with very different virulence and habitat, to survey and characterize their serine proteinases (subtilases and trypsins), and provide an understanding of general processes in fungal gene family evolution. The survey of three families of subtilases in nine fungal genomes (plus ESTs from *M. anisopliae*) revealed that basidiomycetes (*Cryptococcus neoformans*, *Coprinus cinereus*, *Ustilago maydis*) and saprophytic ascomycetes (*Saccharomyces cerevisiae*, *Schizosaccharomyces pombe*, *Aspergillus nidulans*, *Neurospora crassa*) lack the large gene families found in the pathogenic ascomycetes (*M. anisopliae*, *Magnaporthe grisea*, *Fusarium graminearum*). Patterns of intron loss and the degree of divergence between paralogs indicate that the proliferation of subtilisins seen in pathogens mostly predated radiation of ascomycete lineages. This suggests that the early ascomycetes had a lifestyle that selected for multiple proteases (pathogenicity?), while the current disparity in gene numbers between ascomycete lineages results mostly from retention of genes in pathogens that have been lost in saprophytes.[60]

9.9. Genetically Engineering Improved Pathogens

The advanced engineered approach attempts to remedy the perceived deficiencies in biologicals by molecular manipulation to improve virulence (speed of kill), restrict or widen host range and/or reduce inoculum loads and alter saprophytic competence. This could theoretically lead to designing the ideal biocontrol agent for a particular pest. Genetic engineering relies on the power of specificity of molecular biology to identify genes conferring pathogenicity to diverse hosts, and the development of a bank of cloned pathogen genes, each of which controls a different virulence trait.

9.9.1. PRODUCING TRANSGENIC STRAINS

Strain improvement can be achieved in a variety of ways, from random selection of chemically induced mutants to site-directed homologous gene replacement techniques. The technique chosen depends upon the availability of suitable selection markers (e.g., antibiotic resistance), transformation systems, and the desired phenotypic change. Many insect pathogens are naturally resistant to the anti-fungal chemicals commonly used as selectable markers for transformation. The benomyl resistance gene and/or a glufosinate selection procedure can be used to introduce multiple transgenes into either *M. anisopliae* or *B. bassiana.*[19] There are also the options of using expression vectors carrying multiple transgenes and co-transformation with multiple plasmid.[19]

Transformation mediated through the plant pathogenic bacteria *Agrobacterium* is relatively straightforward for both *B. bassiana* and *M. anisopliae* (Bidochka, personnel communication), and may become the preferred method for generating insertional libraries.[59] *Agrobacterium* mediated transformation has been used successfully to transform various fungi including members of the Ascomycetes, Basidiomycetes, Zygomycetes and Oomycetes.[59] The ability of *Agrobacterium* to transfer its DNA to fungi belonging to various classes is indicative of the potential of this transformation system for introducing biotechnology to fungi such as *Erynia* spp. and *Lagenedium* spp. that have so far not been transformed. *Agrobacterium* may therefore provide a simple standardized method for transformation of essentially any entomopathogenic species that would obviously be novel and useful.

The broad classes of pathogenicity genes detailed above suggest that directed changes to alter virulence could result from manipulation of nearly every aspect of fungal developmental biology. An immediate issue of prime importance is how to select those genes that offer the greatest immediate potential in improving the efficacy and reliability of fungi for insect control. The following include some promising candidates:

9.9.2. ADHESINS

We would like to identify genes with the potential to change host range; either increasing it or diminishing it. Adhesins are key virulence factors for many bacterial and fungal pathogens that act by establishing and maintaining interactions with hosts.[65] The molecular interactions of adhesion defines the host range and aggressiveness of several entomopathogens, including *M. anisopliae.*[25,66] *M. anisopliae* produces at least two adhesins: *Mad*1 (for *Metarhizium* adhesin-like protein 1) (DQ338437) and *Mad*2 (DQ338439). *Mad*1 was originally tagged as an adhesin because of sequence similarities with *Candida* ALS (Agglutinin-Like-Sequence) proteins with their characteristic three-domain structure and middle domain containing tandem repeats.[65] *Mad*1 is the third most highly expressed gene in hemolymph (called AAM46085),[15] but is also transcriptionally regulated during germination. Gene knockout confirmed that the protein is involved in specific adhesion of spores to host cuticles during swelling (as distinct from earlier non-specific interactions mediated by the hydrophobins), with a large reduction in virulence in the Δ*Mad*1 mutant. Conversely, the Δ*Mad*2 mutant does not adhere to plant surfaces showing that *M. anisopliae* exploits different subsets of genes to adapt to different environments (Wang and St. Leger, unpublished data).

9.9.3. EMPLOYING PRODUCTS SECRETED BY THE PATHOGEN TO IMPROVE VIRULENCE-SPEED OF KILL

A major deterrent to the development of fungi as pesticides has been that it can take 5–15 days post-infection to kill the targeted pest. This not only makes them poorly competitive, but also limits industrial investment in application and formulation technologies for advanced efficacy. Unfortunately, the host specific strains in particular kill slowly and produce fewer toxins than the generalists.[67] Presumably strains that are not specifically adapted to subvert/avoid/overcome the immune response of a particular insect are best served by achieving a rapid kill with toxins. An adapted strain may optimize utilization of host nutrients and production of infectious propagules by growing within the living host. Adding new genes to the fungus that will allow it to kill the insect host more quickly is a solution. This could also contribute to their escape from environmental hazards. The most attractive initial candidates for this approach include cuticle-degrading enzymes and toxins that are encoded by single genes as they are highly amenable to manipulation by gene transfer.

Many of the cuticle-degrading enzymes that act synergistically to solubilize cuticles are multiple gene products with distinctive activity profiles.[18,68] The variability of molecules with activity against host substrates increases the range of tools naturally available to develop biotechnological procedures for pest control. Furthermore, these molecules possess pathogenic specializations that distinguish them from similar molecules produced by saprophytes. For example, stronger binding, due to the positively charged surface groups on the subtilisin protease Pr1 contribute to increasing Pr1 activity 33-fold against insoluble cuticle proteins compared to proteinase K from a related saprophyte.[21] Pr1 is also resistant to proteinase inhibitors (serpins) in hemolymph and even to being in a rapidly melanizing suspension, mimicking the insect defense response.[69] These pathogenic specializations are suggestive that entomopathogenic fungi have spent millions of years of evolution refining chemicals that subdue their hosts. The toxins they now produce become choice candidates for producing improved transgenic organisms.

Optimal pathogenicity may require manipulation of several genes encoding enzymes and toxins that act additively or synergistically. However, recombinant *Metarhizium* strains that constitutively overexpress the subtilisin protease Pr1a have improved pathogenic qualities at all stages of infection.[18] In contrast to the wild-type, transgenic strains continued to produce Pr1 in the haemocoel of *Manduca sexta* caterpillars following penetration of the cuticle. This caused extensive melanization in the body cavity, and cessation of feeding 40 h earlier than controls infected with wild type. Inhibitors of trypsin that have no effect on Pr1 nevertheless blocked Pr1 induced activation of host prophenoloxidase, indicating that Pr1 acts indirectly by activating an earlier stage in a cascade terminating in prophenoloxidase activation. Insects killed by transgenic strains and extensively melanized were very poor substrates for fungal growth and sporulation. This reduces transmission of the recombinant fungi, which assisted in obtaining permission for the field trial.[45] It is also consistent with the emphasis of using entomopathogenic fungi as "contact insecticides" that achieve a quick kill.[4] In addition, using the multifarious secreted compounds produced by the entomopathogens themselves as a resource for their genetic improvement, albeit under altered regulation, provides an experimental design that seems inherently unlikely to raise public concern.

The availability of these genes raised the possibility of creating novel combinations of insect specificity and virulence by expressing them in other fungi, bacteria or viruses to produce improved pathogens. Thus, the Pr1 gene from *M. anisopliae* has been used to increase virulence of *B. bassiana*[70] and baculoviruses (Huang, Hughes, St. Leger, and Wood, unpublished data). Similar subtilases have improved the biocontrol potential of fungal pathogens of other fungi[71] and nematodes.[72]

9.9.4. INVESTIGATING PATHOGEN GENES THAT LIMIT THE IMMUNE RESPONSE

We are investigating a selection of genes that are differentially expressed in hemolymph and therefore implicated in adaptation to this host environment. However, an insect's greatest defense mechanism may be avoidance of entomopathogenic fungi,[73] and *M. anisopliae* is repellent to many insect species including Japanese beetle (*Popillia japonica*) in turfgrass.[74] Thus *M. anisopliae* in the rhizosphere could provide a repellent barrier around roots that would offer more effective protection to the plant than causing disease, as there is an inevitable time lag following infection before cessation of feeding. The nature of fungal repellency has not been determined but is influenced in termites by the specific strain of entomopathogen.[73]

The effectiveness of pathogens as biological control agents will also be determined by the efficacy of the insect's immune system. Thus, fungal adaptations to host defenses are likely to play an important role in virulence and specificity. *Mcl1* is the most highly expressed gene when strain 2575 is grown in hemolymph (5.6% of total transcripts) and encodes a cell wall protein with a long collagenous domain. Gene knockout confirmed that Mcl1 is required for immune evasion.[17] The mutant is rapidly attacked by hemocytes and has reduced virulence to *Manduca sexta*. RT-PCR confirmed that *Mcl*1 is expressed during growth in the hemolymph of a diverse array of insect species, consistent with the broad host range of 2575. However, it was not expressed in other media, consistent with it being involved in pathogenesis.

9.9.4.1. *The Matter of Promoters*

Specificity is usually controlled by infection events at the level of the cuticle,[46] so altering post-penetration events should not reduce environmental safety derived from species selectivity. We have mostly over-expressed genes in *M. anisoplia*e under control of strong constitutive promoters (e.g., *gpd* and *mtr*). The Seegene DNA Walking SpeedUp™ kit (Rockville, MD) has allowed us to accumulate *M. anisopliae* promoters that are capable of expressing homologous and heterologous genes in a regulated fashion and that vary in levels of expression. The highly expressed *Mcl*1 promoter seems optimal for targeted expression of transgenes. Aside from the possibility of increasing virulence, regulation of toxin expression to growth in the hemolymph has safety considerations, by precluding casual release of the toxin by the fungus living as a saprophyte.

Precise information on the host related signals that induce the promoter is required for engineering purposes, and for regulatory bodies to determine whether the specificity of the promoter can be relied on in field conditions. We transformed 2575 with the jellyfish gene for green fluorescent protein (GFP) fused to the 2,000 bp segment up-stream of the *Mcl*1 coding region to confirm targeted expression to the hemocoel. The procedure worked well with rapid production of GFP in *M. sexta* hemolymph in vitro and in vivo and quick protein decay under repressing conditions. This highlighted the tight control of expression consistent with our RT-PCR and microarray data. We have used the *Mcl*1 promoter to drive expression of the transgenes in *M. anisopliae*, including the scorpion venom gene AaIT.

9.9.5. HYPERVIRULENT PATHOGENS EXPRESSING ADDITIONAL TOXINS

Biocontrol agents expressing multiple toxins targeting different pathways can significantly increase killing speed. The best studied *M. anisopliae* toxin is destruxin.[9] Destruxins are cyclic peptides composed of an alpha-hydroxy acid and five amino acid residues. Destruxin-induced membrane depolarization due to the opening of Ca^{2+} channels has been implicated as a cause of paralysis and death.[75] Destruxins also cause signaling changes, through the phosphorylative activation of certain proteins in lepidopteran and human cell lines. Destruxins cause morphological and cytoskeletal changes in insect plasmatocytes in vitro, and this adversely affects insect cellular immune responses such as encapsulation and phagocytosis.[76]

The mechanisms by which destruxins achieve their varied biological activities have not been studied in vivo, except for their ability to open calcium channels. We used *Drosophila melanogaster* to characterize the range of functions affected by destruxins. We exposed *Drosophila* to pathogen molecules, e.g., *M. anisopliae* cell wall components, secreted enzymes and destruxins, and used *Drosophila* microarrays to identify which of these generate or alter the host defense response. Destruxins suppressed most of the *Drosophila* antimicrobial gene activation program. This included suppression of production of antimicrobial peptides such as drosomycin, metchnokovin and cercropins, but the antifungal peptide attacin was elevated (though attacin has no effect on *M. anisopliae*). Destruxins did not block phagocytosis, but did block maturation of phagocytes. Most interestingly, destruxin was sufficient to turn injected *E. coli* cells into a virulent pathogen that increased exponentially in the hemolymph.[77]

Unfortunately for the purposes of genetic engineering, destruxins are secondary metabolites and encoded by genes that are too large at 20 kb for convenient molecular manipulations. We have supplemented toxic proteins from the generalist *M. anisopliae* strain 2575 with the insect-selective 70 aa AaIT neurotoxin from the scorpion *Androctonus australis*. This has already provided the most promising recombinant baculoviruses,[78] with improved performance against lepidopteran larvae in several field trials.[79] AaIT acts on the neuronal sodium channel causing presynaptic excitatory effects. Interestingly, lepidopterans are relatively tolerant to this toxin compared to locusts, beetles and crickets.[78] Baculoviruses are primarily pathogens of lepidopterans, with some notable exceptions such as *Orcyctes rhinoceros* baculovirus. However, many insects not susceptible to baculoviruses are targeted by *M. anisopliae*. These studies are providing an opportunity: (1) to diversify the deployment of this useful, very well studied toxin, which like *M. anisopliae* has already passed many regulatory hurdles, and (2) to directly compare the efficacy of fungal toxins with the most frequently studied arthropod one. Judging from the literature and our own results we expect fungal and arthropod toxins to have good killing power singly, but synergistic effects derived from combining them in a single strain could produce a large magnitude of hypervirulence. An underlying premise behind this work is that by comparing arthropod and fungal toxins it will increase interest in fungi as a resource of genes for biotechnology. Fungi have been under-exploited to date. This is particularly true for the insect pathogens, even though they are exceptionally rich sources of novel biologically active substances.[80]

One of our principal candidates for genetic enhancement is *M. anisopliae sf. acridum*. Its development as a locust mycoinsecticide is being hindered in China and sub-Saharan Africa by its slow speed of kill.[5] Strain 324 does not express several lytic enzymes/toxins produced by strain 2575, including phospholipases.[12] Thus, we are investigating the extent of increases in virulence that result from appropriate combinations of several genes from *M. anisopliae* strain 2575 encoding enzymes and toxins that act additively or synergistically to quickly kill insects or to prevent them from feeding. To analyze gene interactions, and the comparative efficacy of the AaIT with fungal toxins, we are comparing disease development (particularly speed of kill) by 324 transformed with two or more transgenes with equivalent 324 strains transformed with the Pr1a subtilisin gene or AaIT separately. Changes to LT_{50} values indicate faster kill consistent with toxicosis[18], while reductions in the median lethal dose (LC_{50}) values indicate that inoculum loads and efficiency of infection (attachment and penetration) are improved.[46] We are also determining if any of the transformations broaden the conditions under which 324 or other strains can produce infection structures (i.e., in the absence of locust inducers or against hydrophilic surfaces).[81] Although we do not expect host range to change, we are evaluating the specificity of transgenic 324 against non-hosts compared with the wild-type (including *Apis mellifera, M. sexta, Acheta domestica, D. melanogaster, Galleria mellonella* and *Tenebrio molitor*). The minimum dosage applied to an insect is 100-fold above the LC_{50} for the susceptible grasshopper host. By varying host density, relative humidity, and temperature, we are attempting to optimize the infection level within an insect population. Low infection rates using these procedures would probably translate into virtually undetectable infection rates under natural conditions. Behavior of infected grasshoppers is also being noted. It is possible that neurotoxin-expressing 324 will cause infected insects to fall off plants, which could reasonably be expected to reduce transmission. Over-expression of Pr1 greatly reduced sporulation providing biological containment.[18] We therefore measure yield of spores by recombinant strains and WT to predict the capacity of transgenics to recycle. Conceivably, rapid kill would reduce the ability of the pathogen to access host tissues for nutrition and thereby decrease spore production.

References

1. T. M. Butt, C. Jackson, and N. Magan, Introduction-fungal biological control agents: Progress, problems and potential, in*Fungal Biocontrol Agents: Progress, Problems and Potential,* edited by T. M. Butt, C. Jackson, and N. Morgan (CAB International, Wallingford, UK, 2001), pp. 1–8.
2. R. P. Bateman, Controlled droplet application of mycoinsecticides: An environmentally friendly way to control locusts. *Antenna* 16, 6–13 (1992).
3. R. J. Milner, Selection and characterization of strains of *Metarhizium anisopliae* for control of soil insects in Australia, in *Biological Control of Locusts and Grasshoppers*, edited by C. J. Lomer and C. Prior (CAB International, Wallingford, UK, 1992), pp. 200–207.
4. C. Prior, Discovery and characterization of fungal pathogens for locust and grasshopper control, in *Biological Control of Locusts and Grasshoppers*, edited by C. J. Lomer and C. Prior (CAB International, Wallingford, UK, 1992), pp. 159–180.
5. A. Hajek, S. P. Wraight, and J. D. Vandenberg, Control of arthropods using pathogenic fungi, in *Bio-exploitation of Filamentous Fungi*, edited by S. P. Pointing and K. D. Hyde (Fungal Diversity Press, Hong Kong, 2001).
6. R. L. Harrison and B. C. Bonning, Genetic engineering of biocontrol agents for insects, in *Biological and Biotechnological Control of Insect Pests*, edited by J. E. Rechcigl and N. A. Rechcigl (Lewis Publishers, Boca RAton, FL, 1998), pp. 243–280.
7. S. P. Wraight, M. A. Jackson, and S. L. deKock, Production, stabilization and formulation of fungal bocontrol agents, in*Fungal biocontrol agents: Progress, problems and potential*, edited by T. M. Butt, C. Jackson, and N. Morgan (CAB International, Wallingford, UK, 2001), pp. 253–287.

8. M. G. Klein, P. S. Grewal, and T. A. Jackson, Lawn, turf and grassland pests, in *Field Manual of Techniques in Invertebrate Pathology*, edited by L. A. Lacey and H. K. Kaya (Kluwer Academic Publishers, Dordrecht, 2000), pp. 681–706.
9. D. W. Roberts and R. J. St. Leger, *Metarhizium* spp. Cosmopolitan insect-pathogenic fungi: Mycological aspects, *Adv. Appl. Microbiol.* 54, 1–70 (2004).
10. M. A. Benjamin, E. Zhioua, and R. S. Ostfeld, Laboratory and field evaluation of the entomopathogenic fungus *Metarhizium anisopliae* (Deuteromycetes) for controlling questing adult *Ixodes scapularis* (Acari: *Ixodidae*), *J. Med. Entomol.* 39, 723–728 (2002).
11. M. Wright, A. K. Raina, and A. R. Lax, A strain of the fungus *Metarhizium anisopliae* for controlling subterranean termites, *J. Econ. Entomol.* 98, 1451–1458 (2005).
12. F. M. Freimoser, S. Screen, S. Bagga, G. Hu, and R. J. St. Leger, EST analysis of two subspecies of *Metarhizium anisopliae* reveals a plethora of secreted proteins with potential activity in insect hosts, *Microbiology* 149, 239–247 (2003).
13. F. M. Freimoser, S. Screen, G. Hu, and R. J. St. Leger, EST analysis of genes expressed by the zygomycete *Conidiobolus coronatus* during optimized secretion of proteins, *Microbiology* 149, 1893–1900 (2003).
14. F. M. Freimoser, G. Hu, and R. J. St. Leger, Variation in gene expression patterns as the insect pathogen *Metarhizium anisopliae* adapts to different host cuticles or nutrient deprivation in vitro, *Microbiology* 151, 361–371 (2005).
15. C. Wang, G. Hu, and R. J. St. Leger, Differential gene expression by *Metarhizium anisopliae* growing in root exudate and host (*Manduca sexta*) cuticle or hemolymph reveals mechanisms of physiological adaptation, *Fungal Genet. Biol.* 42, 704–718 (2005).
16. C. Wang, T. M. Butt, and R. J. St. Leger, Colony sectorization of *Metarhizium anisopliae* is a sign of ageing, *Microbiology* 151, 3223–3236 (2005).
17. C. Wang and R. J. St. Leger, A collagenous protective coat enables *Metarhizium anisopliae* to evade insect immunity, *Proc. Natl. Acad. Sci.* 103, 6647–6652 (2006).
18. R. J. St. Leger, L. Joshi, M. J. Bidochka, and D. W. Roberts, Construction of an improved mycoinsecticide over-expressing a toxic protease, *Proc. Natl. Acad. Sci.* 93, 6349–6354 (1996).
19. S. E. Screen, G. Hu, and R. J. St. Leger, Transformants of *Metarhizium anisopliae* sf. *anisopliae* overexpressing chitinase from *Metarhizium anisopliae* sf. *acridum* show early induction of native chitinase but are not altered in pathogenicity to *Manduca sexta*, *J. Invertebr. Pathol.* 78, 260–266 (2001).
20. A. Leclerque, H. Wan, A. Abschultz, S. Chen, G. V. Mitina, G. Zimmerman, and H. U. Scharer, Agrobacterium-mediated insertional mutagenesis (AIM) of the entomopathogenic fungus *Beauveria bassiana*, *Curr. Genet.* 45, 111–119 (2004).
21. R. J. St. Leger, B. May, L. Allee, D. C. Frank, D. W. Roberts, and R. C. Staples, Genetic differences in allozymes and in formation of infection structures among isolates of the entomopathogenic fungus *Metarhizium anisopliae, J. Invertebr. Pathol.* 60, 89–101 (1992).
22. M. J. Bidochka, R. J. St. Leger, and D. W. Roberts, Differentiation of species and strains of entomopathogenic fungi by random amplification of polymorphic DNA (RAPD), *Curr. Genet.* 25, 107–113 (1994).
23. G. H. Sung, J. W. Spatafora, R. Zare, K. T. Hodge, and W. Gams, A revision of *Verticillium* sect. *Prostrata,* II: Phylogenetic analysis of SSU and LSU nuclear rDNA sequences from anamorphs and teleomorphs of the Clavicipitaceae, *Nova Hedwigia* 72, 311–328 (2001).
24. F. Driver, R. J. Milner, and J. W. H. Trueman, A taxonomic revision of *Metarhizium* based on a phylogenetic analysis of rDNA sequence data, *Mycol. Res.* 104, 134–150 (2000).

25. J. Fargues, Adhesion of the fungal spore to the insect cuticle in relation to pathogenicity, in *Infection Processes of Fungi*, Conference Report, edited by D. W. Roberts and J. R. Aist (Rockefeller Foundation, 1984), pp. 90–110.
26. R. J. St. Leger, M. J. Bidochka, and D. W. Roberts, Germination triggers of *Metarhizium anisopliae, Microbiology* 140, 1651–1660 (1994).
27. B. Amiri-Besheli, B. Khambay, S. Cameron, M. L. Deadman, and T. M. Butt, Inter- and intra-specific variation in destruxin production by insect pathogenic *Metarhizium* spp., and its significance to pathogenesis, *Mycol. Res.* 104, 447–452 (2000).
28. T. M. Butt and M. Brownbridge, Increasing the efficacy of entomogenous fungi, in *Enhancing Biocontrol Agents and Handling Risks*, edited by M. Vurro, J. Gressel, T. Butts, G. Harman, A. Pilgeram, R. J. St. Leger, and D. Nuss (IOS Press, Amsterdam, The Netherlands, 2001), pp. 52–63.
29. G. D. Inglis, M. S. Goettel, T. M. Butt, and H. Strasser, Use of hyphomycetous fungi for managing insect pests, in *Fungi as Biocontrol Agents*, edited by T. M. Butt, C. Jackson, and N. Magan (CAB International, Wallingford, UK, 2001), pp. 23–69.
30. J. Drummond, J. B. Heale, and A. T. Gillespie, Germination and effect of reduced humidity on expression of pathogenicity in *Verticillium lecani*, *Ann. Appl. Biol.* 111, 193–201 (1987).
31. M. J. Bidochka, A. M. Kamp, T. M. Lavender, J. Dekoning, and J. N. A. De Croos, Habitat association in two genetic groups of the insect-pathogenic fungus Metarhizium anisopliae: Uncovering cryptic species? *Appl. Environ. Microbiol.* 67, 1335–1342 (2001).
32. D. E. N. Rangel, G. U. L. Braga, A. J. Anderson, and D. W. Robert, Influence of growth environment on tolerance to UV-B radiation, germination speed, and morphology of Metarhizium anisopliae var. acridum conidia, *J. Invetebrate Pathol.* 90, 55–58 (2005).
33. G. U. L. Braga, D. E. N. Rangel, S. D. Flint, A. J. Anderson, and D. W. Roberts, Conidial pigmentation is important to tolerance against solar-simulated radiation in the entomopathogenic fungus *Metarhizium anisopliae*, *Photochem. Photobiol.* 82, 418–422 (2006).
34. H. De Grotte, O. Douro-Kpindou, Z. Ouambama, C. Gbongboui, D. Müller, S. Attignon, and Chris Lomer, Assessing the feasibility of biological control of locusts and grasshoppers in West Africa: Incorporating the farmers' perspective, *Agricu. Human Values* 18, 413–428 (2001).
35. Z. Long and D. M. Hunter, Laboratory and field trials of Green Guard® (*Metarhizium anisopliase* var. *acridium*) (Deuteromycotina: Hyphomycetes) against the oriental migratory locust (*Locusta migratoria manilensis*) (Orthoptera: Acrididae), *China* 14, 27–30 (2005).
36. E. J. Scholte, K. Ng'habi, J. Kihonda, W. Takken, K. Paaijmans, S. Abdulla, G. F. Killeen and B. G. J. Knols, An entomopathogenic fungus for control of adult African malaria mosquitoes, *Science* 308, 1641–1642 (2005).
37. S. Blanford, B. H. K. Chan, N. Jenkins, D. Sim, R. J. Turner, A. F. Read, and M. B. Thomas, Fungal pathogen reduces potential for malaria transmission, *Science* 308, 1638–1641 (2005).
38. A. Flores, Saving bees: Fungus found to attack *Varroa* mites, *Agric. Res. Mag.* 52, 18 (2004).
39. M. S. Goettel, and G. D. Inglis, Fungi: Hyphomycetes, in *Manual of Techniques in Insect Pathology,* edited by L. Lacey (Kluwer Academic Publishers, 1997), pp. 213–249.
40. M. J. Bidochka, Monitoring the fate of biocontrol fungi, in*Fungal Biocontrol Agents: Progress, Problems and Potential,* edited by T. M. Butt, C. Jackson, and N. Morgan (CAB International, Wallingford, UK, 2001), pp. 193–218.

41. K. P. Barley, The configuration of the root system in relation to nutrient uptake, *Adv. Agron.* 22, 159–201 (1970).
42. T. F. Duda and S. R. Palumbi, Molecular genetics of ecological diversification: Duplication and rapid evolution of toxin genes of the venomous gastropod *Conus*, *Proc. Natl. Acad. Sci.* 96, 6820–6823 (1999).
43. G. E. Harman and B. G. G. Dozelli, Enhancing crop performance and pest resistance with genes from biocontrol agents, in *Enhancing Biocontrol Agents and Handling Risks*, edited by M. Vurro, J. Gressel, T. Butts, G. Harman, A. Pilgeram, R. St. Leger, and D. Nuss (IOS Press, Amsterdam, The Netherlands, 2001), pp. 114–125.
44. J. M. Whipps, Microbial interactions and biocontrol in the rhizosphere, *J. Exp. Bot.* 52, 487–511 (2001).
45. G. Hu and R. J. St. Leger, Field studies of a recombinant mycoinsecticide (*Metarhizium anisopliae*) reveal that it is rhizosphere competent. *Appl. Environ. Microbiol.* 68, 6383–6387 (2002).
46. R. J. St. Leger and S. Screen, Prospects for strain improvement of fungal pathogens of insects and weeds, in *Fungal Biocontrol Agents: Progress, Problems and Potential*, edited by T. M. Butt, C. Jackson, and N. Morgan (CAB International, Wallingford, UK, 2001), pp. 219–238.
47. R. J. St. Leger, M. J. Bidochka, and D. W. Roberts, Co-transformation of *Metarhizium anisopliae* by electroporation or the gene gun to produce stable GUS transformants, *FEMS Microbiol. Lett.* 131, 289–294 (1995).
48. B. R. Kerry, Rhizosphere interactions and the exploitation of microbial agents for the biological control of plant-parasitic nematodes, *Annu. Rev. Phytopathol.* 38, 423–441 (2000).
49. R. Baker, Diversity in biological control, *Crop Protection* 10, 85–94 (1991).
50. D. W. Roberts and A. E. Hajek, Entompathogenic fungi as bioinsecticides, in *Frontiers in Industrial Mycology,* edited by G. F. Leatham (Chapman & Hall, New York, 1992), pp. 144–159.
51. J. M. Barea, M. J. Pozo, R. Azcon, and C. Azcon-Aguilar, Microbial co-operation in the rhizosphere, *J. Exp. Bot.* 56, 1761–1778 (2005).
52. E. B. Nelson, Microbial dynamics and interactions in the spermosphere, *Annu. Rev. Phytopathol.* 42, 271–309 (2004).
53. P. Liang and A. B. Pardee, Analyzing differential gene expression in cancer, *Nat. Rev. Cancer* 3, 869–876 (2003).
54. Y. Ben-Shahar, A. Robichon, M. B. Sokolowski, and G. E. Robinson, Influence of gene action across different time scales on behavior, *Science* 296, 741–744 (2002).
55. W. Enard, P. Khaitovich, J. Klose, S. Zollner, F. Heissig, P. Giavalisco, K. Nielset-Struwe, E. Muchmore, A. Varki, R. Ravid, G. M. Doxiadis, R. E. Bontrop, and S. Paabo, Intra and inter specific variation in primate gene expression patterns, *Science* 296(5566), 233–235 (2002).
56. R. B. Brem, G. Yvert, R. Clinton, and L. Kruglyak, Genetic dissection of transcriptional regulation in budding yeast, *Science* 296, 752–755 (2002).
57. T. L. Ferea, D. Botstein, P. O. Brown, and R. F. Rosenzweig, Systematic changes in gene expression patterns following adaptive evolution in yeast, *Proc. Natl. Acad. Sci.* 96, 9721–9726 (1999).
58. J. P. Townsend, D. Cavalieri, and D. L. Hartl, Population genetic variation in genome-wide gene expression, *Mol. Biol. Evol.* 20, 955–963 (2003).
59. M. V. Olson, When less is more: Gene loss as an engine of evolutionary change, *Am. J. Hum. Genet.* 64, 18–23 (1999).

60. G. Hu and R. J. St. Leger, A phylogenomic approach to reconstructing the diversification of serine proteases in fungi, *J. Evol. Biol.* 17, 1204–1214 (2004).
61. R. H. Waterston, K. Lindblad-Toh, E. Birney, et al., Initial sequencing and comparative analysis of the mouse genome, *Nature* 420, 520–562 (2002).
62. D. G. Boucias and J. C. Pendland, Detection of protease inhibitors in the hemolymph of resistant *Anticarsia gemmatalis* inhibitory to the entomopathogenic fungus *Nomuraea rileyi*, *Experientia* 43, 336–3439 (1997).
63. S. Bagga, S. E. Screen, and R. J. St. Leger, Reconstructing the diversification of subtilisins in the pathogenic fungus *Metarhizium anisopliae*. *Gene,* 324, 159–169 (2004).
64. C. B. Michielse, P. J. Hooykaas, C. A. van den Hondel, and A. F. Ram, *Grobacterium*-mediated transformation as a tool for functional genomics in fungi, *Curr. Genet.* 48, 1–17 (2005).
65. L. Hoyer, The ALS gene family of *Candida albicans*, *Trends Microbiol.* 9, 176–180 (2001).
66. K. Al-Aidroos and D. W. Roberts, Mutants of *Metarhizium anisopliae* with increased virulence toward mosquito larvae, *Can. J. Genet. Cytol.* 20, 211–220 (1978).
67. A. K. Charnley, Fungal pathogens of insects: cuticle degrading enzymes and toxins, *Adv. Bot. Res.* 40, 241–321 (2003).
68. R. J. St. Leger, Biology and mechanisms of invasion of deuteromycete fungal pathogens, in: *Parasites and Pathogens of Insects,* edited by N. C. Beckage, S. N. Thompson, and B. A. Federici (Academic Press, New York, 1993), vol. 2, pp. 211–229.
69. R. J. St. Leger, R. M. Cooper, and A. K. Charnley, The effect of melanization of *Manduca sexta* cuticle on growth and infection by *Metarhizium anisopliae, J. lnvertebr. Pathol.* 52, 459–470 (1988).
70. C. E. Gongora, Transformacion de *Beauveria bassiana* cepa Bb9112 con les genes de la proteina verde fluorescente y la protease pr1A de *M. anisopliae*, *Rev. Colomiana Entomol.* 30, 1–5 (2004).
71. A. Flores, I. Chet, and A. Herrera Estrella, Improved biocontrol activity of *Trichoderma harzianum* by overexpression of the proteinase encoding gene, *prb1*, *Curr. Genet.* 31, 30–37 (1997).
72. J. Ahman, T. Johansson, M. Olsson, P. J. Punt, A. M. J. Cees, van den Hondel, and A. Tunlid, Improving the pathogenicity of a nematode-trapping fungus by genetic engineering of a subtilisin with nematotoxic activity, *Appl. Environ. Microbiol.* 68, 3408–3415 (2002).
73. A. C. Rath, The use of entomopathogenic fungi for control of termites, *Biocont. Sci. Technol.* 10, 563–581 (2000).
74. M. G. Villani, S. R. Krueger, P. C. Schroeder, F. Consolie, N. H. Console, L. M. Preston-Wisey, and D. W. Roberts, Soil application effects of *Metarhizium anisopliae* on Japanese beetle (Coleoptera: Scarabaeidae) behavior and survival in turfgrass microcosms, *Environ. Entomol.* 23, 502–513 (1994).
75. R. L. Samuels, S. E. Reynolds, and A. K. Charnley, Calcium-channel activation of insect muscle by destruxins, insecticidal compounds produced by the entomopathogenic fungus Metarhizium anisopliae. *Comp. Biochem. Physiol.* C 90, 403–412 (1988).
76. Vey, A., Matha, V., and Dumas, C. Effects of the peptide mycotoxin destruxin E on insect hemocytes and on dynamics and efficiency of the multicellular immune reaction, *J. Invertebr. Pathol.* 80, 177–187 (1988).
77. S. Pal, R. J. St. Leger, and L. P. Wu, Fungal peptide destruxin A plays a specific role in suppressing the innate immune response in *Drosophila melanogaster*, *J. Biol. Chem.* (2007) in press.
78. E. Zlotkin, Y. Fishman, M. Elazer, AaIT: From neurotoxin to insecticide, *Biochemie* 82, 869–881 (2000).

79. X. Sun, X. Chen, Z. Zhang, H. Wang, F. J. Bianchi, H. Peng, J. M. Vlak, and Z. Hu, Bollworm responses to release of genetically modified *Helicoverpa armigera* nucleopolyhedroviruses in cotton, *J. Invertebre. Pathol.* 81, 63–69 (2002).
80. M. Isaka, P. kittakoop, K. Kirtikra, N. Hywel-Jones, and Y. Thebtaranonth, Bioactive substances from insect pathogenic fungi, *Acc. Chem. Res.* 38, 813–823 (2005).
81. C. Wang and R. J. St. Leger, Developmental and transcriptional responses to host and non hostcuticles by the specific locust pathogen *Metarhizium anisopliae* sf. *Acridum*, *Eukaryotic Cell* 4, 937–947 (2005).

10. *SCLEROTINIA MINOR*—BIOCONTROL TARGET OR AGENT?

Alan Watson*
Department of Plant Science, McGill University, Sainte-Anne-de-Bellevue, Quebec, Canada

Abstract. *Sclerotinia minor* is the causal agent of several important crop diseases including lettuce drop and Sclerotinia blight of peanut. Extensive search for biocontrol of the Sclerotinia diseases has culminated in the commercialization of *Coniothyrium minitans*. *Sclerotinia minor* is also an effective bioherbicide that can be effectively and safely used to control broadleaf weeds in turf environments.

Keywords: host range, crop pathogen, virulence, dissemination

10.1. Sclerotinia Minor—The Crop Pathogen, a Target for Biological Control

Sclerotinia minor Jaggar is a soil borne Discomycetes fungus characterized by small (0.5–2.0 mm), irregular sclerotia that germinate by eruptive growth of mycelium and colonize susceptible plant tissues.[1,2] *S. minor* is closely related to *S. sclerotiorum* (Lib.) de Bary and *S. trifoliorum* Erikss. These species are serologically related with *S. sclerotiorum* a tetraploid form of *S. minor* whilst *S. trifoliorum* a hybrid with part of the genome contributed by *S. minor*.[3] In contrast to *S. sclerotiorum*, apothecia and ascospore production in *S. minor* is very rare in the field and has not been recorded to occur in North America[4] and only one report from New Zealand[5] of their natural occurrence. Various workers[1,6] have concluded that ascospores are unimportant in the epidemiology of *S. minor* caused disease. *Sclerotinia minor* has been known to occur in North America prior to 1900 and has been the target of extensive research with results widely published[1,2,4,6] in the scientific literature.

10.1.1. DISTRIBUTION AND HOST RANGE

S. minor is distributed worldwide, except for the warm tropics, and occurs on many plant species.[7] Most susceptible species are dicotyledonous with

---

*To whom correspondence should be addressed, e-mail: alan.watson@mcgill.ca

M. Vurro and J. Gressel (eds.), Novel Biotechnologies for Biocontrol Agent Enhancement and Management, 205–211.

only three monocotyledonous plants, asparagus, tulip, and banana, reported as hosts. *S. minor* is not a serious pathogen on most plants as economic losses to *S. minor* in the United States have only occurred in lettuce[8] and peanuts[9] and in eastern Canada,[10] only on lettuce. The most economically important diseases caused by *S. minor* are lettuce drop, Sclerotinia blight of peanut, and Sclerotinia stem rot of sunflower. *S. minor* is not pathogenic on any species of the grass family Poaceae.

10.1.2. LIFE CYCLE

Dispersal and transmission of the disease is exclusively by the direct contact with germinating sclerotia to produce infective hyphae which colonize plants and eventually produce more sclerotia which are returned to the soil.[4,11] Sclerotia must be within 2 cm of the taproot of lettuce and 8 cm of the soil surface to cause disease.[12] Plant-to-plant spread between diseased and healthy plants can occur by direct contact with infected tissue.[4] Infection of lettuce with *S. minor* can occur at the soil line through lower senescent leaves or below ground to a depth of 10 cm through root tissues.[8] Favorable conditions for germination and infection include temperatures of 17–21°C and relative humidity greater than 95%. Infection occurs by mycelium arising from sclerotia or infected plant debris. Plants may be attacked at any stage from seedling to maturity. Under moist and cool conditions the fungus rapidly invades the tissues of the host in which a light brown, watery rot develops and a white, cottony-like mycelium grows over the infected tissues. Stunting, premature ripening and sudden collapse of the host are common symptoms of infection. After several days of mycelium growth, small, compact bodies develop either on the surface of the host or in cavities within it. These aggregates of mycelium are young resting vegetative structures (sclerotia). At first they are white, but when mature are black in color. Large numbers of sclerotia accumulate in plant debris and in soil where they can remain dormant for long periods. Alternatively they may germinate after a short resting period.

The *S. minor* inoculum of lettuce drop disease is sedentary and spread between commercial fields is slow and restricted.[8,13] After harvest, lettuce residues infested by *S. minor*, including those with sclerotia already formed are disked into the soil where the inoculum remains dormant until the next planting. Disease is often in clumped or aggregated distribution patterns within infested fields. Several modes of dissemination of *S. minor* have been proposed; mycelia infected seed, seed contaminated with sclerotia, movement of infested soil by machinery, winter annual weed species acting as reservoirs in rotational systems and passage through animals fed or bedded with peanut plant material.[14] Viable sclerotia of *S. minor* were recovered from fecal and ruminal samples of heifers fed infested peanut hay.[15]

10.1.3. SURVIVAL AND PERSISTENCE

Soil moisture and temperature, sclerotial position and duration in the soil, sclerotia shape, soil gases and chemicals, microbial activity, nutrition, and other factors are known to affect survival and germination of sclerotia.[6,11] There have been reports[16] of the sclerotia of *Sclerotinia* species surviving in the soil for 4–5 yearswhile others report[17] rapid decline in sclerotia survival with few surviving into the following year. In soil box and field trials in New Zealand, only 22% of the *S. minor* sclerotia could be recovered after 3 months, and only 2% (which were 50% viable) were recovered after 11 months.[17] In water saturated soil, sclerotia of *S. minor* disintegrate or fail to germinate within 8 weeks.[18]

10.1.4. NATURAL CONTROL OF SCLEROTINIA MINOR

The biological component of the soil is the most important component affecting survival of *S. minor* sclerotia. Forty-six fungi, two bacteria, two insects, a mite, and a snail are reported as antagonists, mycoparasites or predators of *Sclerotinia* spp.[11,16,19]. These organisms are thought to be responsible for the occurrence of "suppressive soils." Several; including *Coniothyrium minitans*, *Trichoderma harzianum*, *Teratosperma oligocladum*, *Talaromyces flavus*, and *Sporidesmium sclerotivorum* have been evaluated as biocontrol agents to deal with *Sclerotinia* spp. in lettuce with varying degrees of success.[4,18,20,21] CONTANS *WG*, a water dispersible granule formulation of *Coniothyrium minitans* is registered for the reduction/control of *Sclerotinia sclerotiorum* and *Sclerotinia minor* in agricultural soils in Europe and the United States (Chapter 12).

10.2. *Sclerotinia spp.* as Bioherbicides

The severe and rapid necrosis caused by *Sclerotinia sclerotiorum* on a wide spectrum of broadleaf weeds has created interest in utilizing *S. sclerotiorum* as a biological agent to control weeds. *S. sclerotiorum* was field tested as a bioherbicide against *Centaurea maculosa* (spotted knapweed) in British Columbia in 1972 (Watson, unpublished). In Montana statewide trials in 1982, *S. sclerotiorum* controlled 20–80% of *Cirsium arvense* (Canada thistle).[22] This work was followed by the selection of non-sclerotia mutants[23] incapable of producing ascospores, but virulence is linked to sclerotia formation. Amino acid auxotroph mutants[24] were also developed in attempts to mitigate reproduction and persistence of a *S. sclerotiorum* bioherbicide. The pathogenicity of amended auxotrophic strains and wild strains of *S. sclerotiorum* were compared in a permanent pasture in New Zealand and the auxotrophic strains were less field fit than the wild strains.[25]

An elaborate risk analysis of using *S. sclerotiorum* for biological control of *Cirsium arvense* simulated dispersal of ascospores in a pasture.[26] A "safety zone" was determined to be the distance from a pasture undergoing biological weed control using *S. sclerotiorum* at which the concentration of dispersing ascospores has declined to that occurring naturally in the air. Regional variation in the width of "safety zones" for sheep and dairy pasture treated with a *S. sclerotiorum* mycoherbicide have been quantified using climatic data and wind direction.[27]

Interest in the weed control potential of *S. minor* was first reported in 1991[28,29] when several *Sclerotinia* species were compared. In one study,[28] a *S. minor* isolate was more virulent on dandelion than the *S. sclerotiorum* and *S. trifoliorum* isolates. Subsequently, this *S. minor* isolate (IMI 344141) became the focus of bioherbicide research.[30–34]

10.3. Sclerotinia Minor IMI 344141 "Sensu Stricto"—The Bioherbicide

S. minor IMI 344141 was obtained from a lettuce field in Sherrington, Québec in 1983. The life cycle, mode of action, moisture and temperature requirements, and host range of *S. minor* IMI 344141 are not different from *S. minor* "*sensu lato*". However, persistence, survival and dissemination are much different when *S. minor* IMI 344141 is employed as an integrated biocontrol product.

10.3.1. THE BIOHERBICIDE PRODUCT

Sclerotinia minor IMI 344141 is the active ingredient of SARRITOR, a bioherbicide proceeding towards registration as a Microbial Pest Control Product (MPCP) in Canada for the control of *Taraxacum officinale* (dandelion) and other broadleaf weeds in turfgrass. The fungus is cultured on ground barley and the bioherbicide granules are broadcast applied to weed infested turf. Favorable conditions for germination and infection include 15–24°C temperatures and 95+% relative humidity. Disease develops quickly and complete kill of dandelion and other broadleaf weeds can be achieved within 7 days, about twice as fast as the standard chemical herbicide Killex™. The product is compatible with normal lawn maintenance operations such as mowing, fertilization and irrigation.

10.3.2. SURVIVAL AND PERSISTENCE OF *S. MINOR* IMI 344141

Questions concerning persistence of the SARRITOR product and effect on turfgrass have been addressed in greenhouse and fields studies.[30,32,33] When applied to turfgrass, *S. minor* IMI 344141 rarely produces sclerotia (melanized survival structures) and these sclerotia do not survive over winter. Sclerotia formation is mainly associated with clumps of inoculum rather than infected weed tissue. Eruptive mycelial growth of *S. minor* IMI 344141 from the inoculum does not persist in the absence of a host and quickly decays within 10 days. Lettuce bioassays of treated field trials have revealed no infectivity of *S. minor* IMI 344141 in the turf environment four months after treatment. Field and greenhouse studies confirmed that turfgrass species are not susceptible to *S. minor* IMI 344141.[32,33] Independent human health and environmental toxicology studies established that *S. minor* IMI 344141 is neither toxic nor pathogenic to non-target organisms. These data support MPCA registration and have been incorporated within the product submission to the Pest Management Regulatory Agency in Ottawa, Canada.

10.3.3. OFF TARGET MOVEMENT OF *S. MINOR* IMI 344141

Sclerotinia minor IMI 344141 does not move off target. When applied, the granules settle down within the turf on or near the soil surface. Granules are not easily dislodged or dispersed from the point of application. SARRITOR granules have been applied to over 250 field research plots in Eastern Canada and there has been no occurrence of off-target movement expressed as disease on plants beyond plot borders.

10.3.4. WEED CONTROL EFFICACY OF SARRITOR

The *Sclerotinia minor* IMI 344141 bioherbicide provides effective control of dandelion and many other broadleaf weeds including broadleaf plantain (*Plantago major*), white clover (*Trifolium repens*), and field bindweed (*Convolvulus arvensis*). Under high weed infestation levels in the field, *S. minor* caused a greater initial reduction of dandelion density than did the herbicide during the 2-weeks-post application period, although reductions were greater in herbicide treated plots by 6 weeks after application.[32] Over the growing season, *S. minor* and the herbicide had similar suppressive effects on dandelion density except under low mowing height (3–5 cm). *Sclerotinia minor* IMI 344141 has no residual activity, thus a vigorous competitive grass sward enhances the efficacy of *S. minor* by minimizing dandelion seedling recruitment in vegetation gaps created by the complete removal of the dandelions.[33,34] All life stages of dandelion from seeds to flowering plants are susceptible to the *Sclerotinia minor* IMI 344141 disease. Disease symptoms were identical on 14 different accessions of dandelion from Europe and North America and the aboveground and belowground biomass were reduced by 94% and 96%, respectively with no difference among accessions.[34]

Unlike most host specific, questionable virulence bioherbicide candidates that are being investigated worldwide, *Sclerotinia minor* IMI 344141 is a highly virulent, and broad spectrum. In addition to being an important crop pathogen, *Sclerotinia minor* can also provide effective broadleaf weed control in turfgrass.

References

1. H. R. Dillard and R. G. Grogan, Relationship between sclerotial spatial pattern and density of *Sclerotinia minor* and the incidence of lettuce drop, *Phytopathology* 75, 90–94 (1985).
2. H. J. Willets and J. A-J. Wong, The biology of *Sclerotinia sclerotiorum, S. trifoliorum, and S. minor* with emphasis on specific nomenclature. *Bot. Rev.* 46, 101–165 (1980).
3. S. W. Scott. Serological relationships of three *Sclerotinia* species. *Trans. Brit. Mycol. Soc.* 77, 674–676 (1981).
4. K. V. Subbarao, Progress toward integrated management of lettuce drop. *Plant Dis.* 82, 1068–1078 (1998).
5. B. T. Hawthorne, Observations on the development of apothecia of *Sclerotinia minor* Jagg. in the field. *N. Z. J. Agr. Res.* 19, 383–386 (1976).
6. J. J. Hao, K. V. Subbarao, and J. M. Duniway, Germination of *Sclerotinia minor* and *S. sclerotiorum* sclerotia under various soil moisture and temperature combinations, *Phytopathology* 93, 443–450 (2003).
7. M. S. Melzer, E. A. Smith, and G. J. Boland, Index of hosts of *Sclerotinia minor, Can. J. Plant Pathol.* 19, 272–280 (1997).
8. S. Abawi and R. G. Grogan, Epidemiology of diseases caused by *Sclerotinia* species, *Phytopathology* 69, 899–904 (1979).
9. D. M. Porter and M. K. Buete, Sclerotinia blight of peanuts, *Phytopathology* 64, 263–264 (1974).
10. W. R. Jarvis, *Sclerotinia minor* as the cause of lettuce drop in southwestern Ontario. *Can. Plant Dis.* Sur. 65, 1 (1985).
11. P. B. Adams, Effects of soil temperature, moisture and depth on survival and activity of *Sclerotinia minor, Sclerotium cepivorum* and*Sporidesmium sclerotivorum, Plant Dis.* 71, 170–174 (1987).
12. J. J. Hao and K. V. Subbarao, Comparative analyses of lettuce drop epidemics caused by *Sclerotinia minor* and *S. sclerotiorum, Plant Dis.* 89, 717–725 (2005).
13. E. D. Imolehin, R. G. Grogan, and J. M. Duniway, Effect of temperature and moisture tension on growth, sclerotial production, germination, and infection by *Sclerotinia minor, Phytopathology* 70, 1153–1157 (1980).
14. J. E. Hollowell, G. G. Shaw, M. A. Cubeta, and J. W. Wilcut, Weed species as hosts of *Sclerotinia minor* in peanut fields, *Plant Dis.* 87, 127–199 (2003).
15. H. A. Melouk, L. L. Singleton, F. N. Owens, and C. N. Akem, Viability of sclerotia of *Sclerotinia minor* after passage through the digestive tract of a crossbred heifer, *Plant Dis.* 73, 68–69 (1989).
16. B. Adams and W. A. Ayers, Ecology of *Sclerotinia* species, *Phytopathology* 69, 896–899 (1979).
17. B. J. R. Alexander and A. Stewart, Survival of sclerotia of *Sclerotinia* and *Sclerotium* spp in New Zealand horticultural soil, *Soil Biol. Biochem.* 26, 1323–1329 (1994).

18. E. D. Imolehin and R. G. Grogan, Factors affecting survival of sclerotia and effects of inoculum density, relative position, and distance of sclerotia from the host on infection of lettuce by *Sclerotinia minor, Phytopathology* 70, 1162–1167 (1980).
19. J. B. Coley-Smith and R. C. Cooke, Survival and germination of fungal sclerotia, *Ann. Rev. Phytopath.* 9, 65–92 (1971).
20. E. E. Jones and A. Stewart, Biological control of *Sclerotinia minor* in lettuce using *Trichoderma* species, in *Proceedings of the 50th N. Z. Plant Protection Conference* 1997, pp. 154–158.
21. H. J. Ridgway, N. Rabeendran, K. Eade and A. Steart, Application timing of *Coniothyriun minitans* A69 influences biocontrol of *Sclerotinia minor* in lettuce, *N. Z. Plant Prot.* 54, 89–92 (2001).
22. B. S. Brosten and D. C. Sands, Field trials of *Sclerotinia sclerotiorum* to control Canada thistle (*Cirsium arvense*), *Weed Sci.* 34, 377–380 (1986).
23. C. Miller, E. F. Ford, and D. Sands, A nonnsclerotial pathogenic mutant of *Sclerotinia sclerotiorum, Can. J. Microbiol.* 35, 517–520 (1989).
24. R. V. Miller, E. J. Ford, N. J. Zidack, and D. C. Sands, A pyrimidine auxotroph of *Sclerotinia sclerotiorum* for use in biological weed control, *J. Gen. Microbiol.* 135, 2085–2091 (1989).
25. I. C. Harvey, G. W. Bourdot, D. J. Saville, and D. C. Sands, A comparison of auxotrophic and wild strains of *Sclerotinia sclerotiorum* used as a mycoherbicide against Californian thistle (*Cirsium arvense*), *Biocontrol Sci. Technol.* 8, 73–81 (1998).
26. M. D. de Jong, G. W. Bourdot, G. A. Hurrell, D. J. Saville, H. J. Erbrink, and J. C. Sadoks, Risk analysis for biological weed control—Simulating dispersal of *Sclerotinia sclerotiorum* (Lib.) de Bary ascospores from a pasture after biological control of *Cirsium arvense* (L.) Scop, *Aerobiologica* 18, 211–111 (2002).
27. G. W. Bourdôt, D. Baird, G. A. Hurrell, and M. D. De Jong, Safety zones for a *Sclerotinia sclerotiorum*-based mycoherbicide: Accounting for regional and yearly variation in climate, *Biocontrol Sci. Technol.* 16, 345–358 (2006).
28. M. Ciotola, L. A. Wymore, and A. K. Watson, *Sclerotinia*, a potential mycoherbicide for lawns, *Weed Abst.* 31, 81 (1991).
29. G. E. Riddle, L. L. Burpee, and G. J. Boland, Virulence of *Sclerotinia sclerotiorum* and *S. minor* on dandelion, *Weed Sci.* 39, 109–118 (1991).
30. S. M. Stewart-Wade, S. Green, G. J. Boland, M. P. Teshler, I. B. Teshler, A. K. Watson, M. G. Sampson, K. Patterson, A. DiTommaso, and S. Dupont, *Taraxacum officinale* (Weber), dandelion (Asteraceae), in *Biological Control Programmes in Canada 1981–2000*, edited by P. G. Mason and J. T. Huber (CABI Publishing, Wallingford, Oxon, UK, 2002), pp. 427–430.
31. M. H. Abu-Dieyeh, J. Bernier, and A. K Watson, *Sclerotinia minor* advances fruiting and reduces germination in dandelion (*Taraxacum officinale*), *Biocontrol Sci. Tech.* 15, 815–825 (2005).
32. M. H. Abu-Dieyeh and A. K. Watson, Effect of turfgrass mowing height on biocontrol of dandelion with *Sclerotinia minor, Biocontrol Sci. Technol.* 16, 509–524 (2006).
33. M. H. Abu-Dieyeh and A. K Watson, Suppression of *Taraxacum officinale* populations by *Sclerotinia minor* and grass over-seeding, *J. App. Ecol.* 44, 115–124 (2007).
34. M. H. Abu-Dieyeh and A. K Watson, Efficacy of *Sclerotinia minor* for dandelion control: effect of dandelion accession, age and grass competition, *Weed Res.* 47, 67–72 (2007).

11. *FUSARIUM OXYSPORUM* F. SP. *STRIGA*, ATHLETES FOOT OR ACHILLES HEEL?

Alan Watson,[1]* Jonathan Gressel,[2] David Sands,[3] Steven Hallett,[4] Maurizio Vurro,[5] and Fenton Beed[6]
[1]*Department of Plant Science, McGill University, Sainte-Anne-de-Bellevue, Quebec, Canada*
[2]*Department of Plant Sciences, Weizmann Institute of Science, Rehovot, Israel*
[3]*Department of Plant Sciences and Plant Pathology, Montana State University, Boseman, Montana, USA*
[4]*Department of Botany & Plant Pathology, Purdue University, West Lafayette, Indiana, USA*
[5]*Istituto di Scienze delle Produzioni Alimentari, Consiglio Nazionale delle Ricerche, Bari, Italy*
[6]*Biological Control Centre for Africa, International Institute of Tropical Agriculture, Cotonou, Benin, West Africa*

Abstract. Parasitic weeds are major contributors to hunger, malnutrition, and food insecurity across sub-Saharan and northern Africa by reducing crop yields in half. Over 20 million hectares of cereal grains in sub-Saharan Africa are infested with *Striga* (witchweed). Food production losses due to *Striga* in African countries range from 20% to 90%, amounting to over 10 million tons of food lost annually. The control options for *Striga* are currently ineffective and management possibilities for these weeds are urgently needed. The research progress with a specific *forma speciales* of *Fusarium oxysporum* as a biological control for *Striga* in Africa illustrates the potential to positively impact many lives and improve the health and livelihood of rural and urban poor. Can *F. oxysporum* wild type be the Achilles heel of *Striga*, or do we need enhanced biocontrol to achieve rapid, safe, cost-effective solutions for this major biotic constraint to food production in Africa?

Keywords: witchweed, chlamydospores, seed coating, rhizosphere competence, hypervirulence

*To whom correspondence should be addressed, e-mail: alan.watson@mcgill.ca

M. Vurro and J. Gressel (eds.), Novel Biotechnologies for Biocontrol Agent Enhancement and Management, 213–222.

11.1. Introduction

Parasitic weeds, the scourge of African farmers, are major intractable biotic constraints to food production in Africa.[1] *Striga* spp. (witchweeds) are obligate parasitic weeds that parasitize the roots of cereal crops and food legumes. After attachment to the crop hosts' roots, they penetrate into the vascular system of the crop, removing water, photosynthates and minerals. Parasitic weeds are major contributors to hunger, malnutrition, and food insecurity across sub-Saharan Africa by having yields in major crops in infested areas. Striga infests 26 million hectares in sub-Saharan Africa. The control options for *Striga* are currently ineffective and novel management strategies for Striga suppression are urgently needed. The development of biological control for *Striga* in Africa illustrates the potential to impact many lives and improve the security of struggling regions. There is a need to ascertain whether biotechnologies can supply rapid, safe, cost-effective solutions to these intractable biotic constraints. Thus, for sustained *Striga* control and management, it is imperative to foster new integrated approaches including biotechnological solutions, with rigorous resource mobilization, wider strategic partnerships, novel multidisciplinary linkages and participatory approaches with farmers.[1]

11.2. Striga in Sub-Saharan Africa

The genus *Striga* (Orobancaceae) contains several obligate hemi-parasitic flowering weeds that are major biotic constraints to cereal and legume production in sub-Saharan Africa (SSA). *Striga* species are hemi-parasites, photosynthesizing about 20% of their needs after emerging from the soil. *Striga hermonthica*, *S. asiatica*, and *S. forbesii* parasitize cereal grain crops, sorghum, millets, maize and upland rice while *S. gesnerioides* parasitizes legume crops, mainly cowpeas. *Striga* species have become a scourge to cereal production and legumes where fertility is low and water/rainfall is low or erratic. The genus is most widespread in western Africa where it infests 17 million hectares or covers 64% of the cereal production area with a potential coverage of almost 100% in the semi-arid and sub-humid tropical zones,[1] In eastern and central Africa, *Striga* infests 3 million hectares (23% of cereal area and 1.6 million hectares in southern Africa (mostly in Mozambique) are infested. The highest infestations are in Nigeria (8.7 million ha), Niger (5.0 million ha), Mali (1.5 million ha) and Burkina Faso (1.3 million ha) (Table I).

In many places in Africa, the *Striga* problem has reached epidemic proportions with the situation being worst in subsistence agriculture. Yield losses from damage by *Striga* are often very significant, ranging from 10% to 70%, depending on the crop cultivar, degree of infestation, rainfall pattern and soil degradation, and estimated at 40% on average.[2] Food production losses due to

TABLE I. Sub-Saharan Africa countries with most *Striga* incidence/infestation

	Striga infested area			
	Sorghum and Millet*		Maize	
Country	('000 ha)	% total	('000 ha)	% total
Botswana	30	30	2	10
Burkina Faso	1,318	50	26	10
Eritrea	64	40	0	0
Ethiopia	528	30	80	5
Kenya	80	53	225	15
Mali	1,513	70	20	10
Mozambique	150	40	122	10
Niger	4,989	70	—	—
Nigeria	8,720	80	904	22
Senegal	411	40	3	0.05
Sudan	1,875	30	17	10
Tanzania	650	90	214	12
Total/mean	20,330	56	1,613	15

Compiled by A.B. Obilana from reports of A.B. Obilana, F. Kanampiu and D. Friesen.
* Includes finger millets in the lake zone of east central Africa.
* Includes both sorghum and pearl millet combined in West African countries only.
Source: Modified from Gressel et al.[1]

TABLE II. Sub-Saharan Africa countries with the highest food production losses due to *Striga**

Country	Estimated yield loss* (%)	Yield loss ('000 tons)
Burkina Faso	35–40	710–820
Eritrea	20–60	30–90
Ghana	35	170
Kenya	35–40	50–60
Mali	40	580
Mozambique	35	40
Niger	40–50	930–1,160
Nigeria	35	3,750
Sudan	30	1,230
Tanzania	Up to 90	550
Togo	35	70
Total/mean	39–45	8,110—8,520

* Loss includes sorghum, millets, and maize. Compiled by A.B. Obilana, from NARS documents, reports and personal records.
Source: Gressel et al.[1]

Striga in the SSA countries range from 20% to 90% (Table II), amounting to over 8 million tons of food lost annually.[1] Although several potential control measures have been developed in the past decades, most of these methods (including the use of chemical herbicides, nitrogen fertilization and soil fumigation) are too costly for poor subsistence farmers that make up about 75–80% of farmers in SSA. Crop rotation is probably one of the most effective ways to reduce *Striga* infestations and increase maize yields and income considering the limited resource base of small-scale subsistence farmers in SSA.[3] Yet most of the rotational crops (forage legumes) do not provide the food needed to sustain the farm families. Land use intensification and increasing cereal mono-cropping, with little or no use of purchased external inputs, have contributed immensely to exacerbate the *S. hermonthica* problem in Africa. The farmers' plight has been compounded by the environmental and policy factors that fostered *Striga* spread.

11.3. *Striga hermonthica*

Striga hermonthica, the most economically important parasitic seed plant in the world,[4] is endemic in the African savanna and the Sahel where it devastates the yields of maize, sorghum, millet, and rice, the major staple foods for over 300 million people in SSA. Annual crop losses in cereals caused by *S. hermonthica* vary from about 10% (at low levels of infestation) to complete crop loss and total abandonment of cereal production in severely infested fields. It causes an annual loss of about US $9 billion. Recent surveys have abundantly confirmed that farmers in these areas urgently and desperately need effective, inexpensive and sustainable control options as components of an integrated *Striga* management (ISM package).

Numerous techniques exist for the management of *Striga*. Each technique has value in certain situations, and limitations in others. For example, a new technique using herbicide treatment of maize seed of a herbicide-resistant maize is highly successful[5] and has been commercialized, but only for east-African short season maize, while it is now being developed for longer season maize, but not for other crops attacked by *Striga*. In many cases, valuable techniques are unavailable to the subsistence farmers who need them the most. The greatest deficiencies in the needs of subsistence farmers are short-term techniques that will enable the effective production of susceptible crops in *Striga* infested land. Techniques that will protect crops from parasitism by *Striga*, and provide remedial control of *Striga* are urgently needed. Thus, for sustained *Striga* control and management, it is imperative to foster new integrated approaches including biotechnological solutions, with concerted resource mobilization, wider strategic partnerships, and novel multidisciplinary linkages in participation with farmers. One potential option, that obviates some of the problems of several of the other options, is the use of *Fusarium oxysporum*

f. sp. *striga* for the biological control of *S. hermonthica*. This solution would be applicable to all varieties of all crops attacked by *Striga*.

11.4. Biological Control of *Striga hermonthica*

Both insects and fungi have been proposed for biocontrol of Striga. The insects attack mainly the seedpods, eating most, but never all of the seeds. Thus, replenishment of the seed bank is sufficient to sustain the weed population while having little yield promotion.[6] Various fungi have been tested both for pathogenicity on *Striga* but none are yet in wide scale field testing.

Fusarium species are the most prevalent fungi associated with diseased *Striga* plants. Controlled environment chamber evaluation of 81 fungal isolates from three countries (Burkina Faso, Mali and Niger) found an isolate of *Fusarium oxysporum* from Mali (isolate M12-4A), grown on sorghum straw and incorporated into pots, that successfully prevented emergence of *S. hermonthica*. This resulted in a fourfold increase of sorghum dry matter.[7] Subsequent evaluation of efficacy of the M12-4A isolate in the field in Mali, using chopped or ground sorghum straw inoculum, resulted in 60% reduction of emerged *Striga* at 82 days after sowing, while sorghum biomass was doubled[8] compared with the control. Further work with isolate M12-4A has reported complete inhibition of *S. hermonthica* emergence when the fungal spore (chlamydospore) powder was added to the soil with sorghum seeds or by sowing sorghum seeds that were also coated[9] with the chlamydospores. Chlamydospore powder treatments reduced *S. hermonthica* emergence by 78% to 92% (Table III). In related studies from Nigeria and Burkina Faso, other isolates of

TABLE III. Effect of *Fusarium oxysporum* M12-4A on *Striga hermonthica* emergence in the field

Inoculum treatments per seed pocket	Striga plants/plot		
Control (no straw incorporated)	32.1*	(17.3)†	a‡
Sterilized straw control (10 g)	16.8	(6.2)	ab
Sterilized ground straw control (2.6 g)	21.3	(12.5)	ab
Solid substrate ground inoculum (2.6 g)	7.9	(4.5)	b
Chlamydospore powder (0.5 g)	6.9	(4.9)	b
Chlamydospore powder (0.5 g) + sterilized straw (10 g)	3.6	(1.9)	b
Chlamydospore powder (1.0 g)	2.7	(1.7)	b
Chlamydospore powder (1.0 g) + sterilized straw (10 g)	2.5	(1.4)	b

Source: Ciotola et al.[11]

* Mean number of *S. hermonthica* in plots.

† Values in parentheses are standard errors.

‡ Values having the same letter are not significantly different at = 0.05 according to the Student–Neuman–Keuls multiple range test.

F. oxysporum (PSM197, 4-3-B) inhibited *Striga* seed germination and reduced the number of emerged *S. hermonthica* plants in pot[10] and field[11,12] trials.

F. oxysporum f. sp. *striga* is host limited. Several crop species (sorghum, pearl millet, maize, rice, fonio, cotton, groundnut, cowpea and okra) were immune to isolate M12-4A.[7] These and other crops are also immune to isolates from Ghana, Sudan, and Nigeria.[13,14] All *S. hermonthica* isolates of *F. oxysporum* f. sp. *striga* are pathogenic only to *S. hermonthica*, and possibly *S. asiatica*.[14] Isolate M12-4A does not produce mycotoxins under all conditions tested, and hence it does not constitute a known health hazard to humans or livestock.[15]

Mass production and delivery of the biocontrol agent to its target are critical phases in biocontrol projects. Techniques that have been suggested for mass production of *F. oxysporum* inoculum include on-farm models, cottage-industry models and small entrepreneur industry models. *Fusarium* can be grown on a range of cheap, crude agricultural products, including sorghum stubble. Several methods for mass production of the fungus on sterilized sorghum straw have been developed.[8,9]Effective biological control of *S. hermonthica* with M12-4A was achieved with inoculum produced using a simple fermentation system with sorghum straw as the growth substrate for inocula. Sorghum seeds were coated with inoculum using gum arabic as the adhesivefor inoculum delivery at farmers' fields in researcher-managed trials.[9] When applied as a seed coat, only 80 g of the chlamydospore powder are required per hectare. To facilitate broad usage of the *F. oxysporum* isolate M12-4A, an inoculum production strategy based on cottage industry model was suggestedthat utilizes a liquid fermentation process and inexpensive locally available substrates (including sorghum straw and gum arabic).[11] Four villages in Mali participated in 2000 in liquid mass production of M12-4A in cooking pots and in coating seed. Seed coating activities were highly successful, but all production vessels became contaminated and no viable inoculum was produced.[16] Other, more rigorous production systems need to be critically evaluated.

In addition to the above powder formulation, inocula have been applied directly into the seeding holes and several granular formulations, including sodium-alginate and wheat flour-kaolin "Pesta" granules have been evaluated.[17]

11.5. Molecular Characterization of *F. oxysporum* Wild Types

The genetic diversity among the various isolates of *Fusarium oxysporum* from *Striga hermonthica* has indicated a high degree of genetic similarity (Ciotola et al., unpublished data). Vegetative compatibilities of 14 isolates of *F. oxysporum* from diseased *S. hermonthica* were determined using nitrate non-utilizing mutants. All *F. oxysporum* f. sp. *striga* collected from Mali, Niger and Kenya were in one vegetative compatibility group (VCG) and thus genetically similar. Random amplified polymorphic DNA assays were carried out on a large range of isolates of *Fusarium oxysporum* to identify markers only common to *F. oxysporum* strains isolated from *Striga*. One fragment of 3500 bp was cloned and used to probe Southern blots of DNA from *Fusarium oxysporum* isolates as well as various heterogeneous organisms and plant tissue. The fragment hybridized only to DNA from *Striga* isolates and two *F. oxysporum* isolates that originated from sorghum. The amplified product (600 bp) was sequenced and two pairs of SCAR (sequence characterized amplified region) primers (M12-4A/R and M12-4A/F) were generated for use in polymerase chain reaction (PCR). One fragment of 600 bp was generated following PCR of all *F. oxysporum* f. sp. *striga* isolates and from one *F. oxysporum* from sorghum. Two new SCAR primers (FUN001 and FUN002) were designed containing the most sequence differences between the target isolate (M12-4A) and the sorghum *F. oxysporum* isolate O-1202 and tested in conventional PCR assays. FUN001 and FUN002 amplified only one band of 157 bp in all isolates from *Striga*. No amplified product was detected in the sorghum *F. oxysporum* isolate. The same primers were used in real-time PCR assays to reconfirm their specificity and determine their sensitivity detection level. PCR assays confirmed the VCG results indicating *F. oxysporum* isolates from *Striga* from west and east Africa are genetically similar suggesting co-existence of *F. oxysporum* f. sp. *striga* with its host across SSA.

11.6. Enhancement of *Fusarium oxysporum* f. sp *striga*?

Different *F. oxysporum* isolates have reduced *S. hermonthica* by 40% to 100% in laboratory, pot and field trials. However, extensive field trials to ascertain field efficacy of *F. oxysporum* to control *S. hermonthica* have not yet been conducted. Will the *F. oxysporum* wild type be sufficiently virulent and competent to achieve the desired level of *Striga* reduction? The *Striga* problem in Africa is critical and it behooves us to examine all means to find a solution for this problem. Perhaps one or more of the following biotechnological approaches may improve the virulence, deployment, and success of *F. oxysporum*.

11.6.1. OVER EXPRESSION OF AMINO ACIDS

Amino acid toxicity has long been observed in plants, with different amino acids effecting different species of plants. It is not surprising then that single amino acids, when applied externally to a plant, can inhibit plant growth and development.[18] Examples are the severe seedling inhibitions when valine is applied to germinating seeds of *Papaver somniferum* and *Cannabis sativa*, methionine inhibition of *Cirsium arvense*, and lysine inhibition of *Centaurea diffusa*. These amino acid inhibitions can be as a result of direct application of specific amino acids to the soil, or by plant pathogens that excrete unusually high amounts of these amino acids.[19] Recently, Vurro et al.[20] have reported that 2 mM methionine was able to almost completely inhibit the germination of *Orobanche ramose*, a related parasitic weed. When methionine was applied to tomato roots, the number of developing tubercles of the parasite was reduced. Preliminary work indicated that *Striga* was sensitive to leucine, threonine and tyrosine.

11.6.2. GENERATING TRANSGENIC HYPERVIRULENT FUSARIUM STAINS

Several strains of *F. oxysporum* and *F.* sp. CNCM I-1621 that attack *Orobanche* spp.[21] have not been successful in providing near the level of control desired by farmers when tested in the field. Transgenes encoding auxin production were introduced into an *Orobanche*-attacking fungal species, doubling virulence,[21] although this is still far less efficacy than farmers need. Far stronger toxic genes are needed to enhance virulence, and the NEP1 gene, used to enhance a different mycoherbicide[22] also was active in enhancing the virulence of a *F.* sp. CNCM I-1621 that is specific to *Orobanche* spp. (Chapter 16).

A variety of hypervirulence genes are being transforming into two strains of *Fusarium* that attack *Orobanche.* These same constructs could be transformed into the *Fusarium* isolates used as biocontrol agents against *Striga*. The biosafety aspects of using transgenically hypervirulent biocontrol agents are specifically addressed in references 23, 24 and Chapter 19.

11.6.3. TECHNIQUES FOR OPTIMAL APPLICATION OF THE BIOCONTROL

The current state of the art is to apply the *Fusarium* biocontrol agent as a seed dressing using gum arabic[9] as a sticker. Alternative approaches may include various granular or pelletized formulations placed in the planting hole or applied during weeding operations. The *Striga* infestations in Africa cover vast areas and the idea of aerial dispersal[25] and soil penetration on seed is most intriguing. One suggestion is to deliver the biocontrol agent on the seed of a non-host reclamation plant species. In this method, it is hypothesized that the biocontrol agent could saprophytically colonize the roots of the seedlings of the reclamation plant as it establishes and take up residence in the soil profile where it could then come in contact with *Striga*. Selection of the carrier plant species will be an interesting challenge.

11.6.4. RHIZOSPHERE COMPETENCE AND PERSISTENCE

The biology of *Fusarium* spp. in the soil, root, and rhizosphere is extremely varied. *Fusarium* spp. can be persistent in the soil as saprophytes, can develop large amounts of mycelium on the rhizoplane and in the rhizosphere, and can invade root epidermal and cortical tissues either pathogenically or non-pathogenically.[26] The activity of *F. oxysporum striga* isolates in each of these regards is not fully understood.

How far will *F. oxysporum* f. sp. *striga* move through the soil/rhizosphere? How long will *F. oxysporum* persist in the rhizosphere? What level of rhizosphere colonization is required for effective control of *S. hermonthica*? How is the biology of *F. oxysporum* in the rhizosphere affected by abiotic and biotic factors? How will selected and transformed strains respond? Answers to these questions will be critical for designing the most effective strategies for the deployment of *F. oxysporum*.

11.6.5. COMPARE BIOLOGICAL EFFICACY OF THE BICONTROL AGENT WITH CONTROL OPTIONS BEING PRACTICED BY FARMERS

Additional on farm field trials with several *F. oxysporum* isolates are presently ongoing in Benin and others are planned. These trials should help answer the question of virulence and biocontrol efficacy of *F. oxysporum*. Certainly, enhanced biocontrol would be an additional benefit in the struggle against *Striga*.

References

1. J. Gressel, A. Hanafi, G. Head, W. Marasas, A. B. Obilana, J. Ochanda, T. Souissi, and G. Tzotos, Major heretofore intractable biotic constraints to African food security that may be amenable to novel biotechnological solutions, *Crop Prot.* 23, 661–689 (2004).
2. K. Elemo, S. T. O. Lagoke, A. Awad, and S. Oikeh, Population dynamics and determinants of *Striga hermonthica* on maize and sorghum in Savanna farming systems, *Crop Prot.* 14, 283–290 (1995).
3. A. Oswald and J. K. Ransom, *Striga* control and improved farm productivity using crop rotation, *Crop Prot.* 20, 113–120 (2001).
4. C. Parker and C. R. Riches, *Parasitic Weeds of the World: Biology and Control* (CAB International. Wallingford, Oxon, UK, 1993).
5. F. K Kanampiu, V. Kabambe, C. Massawe, L. Jasi, J. K. Ransom, D. Friesen, and J. Gressel, Multisite, multi-season field tests demonstrate that herbicide seed-coating herbicide-resistance maize controls *Striga* spp. and increases yields, *Crop Prot.* 22, 697–706 (2003).
6. M. C. Smith and M. Webb, Estimation of the seed bank of *Striga* spp. (Scrophulariaceae) in Malian fields and the implications for a model of biocontrol of *S. hermonthica*, *Weed Res.* 36, 85–92 (1996).
7. M. Ciotola, S. G. Hallett, and A. K. Watson, Discovery of an isolate of *Fusarium oxysporum* with potential to control *Striga hermonthica* in Africa, *Weed Res.* 35, 303–309 (1995).
8. C Diarra, M., Ciotola, S. G. Hallett. D. E. Hess, and A. K. Watson, Field efficacy of *Fusarium oxysporum* for the control of *Striga hermonthica*, *Nuis. Pests Prag.* 4, 257–263 (1996).

9. M. Ciotola, A. DiTommaso, and A. K. Watson, Chlamydospore production, inoculation methods and pathogenicity of *Fusarium oxysporum* M12-4A, a biocontrol for *Striga hermonthica, Biocontrol Sci. Technol.* 10, 129–145 (2000).
10. P. S. Marley, S. M. Ahmed, J. A. Y. Shebayan, and S. T. O. Lagoke, Isolation of *Fusarium oxysporum* with potential for biological control of the witchweed (*Striga hermonthica*) in the Nigerian savanna, *Biocontrol Sci. Technol.* 9, 159–163 (1999).
11. D. Yonli, H. Traore, D. E. Hess, A. A. Abbasher, and I. J. Boussim, Effect of growth medium and method of application of *Fusarium oxysporum* on infestation of sorghum by *Striga hermonthica* in Burkina Faso, *Biocontrol Sci. Technol.* 14, 417–421 (2004).
12. P. S. Marley and J. A. Y. Shebayan, Field assessment of *Fusarium oxysporum* based mycoherbicide for control of *Striga hermonthica* in Nigeria, *BioControl* 50, 389–399 (2005).
13. A. A. Elzein, Development of a granular mycoherbicidal formulation of *Fusarium oxysporum* Foxy 2 for the biological control of *Striga hermonthica,* in *Tropical Agriculture—Advances in Crop Research,* vol. 12 (pt. 2), edited by J. Kroschel (Margraf Verlag, Weikersheim, Germany, 2003).
14. M. E. Savard, J. D. Miller, M. Ciotola, and A. K. Watson, Secondary metabolites produced by a strain of *Fusarium oxysporum* used for *Striga* control in West Africa, *Biocontrol Sci. Technol.* 7, 61–64 (1997).
15. A. A. Elzein and J. Kroschel, *Fusarium oxysporum* Foxy 2 shows potential to control both *Striga hermonthica* and *S. asiatica, Weed Res.* 44, 433–438 (2004).
16. C. Bastian, Seed coating with *Fusarium oxysporum* M12-4A for the biocontrol of *Striga hermonthica,* M.Sc. thesis (McGill University, 2002).
17. A. A. Elzein, J. Kroschel, and D. Muller-Stover, Effects of inoculum type and propagule concentration on shelf life of Pesta formulations containing *Fusarium oxysporum* Foxy 2, a potential mycoherbicide agent for *Striga* spp, *Biol. Control* 30, 203–211 (2004).
18. D. C. Sands, A. L. Pilgeram, T. W. Anderson, and K. S. Tiourebaev, Virulence enhancement of bioherbicides, US Patent no. 6,673,746 (2004).
19. D. C. Sands, A. L. Pilgeram, and K. S. Tiourebaev, Enhancing the efficacy of bicontrol agents of weeds, in *Enhancing Biocontrol Agents and Handling Risks,* edited by M. Vurro et al. (IOS Press, Amsterdam, 2001), pp. 3–10.
20. M. Vurro, A. Boari, A. L. Pilgeram, and D. C. Sands, Exogenous amino acids inhibit seed germination and tubercle formation by *Orobanche ramosa* (Broomrape): Potential application for management of parasitic weeds, *Biol. Control* 36, 258–265 (2006).
21. Z. Amsellem, Y. Kleifeld, Z. Kerenyi, L. Hornok, Y. Goldwasser, and J. Gressel, Isolation, identification, and activity of mycoherbicidal pathogens from juvenile broomrape plants, *Biol. Control* 21, 274–284 (2001).
22. Z. Amsellem, B. A. Cohen, and J. Gressel, Transgenically conferring sufficient hypervirulence on an inundative mycoherbicidal fungus for efficient weed control, *Nat. Biotechnol.* 20, 1035–1039 (2002).
23. J. Gressel, Potential failsafe mechanisms against the spread and introgression of transgenic hypervirulent biocontrol fungi, *Trends Biotechnol.* 19, 149–154 (2001).
24. J. Gressel, *Molecular Biology of Weed Control* (Taylor & Francis, London, 2002).
25. D. C. Sands, K. S. Tiourebaev, A. L. Pilgeram, and T. W. Anderson, Carrier methodology for aerial dispersal and soil penetration of bioactive agents, US Patent no. 6,403,530 (2002)
26. P. E. Nelson, T. A. Toussoum, and R. J. Cook (editors), *Fusarium: Disease, Biology and Taxonomy* (Pennsylvanian State University Press. University Park, PA, 1981).

12. CONTROL OF SCLEROTIAL PATHOGENS WITH THE MYCOPARASITE *CONIOTHYRIUM MINITANS*

John M. Whipps,[1]* Amanda Bennett,[1] Mike Challen,[1] John Clarkson,[1] Emma Coventry,[1] S. Muthumeenakshi,[1] Ralph Noble,[1] Chris Rogers,[1] S. Sreenivasaprasad,[1] and E. Eirian Jones[2]
[1] *Warwick HRI, University of Warwick, Wellesbourne, Warwick, CV35 9EF, UK*
[2] *National Centre for Advanced Bio-Protection Technologies, PO Box 84, Lincoln University, Canterbury, New Zealand*

Abstract. Pressure to reduce the use of chemicals in the environment has led to the search for alternative sustainable methods to control soil-borne pathogens, especially those plant pathogens that form long-lived resting bodies (sclerotia). Mycoparasites that attack sclerotia have been explored as biocontrol agents of these pathogens and some mycoparasites such as *Coniothyrium minitans* and *Trichoderma* species have been the focus of particular study. This paper reviews recent developments in the use, ecology, impact and modes of action of *C. minitans* especially against *Sclerotinia sclerotiorum* that may be influential in improving reproducibility of disease control in the future. Some studies of the use of *Trichoderma viride* to control *Allium* white rot caused by *Sclerotium cepivorum* are also discussed.

Keywords: biological disease control; *Coniothyrium minitans*; impact; integrated control; inoculum; mode of action; mycoparasite; pathogenicity genes; sclerotia; *Sclerotinia*; *Sclerotium*; *Trichoderma*

12.1. Introduction

The pressure to reduce chemical use in the environment has led to a decrease in the number of active ingredients available for control of plant pathogens. The withdrawal of the soil sterilant methyl bromide has had a particularly detrimental effect for control of soil-borne plant pathogens as it was so widely used and cost-effective despite its price. Those chemicals remaining have been subject to rigorous, costly, and time-consuming scrutiny by regulatory authorities, and although they are generally more targeted in their mode of

*To whom correspondence should be addressed, e-mail: john.whipps@warwick.ac.uk

M. Vurro and J. Gressel (eds.), Novel Biotechnologies for Biocontrol Agent Enhancement and Management, 223–241.

TABLE I. Examples of mycoparasites of sclerotial pathogens (from[21])

Mycoparasite	Host sclerotial pathogen
Coniothyrium minitans, *Sporidesmium sclerotivorum*, *Teratosperma oligocladum*	*Botrytis* spp., *Sclerotinia* spp., *Sclerotium cepivorum*
Gliocladium spp. including *G. catenulatum*, *G. roseum*	*Botrytis* spp., *Rhizoctonia solani*, *Sclerotinia* spp., *Sclerotium* spp., *Verticillium dahliae*
Laetisaria arvalis	*Rhizoctonia solani*
Microsphaeropsis spp.	*Rhizoctonia solani*
Pythium oligandrum	*Sclerotinia sclerotiorum*, *Verticillium dahliae*
Stachybotrys elegans	*Rhizoctonia solani*
Talaromyces flavus	*Rhizoctonia solani*, *Sclerotium rolfsii*, *Sclerotinia sclerotiorum*, *Verticillium dahliae*
Trichoderma spp. including *T. aureoviride*, *T. hamatum*, *T. harzianum*, *T. koningii*, *T. pseudokoningii*, *T. viride*, *T. virens*	*Botrytis* spp., *Rhizoctonia solani*, *Sclerotinia* spp., *Sclerotium* spp., *Verticillium dahliae*
Verticillium biguttatum	*Rhizoctonia solani*

action, there are still problems with pathogens evolving resistance to them. Clearly, there are needs not being met by conventional fungicides, especially with niche crops, where companies are now unwilling to register fungicides. Consequently, there is a need to find alternative, sustainable control measures for soil-borne plant pathogens, especially those that form resting bodies such as sclerotia, which can survive in soil for many years and for which there are few or no resistant plant cultivars available. Indeed, in the UK, some growers have abandoned growing onions on some land in Kent due to the presence of *Sclerotium cepivorum*, the causal agent of *Allium* white rot disease[1] and others in Lancashire have ceased growing lettuce on land due to *Sclerotinia sclerotiorum*, the causal agent of Sclerotinia drop or rot. The sclerotia of these pathogens survive in soil for more than 20 and 5 years, respectively,[2,3] meaning that normal crop rotation is not effective. The problem is exacerbated with *S. sclerotiorum*, as its host range extends to over 400 species of plants.[4]

One approach has been to develop biological disease control agents (BCAs) for sclerotium forming pathogens and over the last 30 years there have been numerous examples of fungi that have been demonstrated to attack or control sclerotial pathogens (Table I). These have largely been identified for their ability to destroy sclerotia directly, acting as mycoparasites. Some such as *Coniothyrium minitans* and *Sporidesmium sclerotivorum* are highly specialized mycoparasites, lacking a free-living saprotrophic stage of growth in soil, whereas others such as *Gliocladium* and *Trichoderma* species are facultative mycoparasites, capable of utilizing a range of dead and dying plant material.[5] There is also increasing evidence that some *Trichoderma* species can live within plants without causing disease.[6] Perhaps not surprisingly, the literature is dominated by studies of *Gliocladium* and *Trichoderma* which are easy to grow in culture and sporulate well. However, a few examples of commercial BCA products based on both specialized and non-specialized sclerotial mycoparasites are known including Contans, Intercept and KONI (*C. minitans*) and BioTrek, Harzian-10, Rootshield, T-22G and Trichoderma 2000 (*Trichoderma* species).[7,8] There are some reports of biological control of sclerotial pathogens with bacteria, either inhibiting mycelial growth, ascospore germination or sclerotium germination,[9–13] as well as bacteria, yeasts and filamentous fungi inhibiting foliar infection from ascospores,[14–20] but these are not considered further here.

The relatively few disease biocontrol products for control of sclerotial pathogens reflects a combination of difficulties associated with commercialization. Firstly, the registration process for a BCA can be essentially the same as that for a chemical pesticide. This means that the costs are extremely high despite some attempts to encourage registration of BCAs by some regulatory authorities. For example, the US Environmental Protection Agency has a fast-track system for some BCAs, and in the UK, the Pesticides Safety Directorate has a reduced cost for dossier assessment of a biopesticide compared with a chemical, and interactive discussions on the registration process are encouraged. However, as many BCAs are selective in their target pathogen range (which can be seen as an environmental plus) such niche markets are often too small to ensure that the cost and time involved in the registration process is worthwhile, although schemes such as the IR-4 in the USA and the Off-label Approval procedure in the UK attempt to help provide crop management tools for minor crop use. Nevertheless, several known BCAs are marketed in some countries as plant growth strengtheners, plant growth promoters, or soil conditioners rather than as BCAs to avoid the need for registration. It also explains why when registration has been obtained for control of one pathogen, that some BCAs are tested for activity against a range of other pathogens to increase potential sales.

A second problem is the difficulty in obtaining a BCA that works reproducibly in commercial field or glasshouse trials. It is relatively easy to demonstrate some form of activity in small-scale laboratory tests but when scale-up and use in a range of different environments is attempted, control is often lost. To optimize efficacy it is essential to characterize the BCA in an overlapping series of studies including: inoculum production, formulation, downstream processing and application technology; physiology; ecology; and mode of action.[22] It also requires an understanding of the etiology and epidemiology of the pathogen, the crop and culture system and the environment of use.

Consequently, this paper will examine some of the approaches used to enhance the reproducibility of control of sclerotial pathogens with special reference to the work done on *C. minitans* as a biological control agent of *S. sclerotiorum* over the last 15 years. The general applicability of some of the approaches considered with *C. minitans* has also been extended to the control of other sclerotial pathogens and this will be illustrated with the use of *Trichoderma viride* to control *Sclerotium cepivorum* on onion.

12.2. *Coniothyrium minitans* As a Biocontrol Agent

Coniothyrium minitans is an ecologically obligate mycoparasite of sclerotia of *Sclerotinia sclerotiorum, Sclerotinia minor, Sclerotinia trifoliorum* and some strains of *Sclerotium cepivorum* and *Botrytis* species.[23] It was first isolated from sclerotia of *S. sclerotiorum* in California in 1947 and its potential as a biological control agent was appreciated even then.[24] Subsequently, it has been recovered from more than 30 countries, on all continents except Antarctica.[25,26] When applied to soil, the mycoparasite has been shown to control *S. sclerotiorum* in numerous glasshouse and field trials involving lettuce, celery, sunflower, bean and oilseed rape[27–31] and *Sclerotium cepivorum* on onion.[32] It survives well in soil for several years after application[27,33,34] and has been associated with the development of sclerotinia suppressive soils.[35] It has also been applied to foliage to prevent ascospore infection and disease development in alfalfa and beans,[36–40] to foliage to decrease sclerotial production and survival in rotations of several crops[29] as well as to crop debris to decrease sclerotia carryover.[41] However, there is evidence that the ability of *C. minitans* to control *S. sclerotiorum* is diminished at high pathogen inoculum levels.[27,41,42] Consequently, recent experiments have sought to understand more about the influence of inoculum quality and quantity, timing of application and sources of *C. minitans* in an attempt to improve effectiveness.

Numerous studies over the years at Warwick HRI have demonstrated reproducible control of *S. sclerotiorum* on lettuce in standardized glasshouse trials when *C. minitans* isolate Conio was applied to soil as a maizemeal-perlite (MP) preparation at 10^{11} colony forming units (cfu) m^{-2}. [27,41,42] However, the recommended application rate of the commercial *C. minitans* product Contans is 10^8 spores m^{-2} (10^{12} spores ha^{-1}) reflecting the balance between costs of production, formulation and acceptable shelf-life relative to chemical pesticides, and the level of reproducibility of control in large-scale field trials, which may not be the same for all crops.[43] Thus, when three isolates of *C. minitans* (Conio, IVT1 and Contans) were applied in standardized glasshouse trials at 10^8 spores m^{-2} none controlled *S. sclerotiorum* in lettuce although the standard MP preparation did at 1000 times higher inoculum level, suggesting that inoculum level was a key factor in control using this mycoparasite under these conditions (Table II). Further work demonstrated that MP preparation

TABLE II. Only a high inoculum rate (10^{11} cfu m^{-2}) of *C. minitans* MP preparation incorporated into soil reduces % *Sclerotinia*-diseased lettuce, number of sclerotia recovered and sclerotial viability in the third of three sequential crops (adapted from[34])

Treatment	Cfu m^{-2}	% disease	No. of sclerotia 2500 cm^{-2}	% Viability
Control (nil)		57a*	260a	95a
Fungicide		16b	19b	96a
Conio MP full	10^{11}	25b	31b	72b
Conio MP reduced	10^{8}	44a	193a	91a
Conio spore	10^{8}	38a	116c	97a
IVT1 spore	10^{8}	47a	101c	96a
Contans spore	10^{8}	51a	176c	97a

*Numbers in same columns followed by the same letters are not significantly different ($P < 0.05$) based on LSD from restricted maximum likelihood analysis.

was consistently more effective at reducing apothecial (fruit body) production than spore suspensions when applied at the same rate.[44] More economical ways of producing inoculum of *C. minitans* allowing greater application rates than currently cost-effective may be a way that its use could be enhanced.

In New Zealand, different isolates of *C. minitans* exhibited differences in ability to infect sclerotia of *S. sclerotiorum*[45] suggesting that selection of more virulent strains of *C. minitans* could be worthwhile to improve efficacy of this mycoparasite. However, such differences were not clearly shown using the three European isolates at Warwick HRI over several years.[34,44,46,47] Importantly, infection of sclerotia by *C. minitans* can be achieved by just 1–2 conidia[36,37] demonstrating the high efficiency of this mycoparasite providing it comes into contact with the sclerotia under appropriate temperatures and levels of water availability, i.e., between 5–25°C and greater than 95% humidity (–7.0 MPa).

A period of 8 weeks was recommended between inoculum application and planting for Contans and this was not carried out in our early work.[34] Subsequently, an investigation was carried out with sclerotia placed in bags in plots in the glasshouse with the same treatments and isolates applied as before, but with the 8 week period before sclerotial recovery, viability and infection by *C. minitans* were assessed.[46] Once again, only the standard MP treatment had any effect but interestingly, different results were found depending on the time of the year. For example, when MP was applied between August–October there was 87% recovery (defined as the proportion of the original number of sclerotia recovered in bags after retrieval), 76% viability and 32% infection of sclerotia; in February–April there was 99% recovery, 9% viability and 92% infection of sclerotia; and in June–August there was 26% recovery, 17% viability and 3%

infection of sclerotia. In August–October, *C. minitans* infection was relatively low, reflecting high temperatures during this period. The optimal temperature for *C. minitans* growth and sclerotia infection is approximately 20°C,[23,48] which may limit its use to control sclerotinia disease in the tropics unless a temperature tolerant isolate is found. During February–April, soil temperatures were lower, allowing high infection but the sclerotia did not degrade. In contrast, in June–August, *C. minitans* infection occurred early but secondary infection by other microorganisms may have resulted in complete degradation of the sclerotia, masking recovery of *C. minitans* from sclerotia. Such secondary colonization following *C. minitans* infection has been demonstrated recently[49] and may be an important part of the control process. Box-bioassays, involving sclerotia placed in soil with the same treatments as the glasshouse trial, carried out at the same time in an adjacent glasshouse, confirmed the ability of *C. minitans* to reduce apothecial production and again highlighted the effect of temperature on the ability of *C. minitans* to prevent apothecial production and the effect of temperature on apothecial production per se (Figure 1). The inhibitory effects of high temperature on *C. minitans* and the need for a period of time for sclerotial colonization and subsequent reduction in apothecial production is reflected in the recommendations for field applications of this mycoparasite. Thus for countries with dry and hot summers and moderate winter temperatures applications should be made in the autumn.[43]

12.3. Integrated Control of Sclerotial Fungi

One approach to enhance the level of control obtained with BCAs is to use some form of additional control measure, ideally achieving synergistic integrated control effects. This may take the form of combination with a fungicide, perhaps with an increased application interval or reduced application rate, combination with another BCA or combination with some other form of cultural measure, such as additions of organic matter or composts, or soil steaming.[50–52]

12.3.1. INTEGRATED CONTROL WITH *CONIOTHYRIUM MINITANS*

The first studies of integrated control of fungicides with *C. minitans* were carried out in a glasshouse experiment involving sclerotinia disease in three sequential crops of lettuce.[53] Disease levels were low in the first crop and increased with crops 2 and 3. Control of *Sclerotinia sclerotiorum* in the third crop was achieved by a combination of *C. minitans* applied to soil and a single application of iprodione applied to foliage and this level of control was equivalent to that achieved with standard prophylactic sprays with iprodione every

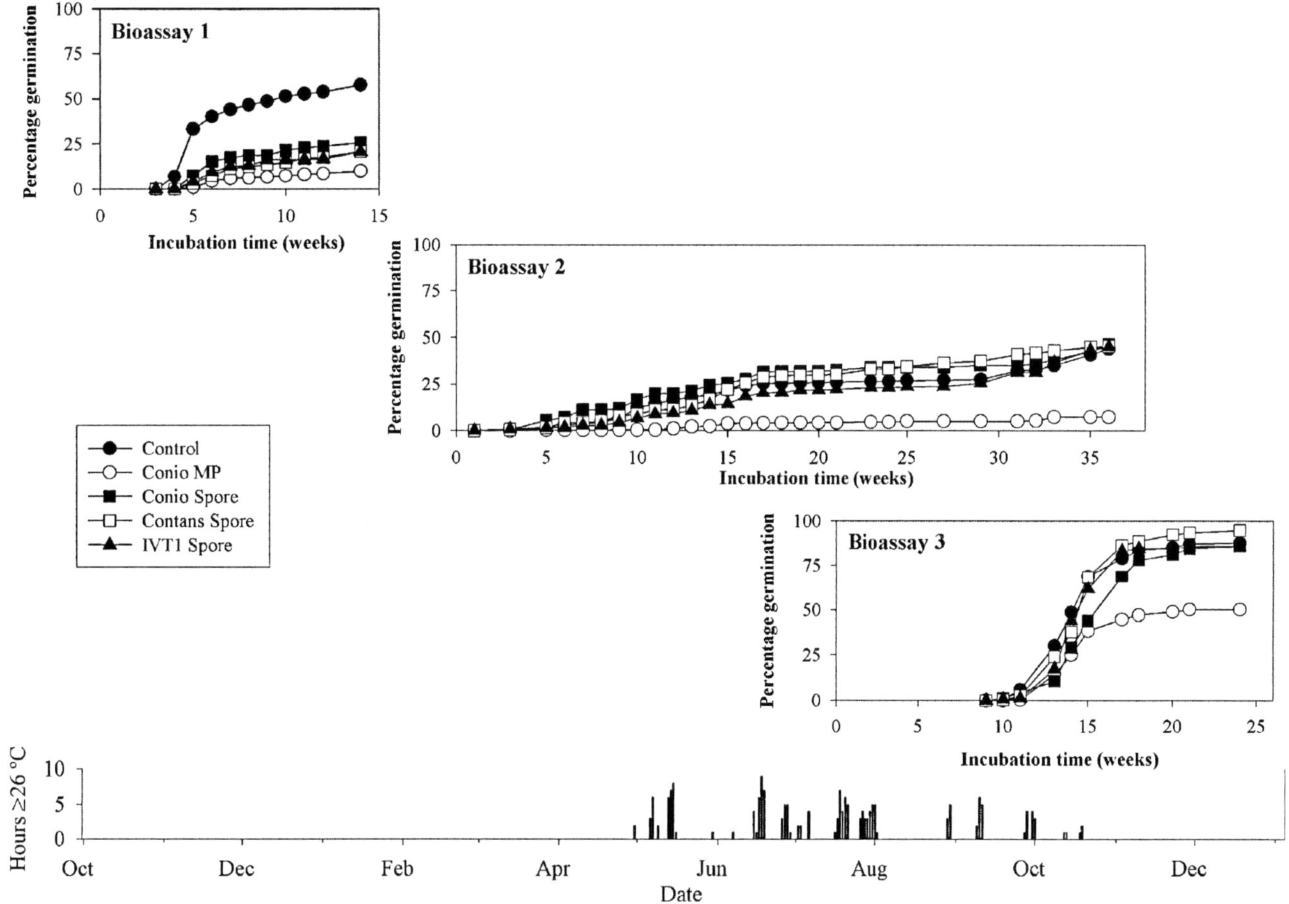

Figure 1. High temperatures (≥26°C) inhibit both germination of *S. sclerotiorum* sclerotia and ability of *C. minitans* to reduce sclerotial germination in soil amended with *C. minitans* as MP (10^{11} cfu m^{-2}) or spore suspension (10^{8} cfu m^{-2}) (adapted from[47])

two weeks. This clearly demonstrated that integrated control of *S. sclerotiorum* was possible. Additional studies also showed that a fungicide tolerant isolate of *C. minitans* was not required because the mycoparasite was essentially protected from the fungicide in the soil and the fungicide targeted the ascospore stage of the life cycle of *S. sclerotiorum* on the foliage rather than the sclerotia in the soil. Nevertheless, stable fungicide tolerant isolates of *C. minitans* may be useful if the mycoparasite is applied to foliage in combination with fungicides.

Combinations of *C. minitans* with *Trichoderma* species for improved control of *S. sclerotiorum* have also been explored under standardized glasshouse trials in repeated lettuce crops.[41] Control was obtained with *C. minitans* but no additional control was achieved when combined with *T. virens*. This was despite the fact that this *T. virens* was originally isolated from a sclerotium of *S. minor*. These results were related to the temperature optima of the two mycoparasites. Separate laboratory tests involving combinations of the two mycoparasites showed that *C. minitans* was active below 20–25°C, effectively infecting and destroying sclerotia at temperatures characteristic of UK glasshouse soil temperatures. However, above 25°C, *C. minitans* became inactive and *T. virens* became more active and dominated the infection of sclerotia, suggesting that combinations of these fungi could be used to enhance control over wider environmental ranges. Recent experiments have repeated the glasshouse trials of [41] with *C. minitans* combined with *Trichoderma harzianum* isolates A6 and T22, both of which have activity against *S. sclerotiorum* in the laboratory (Couper and Whipps, unpublished data). Once again, under these UK glasshouse conditions, only treatments with *C. minitans* provided disease control and both *Trichoderma* isolates were ineffective. As with the experiments concerned with inoculum application timing discussed above (Section 9.2), understanding temperature optima, ecology and efficacy of BCAs is of key importance for achieving successful biocontrol activity.

Some interest has been shown in the concept of integrating partial soil sterilization (pasteurization) with a subsequent *C. minitans* application when the soil has cooled, for more sustainable control of *S. sclerotiorum*. Here, *C. minitans* would kill any sclerotia surviving the steaming process. Low-temperature steaming at 80°C carried out for 3 min, mimics the potential exposure of sclerotia to novel more rapid steaming procedures, and effectively kills all sclerotia of *S. sclerotiorum*.[52] Introduction of *C. minitans* into two steamed soils containing pasteurized sclerotia resulted in a more rapid colonization of pasteurized sclerotia by *C. minitans* than that found on non-pasteurized sclerotia in non-sterile soil[54] indicating that live sclerotia of *S. sclerotiorum* are more resistant to *C. minitans* infection than those weakened by pasteurization. *C. minitans* also increased in cfu following application to pasteurized soil whereas no proliferation occurred in non-pasteurized soil illustrating that *C. minitans* is subject to inhibition by microorganisms present in non-sterile soil. Application of *C. minitans* to pasteurized soil on the day of pasteurization resulted in greater colonization of pasteurized sclerotia by *C. minitans* than when the mycoparasite was introduced after 7 days, and may reflect that *C. minitans* cannot colonize sclerotia already occupied by other fungi although the possibility exists that the mycoparasite is simply masked by the presence of other fungi which colonized the sclerotia first.

Another area that could be worth exploring for integrated control is the combination of BCAs with soil fauna. *C. minitans* has been shown to be dispersed by collembolans, mites, sciarid larvae and slugs[23,55–59] allowing the possibility of improved dispersal than when applied in the absence of such soil fauna. Indeed, collembolans have been maintained for several years on agar or grain cultures of *C. minitans*[55](Whipps, unpublished data) and offer the opportunity to produce simultaneously a combined soil treatment comprising a BCA with dispersal agent. This approach deserves further investment of effort.

12.3.2. INTEGRATED CONTROL *OF ALLIUM* WHITE ROT WITH *TRICHODERMA VIRIDE*

Obtaining BCAs that reproducibly control *Allium* white rot (AWR) caused by *Sclerotium cepivorum* in the field has been notoriously difficult.[60] Following 5 years of screening involving sclerotial degradation assays and glasshouse based pot bioassays, two isolates of *Trichoderma viride*, S17a and L4, provided control of AWR in 2 years of field trials when applied as a fluid-drilled bran preparation at the same time as seeding.[1] Direct seed application of *T. viride* was not done in these trials but is currently being examined (Whipps unpublished). Stem base spray applications of *T. viride* conidia in the field trials failed to provide any control reflecting the need to get actively growing mycelium of this BCA to the zone of sclerotial germination and plant infection. Nevertheless, disease control was not complete and attempts were subsequently made to integrate these two BCAs with either the fungicide tebuconazole or onion waste compost (OWC).[61] OWC alone has been shown to provide reproducible partial control of AWR in both glasshouse bioassays and field trials.[62–64] Combination treatments of *T. viride* with either tebuconazole or OWC in glasshouse tests enhanced control in comparison with the individual treatments used alone and, in some cases, disease was almost eliminated.[61] However, in associated field trials, control of AWR by *T. viride* S17a was more variable than found in previous years and when combined with tebuconazole, a similar level of AWR control was achieved as with tebuconazole used alone. This may reflect different environmental conditions between field trials as efficacy of S17a and two other *Trichoderma* isolates was dependent on appropriate temperature and water potential of soils, with degradation of sclerotia occurring optimally between 10–25°C and at or above –0.00012 MPa.[65] Nevertheless, taken overall, these results still suggest that the use of *T. viride* is potentially compatible with tebuconazole, but additional work on fungicide compatibility with these BCAs is required.[61]

Interestingly, spent mushroom compost alone had no effect on AWR but when combined with *T. viride* S17a, significant control was obtained.[64] This suggests that it may be possible to enhance the efficacy of BCAs by combining them with compost or other organic matter that alone may have no effect. Whether this is due solely to provision of substrate for the BCA or that some more specific interaction is involved is unknown. Certainly, *Trichoderma koningii* TD_{22}, a known BCA of *S. minor* and *S. cepivorum,* was grown in a mixture of wood fiber waste compost:millet seed (80:20 w/w) and, when incorporated at 10–20% in soil, provided almost complete control of *S. minor* in glasshouse pot tests using lettuce.[66] Compost enriched with *T. koningii* TD_{22} for AWR control has now been developed in Tasmania (Dean Metcalf, personal communication).

12.4. Ecology of *Coniothyrium minitans* and Impact on Non-target Microorganisms

The relative paucity of ecological information concerning *C. minitans*[67] has gradually been addressed over the last 15 years and many new developments in this area have recently been reviewed[68] or mentioned in the sections above. Nevertheless, ecological studies with *C. minitans* are not straightforward as there is no selective isolation medium available and it grows very slowly in comparison with many other soil fungi, and so consequently, is frequently overgrown on soil dilution plates. Indeed, its presence in soil is often indicated by its recovery from sclerotia used as bait[26,69] but quantification under these circumstances is not possible. This limits the level of detection normally to circa 10^3–10^4 cfu cm^{-3} substratum using standard plating procedures.[34] Strains genetically marked with GUS (β-glucuronidase (*uid A*)) and hygromycin resistance (hygromycin phosphotransferase (*hph*)) are now available enabling more detailed work on survival, spread and establishment to be undertaken within the constraints of working with genetically modified microorganisms in the environment.[70] Using such a marked strain of *C. minitans*, it has been demonstrated that the mycoparasite infects a large proportion of sclerotia placed in a range of different soils in a short time from an initially low population.[49] These studies also showed for the first time that fungi colonizing sclerotia already infected by *C. minitans* normally mask the detection of *C. minitans* in sclerotia rather than displacing the mycoparasite. This could be viewed as an augmentation of the biocontrol effect of *C. minitans* by saprotrophic fungi killing sclerotia more rapidly.

There is good evidence that *C. minitans* survives well in soil for several years after application,[27,33,34] but is incapable of growth through raw soil.[71] This raises the question of the mechanisms of survival of this mycoparasite between hosts. The conidia are pigmented, presumably melanized, which aids survival of many fungal propagules[72] and so this may be one factor. Survival within infected sclerotia has been suggested as another survival mechanism following recovery of *C. minitans* from the rind of a disintegrated sclerotium after 15 months in soil.[73] Nevertheless, these studies were non-quantitative and it was unclear whether the mycoparasite survived as free conidia or conidia within pycnidia. Recent survival studies in soil demonstrated that *C. minitans* rapidly colonized sclerotia of *S. sclerotiorum*, producing pycnidia in the sclerotial cortex and conidial droplets on the rind surface.[74] The majority of sclerotial medulla had been converted to pycnidia by 30 days after inoculation, with the sclerotial rind remaining largely intact. The pycnidia and dried conidial droplets were still observed after 6 months and by 10 months circa 13% of conidia in dried droplets were still viable. This clearly demonstrated the potential for *C. minitans* infected sclerotia of *S. sclerotiorum* to act as reservoirs for the survival of *C. minitans* in soil.

The impact that large-scale introductions into the soil may have on the existing microbial population is another feature of the ecology of *C. minitans* that has been overlooked. It could be argued that because of its ecologically obligate lifestyle and lack of competitive ability in soil that its introduction to soil would have negligible effect and can be ignored. However, there is now greater concern about the impact of the introduction of BCAs into soil than previously, especially for BCAs that produce antibiotics.[75,76] In view of the recent finding that *C. minitans* can produce antimicrobial metabolites in culture[77–79] (although whether they are important in sclerotial infection is unknown), some studies to examine the impact of *C. minitans* on the soil microbial population have commenced. Initial studies examined the influence on culturable bacterial and fungal populations of introductions of *C. minitans* into three soil types at 10^3 and 10^6 cfu g^{-1} soil.[80] Over the 30 day period of the experiment, *C. minitans* survived at the level of introduction and neither application rate had an effect on bacterial numbers. There was a significant decrease in indigenous fungi at the higher rate of *C. minitans* application but this was small (0.1 $\log_{10}$ cfu g^{-1} soil). These studies used a wild-type *C. minitans* and monitored impact solely through culturable bacteria and fungi. Less than 5% of soil bacteria are culturable on normal lab media used for isolations,[81,82] so recently, experiments have commenced using a marked strain of *C. minitans* for increased sensitivity of detection and using direct DNA extraction from soil and PCR denaturing gradient gel electrophoresis (DGGE) for microbial fingerprinting of the 16S rRNA and 18S rRNA gene sequences for bacteria and fungi, respectively,[83] (Rogers and Whipps, unpublished data). Preliminary data indicate once again that *C. minitans* survives in soil for 6 months with little loss in cfu and that there is little impact on the microbial populations. As *C. minitans* does not grow in non-sterile soil it would seem unlikely that bioactive levels of antibiotic are released from *C. minitans* when conidia are simply applied to soil but this has not been determined chemically.

12.5. Pathogenicity Traits of *Coniothyrium minitans*

The use of molecular techniques to understand more about modes of action is a major advance in mycoparasitism studies. Particular emphasis has been placed on mycelium–mycelium interactions in vitro that allow both host and mycoparasite gene expression to be identified and quantified. Most work has been done with *Trichoderma*, which has the advantage that the genome of one species has been sequenced, that libraries of expressed sequence tags (ESTs) are available and experimental systems and tool-kits are well-established.[84–90] However, little work has been done to understand the molecular interactions between mycoparasites and sclerotia, which may be markedly different to mycelium–mycelium interactions. This has formed a focus of our work over the last few years (Challen, Muthumeenakshi, Rogers, Sreenivasaprasad, and Whipps, unpublished data). The recent sequencing of the genome of *S. sclerotiorum*[91] has considerably assisted this work.

Three approaches have been adopted. The first approach has been to isolate genes putatively involved in signaling and colonization from sequence information available in other pathogenic systems. Using degenerate PCR methods and a macroarrayed cosmid library of *C. minitans*, *pkaC*, *pmk1*, and *cmg1* genes have been obtained.

The second approach involved insertional mutagenesis of *C. minitans* using REMI and T-DNA tagging. Nine pathogenicity mutants were obtained from a panel of over 4000 transformants.[92] T-DNA tagging was also successfully used to obtain two sclerotial pathogenicity mutants in another isolate of *C. minitans* but these were not characterized further.[79] Amongst the genes identified using these techniques, we have found one with similarity to the *PIF1* helicase of *Neurospora crassa*, which may be important in energy generation, or repair and recombination of DNA. Molecular characterization and analysis of these pathogenicity mutants is now underway.

The third approach used suppression subtraction hybridization (SSH) between cDNA from *C. minitans* grown in culture and *C. minitans* colonizing sclerotia of *S. sclerotiorum.* A subtracted library of 672 clones containing cDNA fragments of putative upregulated genes was established. Sequencing

TABLE III. Putative functions and percentage distribution of unisequences derived from SSH between cDNA from *C. minitans* grown in culture and *C. minitans* colonizing sclerotia of *S. sclerotiorum*

Putative function	% of sequences
Metabolism	25
Energy	5
Transcription	3
Protein synthesis	6
Protein destination	4
Transport facilitation	7
Cell communication and signal transduction	2
Cell rescue, defense, death and aging	25
Cellular organization	1
Retro elements	1
Transmembrane proteins	1
Unknown/hypothetical	12
Poor or no hit	12

of these cDNA clones and bioinformatics analysis led to the putative identification of 251 ESTs and assignment of putative functions (Table III).

Dot blot and virtual northern analysis showed different levels of upregulation of various *C. minitans* genes during sclerotial colonization. Characterization of some potentially key genes has now begun and gene silencing and complementation studies to investigate their role in sclerotial parasitism have been initiated. Currently, results indicate a role for various genes associated with overcoming stress during the sclerotial mycoparasitism process by *C. minitans*. Having isolated genes associated with pathogenicity it may be possible to enhance activity of these genes by increasing copy number or overexpressing directly in *C. minitans*. Alternatively, these genes could be transferred to other mycoparasites such as *Trichoderma* which may exhibit different ecological attributes to enhance their effectiveness and, potentially, host range. These ecological attributes could include ability to grow through soil and compete saprotrophically with other microorganisms, and growth at higher temperature ranges which *C. minitans* is unable to do. Nevertheless, use of any genetically modified microorganism in the environment will need to be assessed for impact and safety and could restrict commercial development.

12.6. Conclusions and Future

Over the last 15 years considerable progress has been made with the use of *C. minitans* as a BCA of sclerotial pathogens, particularly *S. sclerotiorum*.

Several commercial products are now available for use in both the greenhouse and field. Much information has been generated concerning its ecology in relation to optimizing timing of application and quantities of mycoparasite preparation applied. However, understanding the basis of sclerotial mycoparasitism by *C. minitans* is still largely in its infancy but an excellent platform for research in this area has been laid. We hope that this will lead to new insights into fungal–fungal interactions in the future and identify genes that could be used to enhance biological disease control in general.

In addition, new developments in the integrated control of sclerotial pathogens have been made, especially the combination of *Trichoderma* with composts for control of AWR. This may lead to new commercial procedures for disease control, especially with pressure to reduce the amount of organic material going to landfill.

Acknowledgements

We would like to thank the following for financial support: the BBSRC, Defra, the EU (Projects: 2E-BCAs in crops; RECOVEG; CT95-0250; CT98-0083), and the UK Horticulture LINK Programme.

References

1. J. P. Clarkson, T. Payne, A. Mead, and J. M. Whipps, Selection of fungal biological control agents of *Sclerotium cepivorum* for control of white rot by sclerotial degradation in a UK soil, *Plant Pathol.* 51, 735–745 (2002).
2. P. B. Adams and W. A. Ayers, Ecology of *Sclerotinia* species, *Phytopathology* 69, 896–899 (1979).
3. J. R. Coley-Smith, C. M. Mitchell, and C. E. Sansford, Long-term survival of sclerotia of *Sclerotium cepivorum* and *Stromatinia gladioli*, *Plant Pathol.* 39, 58–69 (1990).
4. G. J. Boland and R. Hall, Index of plant hosts of *Sclerotinia sclerotiorum*, *Can. J. Plant Pathol.* 16, 93–108 (1994).
5. J. M. Whipps, Effects of mycoparasites on sclerotia-forming fungi, in *Biotic Interactions and Soil-borne Diseases*, edited by A. B. R. Beemster, G. J. Bollen, M. Gerlagh, M. A. Ruissen, B. Schippers, and A. Tempel (Elsevier, Amsterdam, 1991), pp. 129–140.
6. G. E. Harman, C. R. Howell, A. Viterbo, I. Chet, and M. Lorito, *Trichoderma* species—Opportunistic, avirulent plant symbionts, *Nature Rev. Microbiol.* 2, 43–56 (2004).
7. J. M. Whipps and K. G. Davies, Success in biological control of plant pathogens and nematodes by microorganisms, in *Biological Control: Measures of Success*, edited by G. Gurr and S. Wratten (Kluwer, Dordrecht, 2000), pp. 231–269.
8. APS Biological Control Committee, Commercial products available in the U.S.A. for use against plant pathogens, available at http://www.oardc.ohio-state.edu/apsbcc/ (2006).
9. K. A. El-Tarabily, M. H. Soliman, A. H. Nassar, H. A. Al-Hassani, K. Sivasithamparam, F. McKenna, and G. E. St. J. Hardy, Biological control of *Sclerotinia minor* using a chitinolytic bacterium and actinomycetes, *Plant Pathol.* 49, 573–583 (2000).

10. P. Hebbar, O. Berge, T. Heulin, and S. P. Singh, Bacterial antagonists of sunflower (*Helianthus annuus,* L) fungal pathogens, *Plant Soil* 133, 131–140 (1991).
11. T. J. McLoughlin, J. P. Quinn, A. Bettermann, and R. Bookland, *Pseudomonas cepacia* suppression of sunflower wilt fungus and role of antifungal compounds in controlling the disease, *Appl. Environ. Microbiol.* 58, 1760–1763 (1992).
12. C. Thaning, C. J. Welch, J. J. Borowicz, R. Hedman, and B. Gerhardson, Suppression of *Sclerotinia sclerotiorum* apothecial formation by the soil bacterium *Serratia plymuthica*: Identification of a chlorinated macrolide as one of the causal agents, *Soil Biol. Biochem.* 33, 1817–1826 (2001).
13. J. J. Levenfors, R. Hedman, C. Thaning, B. Gerhardson, and C. J. Welch, Broad-spectrum antifungal metabolites produced by the soil bacterium *Serratia plymuthica* A 153, *Soil Biol. Biochem.* 36, 677–685 (2004).
14. D. J. Hannusch and G. J. Boland, Interactions of air temperature, relative humidity and biological control agents on grey mold of bean, *Eur. J. Plant Pathol.* 102, 133–142 (1996).
15. D. J. Hannusch and G. J. Boland, Influence of air temperature and relative humidity on biological control of white mold of bean (*Sclerotinia sclerotiorum*), *Phytopathology* 86, 156–162 (1996).
16. H. C. Huang, E. G. Kokko, L. J. Yanke, and R. C. Phillippe, Bacterial suppression of basal pod rot and end rot of dry peas caused by *Sclerotinia sclerotiorum, Can. J. Microbiol.* 39, 227–233 (1993).
17. M. Kamensky, M. Ovadis, I. Chet, and L. Chernin, Soil-borne strain IC14 of *Serratia plymuthica* with multiple mechanisms of antifungal activity provides biocontrol of *Botrytis cinerea* and *Sclerotinia sclerotiorum* diseases, *Soil Biol. Biochem.* 35, 323–331 (2003).
18. G. Q. Li, H. C. Huang, and S. N. Acharya, Antagonism and biocontrol potential of *Ulocladium atrum* on *Sclerotinia sclerotiorum, Biol. Control* 28, 11–18 (2003).
19. R. D. Reeleder, The use of yeasts for biological control of the plant pathogen *Sclerotinia sclerotiorum, Biocontrol* 49, 583–594 (2004).
20. S. Savchuk and W. G. D. Fernando, Effect of timing of application and population dynamics on the degree of biological control of *Sclerotinia sclerotiorum* by bacterial antagonists, *FEMS Microbiol. Ecol.* 49, 379–388 (2004).
21. J. P. Clarkson and J. M. Whipps, Control of sclerotial pathogens in horticulture. *Pesticide Outlook* June 2002, 97–101 (2002).
22. J. M. Whipps, Developments in the biological control of soil-borne plant pathogens, *Adv. Bot. Res.* 26 1–134 (1997).
23. G. J. Turner and H. T. Tribe, *Coniothyrium minitans* and its parasitism of *Sclerotinia* species, *Trans. Brit. Mycol. Soc.* 66, 97–105 (1976).
24. W. A. Campbell, A new species of *Coniothyrium* parasitic on sclerotia, *Mycologia* 39, 190–195 (1947).
25. C. Monaco, Evaluation of the efficiency of mycoparasites on *Sclerotinia sclerotiorum* "in vitro," *Rev. Fac. Agronomia* 65, 67–73 (1989).
26. C. Sandys-Winsch, J. M. Whipps, M. Gerlagh, and M. Kruse, World distribution of the sclerotial mycoparasite *Coniothyrium minitans, Mycol. Res.* 97, 1175–1178 (1993).
27. S. P. Budge and J. M. Whipps, Glasshouse trials of *Coniothyrium minitans* and *Trichoderma* species for the biological control of *Sclerotinia sclerotiorum* in celery and lettuce, *Plant Pathol.* 40, 59–66 (1991).
28. H. C. Huang and J. A. Hoes, Importance of plant spacing and sclerotial position to development of Sclerotinia wilt of sunflower, *Plant Dis.* 64, 81–84 (1980).

29. M. Gerlagh, H. M. Goossen-van de Geijn, N. J. Fokkema, and P. F. G. Vereijken, Long-term biosanitation by application of *Coniothyrium minitans* on *Sclerotinia sclerotiorum* infected crops, *Phytopathology* 89, 141–147 (1999).
30. P. Lüth, The control of *Sclerotinia* spp. and *Sclerotium cepivorum* with the biological fungicide Contans® WG—experiences from field trials and commercial use, in *Proceedings of the XI International Sclerotinia Workshop,* York, England, July 8–12, 2001, edited by C. S. Young and K. J. D. Hughes (Central Science Laboratory, York, 2001), pp. 37–38.
31. D. L. McLaren, H. C. Huang, G. C. Kozub, and S. R. Rimmer, Biological control of Sclerotinia wilt of sunflower with *Talaromyces flavus* and *Coniothyrium minitans, Plant Dis.* 78, 231–235 (1994).
32. A. H. M. Ahmed and H. T. Tribe, Biological control of white rot of onion (*Sclerotium cepivorum*) by *Coniothyrium minitans, Plant Pathol.* 26, 75–78 (1977).
33. M. P. McQuilken, S. J. Mitchell, S. P. Budge, J. M. Whipps, J. S. Fenlon, and S. A. Archer, Effect of *Coniothyrium minitans* on sclerotial survival and apothecial production of *Sclerotinia sclerotiorum* in field-grown oilseed rape, *Plant Pathol.* 44, 883–896 (1995).
34. E. E. Jones and J. M. Whipps, Effect of inoculum rates and sources of *Coniothyrium minitans* on control of *Sclerotinia sclerotiorum* disease in glasshouse lettuce, *Eur. J. Plant Pathol.* 108, 527–538 (2002).
35. H. C. Huang and G. C. Kozub, Monocropping to sunflower and decline of Sclerotinia wilt, *Bot. Bull. Acad. Sinica* 32, 163–170 (1991).
36. M. Gerlagh, H. M. Goossen-van de Geijn, A. E. Hoogland, and P. F. G. Vereijken, Quantitative aspects of infection of *Sclerotinia sclerotiorum* sclerotia by *Coniothyrium minitans*—Timing of application, concentration and quality of conidial suspension of the mycoparasite, *Eur. J. Plant Pathol.* 109, 489–502 (2003).
37. M. Gerlagh, H. M. Goossen-van de Geijn, A. E. Hoogland, P. F. G. Vereijken, P. F. M. Horsten, and B. H. de Haas, Effect of volume and concentration of conidial suspensions of *Coniothyrium minitans* on infection of *Sclerotinia sclerotiorum* sclerotia, *Biocontrol Sci. Technol.* 14, 675–690 (2004).
38. H. C. Huang, E. Bremer, R. K. Hynes, and R. S. Erickson, Foliar application of fungal biocontrol agents for the control of white mold of dry bean caused by *Sclerotinia sclerotiorum, Biol. Control* 18, 270–276 (2000).
39. G. Q. Li, H. C. Huang, and S. N. Acharya, Importance of pollen and senescent petals in the suppression of alfalfa blossom blight (*Sclerotinia sclerotiorum*) by *Coniothyrium minitans, Biocontrol Sci. Technol.* 13, 495–505 (2003).
40. P. Trutmann, P. J. Keane, and P. R. Merriman, Biological control of *Sclerotinia sclerotiorum* on aerial parts of plants by the hyperparasite *Coniothyrium minitans, Trans. Brit. Mycol. Soc.* 78, 521–529 (1982).
41. S. P. Budge, M. P. McQuilken, J. S. Fenlon, and J. M. Whipps, Use of *Coniothyrium minitans* and *Gliocladium virens* for biological control of *Sclerotinia sclerotiorum* in glasshouse lettuce, *Biol. Control* 5, 513–522 (1995).
42. M. P. McQuilken, S. P. Budge, and J. M. Whipps, Production, survival and evaluation of liquid culture-produced inocula of *Coniothyrium minitans* against *Sclerotinia sclerotiorum, Biocontrol Sci. Technol.* 7, 23–36 (1997).
43. T. de Vrije, N. Antoine, R. M. Buitelaar, S. Bruckner, M. Dissevelt, A. Durand, M. Gerlagh, E. E. Jones, P. Lüth, J. Oostra, W. J. Ravensberg, R. Renaud, A. Rinzema, F. J. Weber, and J. M. Whipps, The fungal biocontrol agent *Coniothyrium minitans*: production by solid-state fermentation, application and marketing, *Appl. Microbiol. Biotechnol.* 56, 58–68 (2001).

44. E. E. Jones, A. Mead, and J. M. Whipps, Evaluation of different *Coniothyrium minitans* inoculum sources and application rates on apothecial production and infection of *Sclerotinia sclerotiorum* sclerotia, *Soil Biol. Biochem.* 35, 409–419 (2003).
45. E. E. Jones and A. Stewart, Selection of mycoparasites of sclerotia of *Sclerotinia sclerotiorum* isolated from New Zealand soils, *N. Z. J. Crop Hort. Sci.* 28, 105–114 (2000).
46. E. E. Jones, A. Mead, and J. M. Whipps, Effect of inoculum type and timing of application of *Coniothyrium minitans* on *Sclerotinia sclerotiorum*: control of sclerotinia disease in glasshouse lettuce. *Plant Pathol.* 53, 611–620 (2004).
47. E. E. Jones, J. P. Clarkson, A. Mead, and J. M. Whipps, Effect of inoculum type and timing of application of *Coniothyrium minitans* on *Sclerotinia sclerotiorum*: Influence on apothecial production, *Plant Pathol.* 53, 621–628 (2004).
48. M. P. McQuilken, S. P. Budge, and J. M. Whipps, Effects of culture media and environmental factors on conidial germination, pycnidial production and hyphal extension of *Coniothyrium minitans, Mycol. Res.* 101, 11–17 (1997).
49. E. E. Jones, A. Stewart, and J. M. Whipps, Use of *Coniothyrium minitans* transformed with the hygromycin B resistance gene to study survival and infection of *Sclerotinia sclerotiorum* sclerotia in soil, *Mycol. Res.* 107, 267–276 (2003).
50. J. M. Whipps, Microbial interactions and biocontrol in the rhizosphere, *J. Exp. Bot.* 52, 487–511 (2001).
51. R. Noble and E. Coventry, Suppression of soil-borne plant diseases with composts: A review, *Biocontrol Sci. Technol.* 15, 3–20 (2005).
52. M. C. A. van Loenen, Y. Turbett, C. E. Mullins, N. E. H. Feilden, M. J. Wilson, C. Leifert, and W. E. Seel, Low temperature-short duration steaming of soil kills soil-borne pathogens, nematode pests and weeds, *Eur. J. Plant Pathol.* 109, 993–1002 (2003).
53. S. P. Budge and J. M. Whipps, Potential for integrated control of *Sclerotinia sclerotiorum* in glasshouse lettuce using *Coniothyrium minitans* and reduced fungicide application, *Phytopathology* 91, 221–227 (2001).
54. A. J. Bennett, C. Leifert, and J. M. Whipps, Effect of combined treatment of pasteurisation and *Coniothyrium minitans* on sclerotia of *Sclerotinia sclerotiorum* in soil, *Eur. J. Plant Pathol.* 113, 197–209 (2005).
55. R. H. Williams, J. M. Whipps, and R. C. Cooke, Role of soil mesofauna in dispersal of *Coniothyrium minitans*: Transmission to sclerotia of *Sclerotinia sclerotiorum, Soil Biol. Biochem.* 30, 1929–1935 (1998).
56. R. H. Williams, J. M. Whipps, and R. C. Cooke, Role of soil mesofauna in dispersal of *Coniothyrium minitans*: Mechanisms of transmission, *Soil Biol. Biochem.* 30, 1937–1945 (1998).
57. P. Trutmann, P. J. Keane, and P. R. Merriman, Reduction of sclerotial inoculum of *Sclerotinia sclerotiorum* with *Coniothyrium minitans, Soil Biol. Biochem.* 12, 461–465 (1980).
58. J. M. Whipps, Growth of the collembolan *Folsomia candida* on cultures of the mycoparasite *Coniothyrium minitans* and sclerotia of *Sclerotinia sclerotiorum, Mycol. Res.* 97, 1277–1280 (1993).
59. J. M. Whipps and S. P. Budge, Transmission of the mycoparasite *Coniothyrium minitans* by collembolan *Folsomia candida* (Collembola, Entomobryidae) and glasshouse sciarid-*Bradysia* sp. (Diptera, Sciaridae), *Ann. Appl. Biol.* 123, 165–171 (1993).
60. S. J. Kay and A. Stewart, Evaluation of fungal antagonists for control of onion white-rot in soil box trials, *Plant Pathol.* 43, 371–377 (1994).
61. J. P. Clarkson, A. Scruby, A. Mead, C. Wright, B. Smith, and J. M. Whipps, Integrated control of *Allium* white rot with *Trichoderma viride*, tebuconazole and composted onion waste, *Plant Pathol.* 55, 375–386 (2006).

62. E. Coventry, R. Noble, A. Mead, and J. M. Whipps, Control of *Allium* white rot (*Sclerotium cepivorum*) with composted onion waste, *Soil Biol. Biochem.* 34, 1037–1045 (2002).
63. E. Coventry, R. Noble, A. Mead, and J. M. Whipps, Suppression of *Allium* white rot (*Sclerotium cepivorum*) in different soils using vegetable wastes, *Eur. J. Plant Pathol.* 111, 101–112 (2005).
64. E. Coventry, R. Noble, A. Mead, F. R. Marin, J. A. Perez, and J. M. Whipps, *Allium* white rot suppression with composts and *Trichoderma viride* in relation to sclerotia viability, *Phytopathology* 96, 1009–1020 (2006).
65. J. P. Clarkson, A. Mead, T. Payne, and J. M. Whipps, Effect of environmental factors and *Sclerotium cepivorum* isolate on sclerotial degradation and biological control of white rot by *Trichoderma, Plant Pathol.* 53, 353–362 (2004).
66. Y. Ramona and M. A. Line, Potential for the large-scale production of a biocontrol fungus—In raw and composted paper mill waste, *Compost Sci. Utiliz.* 10, 57–62 (2002).
67. J. M. Whipps and M. Gerlagh, Biology of *Coniothyrium minitans* and its potential for use in disease biocontrol, *Mycol. Res.* 96, 897–907 (1992).
68. J. M. Whipps, Ecological and biotechnological considerations in enhancing disease biocontrol, in *Enhancing Biocontrol Agents and Handling Risks,* edited by M. Vurro, J. Gressel, T. Butt, G. E. Harman, A. Pilgeram, R. J. St. Leger, D. L. Nuss (IOS Press, Ohmsha, 2001), pp. 43–51.
69. H. C. Huang, Distribution of *Coniothyrium minitans* in Manitoba sunflower fields, *Can. J. Plant Pathol.* 3, 219–222 (1981).
70. E. E. Jones, M. Carpenter, D. Fong, A. Goldstein, A. Thrush, A. Crowhurst, and A. Stewart, Co-transformation of the sclerotial mycoparasite *Coniothyrium minitans* with hygromycin B resistance and β-glucuronidase markers, *Mycol. Res.* 103, 929–937 (1999).
71. R. H. Williams, Dispersal of the mycoparasite *Coniothyrium minitans,* PhD thesis (Department of Animal and Plant Sciences, University of Sheffield, Sheffield, 1996), p. 144.
72. M. J. Butler and A. W. Day, Fungal melanins: A review, *Can. J. Microbiol.* 44, 1115–1136 (1998).
73. H. T. Tribe, On the parasitism of *Sclerotinia trifoliorum* by *Coniothyrium minitans, Trans. Brit. Mycol. Soc.* 40, 489–499 (1957).
74. A. J. Bennett, C. Leifert, and J. M. Whipps, Survival of *Coniothyrium minitans* associated with sclerotia of *Sclerotinia sclerotiorum* in soil, *Soil Biol. Biochem.* 38, 164–172 (2006).
75. T. A. Brimner and G. J. Boland, A review of the non-target effects of fungi used to biologically control plant diseases, *Agr. Ecosyst. Environ.*100, 3–16 (2003).
76. A. Winding, S. J. Binnerup, and H. Pritchard, Non-target effects of bacterial biological control agents suppressing root pathogenic fungi, *FEMS Microbiol. Ecol.* 47, 129–141 (2004).
77. M. P. McQuilken, J. Gemmell, and J. M. Whipps, Some nutritional factors affecting production of biomass and antifungal metabolites of *Coniothyrium minitans, Biocontrol Sci. Technol.* 12, 443–454 (2002).
78. M. P. McQuilken, J. Gemmell, R. A. Hill, and J. M. Whipps, Production of macrosphelide A by the mycoparasite *Coniothyrium minitans, FEMS Microbiol. Lett.* 219, 27–31 (2003).
79. M. Li, X. Gong, J. Zheng, D. Jiang, Y. Fu, and M. Hou, Transformation of *Coniothyrium minitans*, a parasite of *Sclerotinia sclerotiorum*, with *Agrobacterium tumefaciens, FEMS Microbiol. Lett.* 243, 323–329 (2005).
80. A. J. Bennett, C. Leifert, and J. M. Whipps, Survival of the biocontrol agents *Coniothyrium minitans* and *Bacillus subtilis* MBI 600 introduced into pasteurised, sterilised and non-sterile soils, *Soil Biol. Biochem.* 35, 1565–1573 (2003).

81. P. Garbeva, J. A. van Veen, and J. D. van Elsas, Microbial diversity in soil: Selection of microbial populations by plant and soil type and implications for disease suppressiveness, *Annu. Rev. Phytopathol.* 42, 243–270 (2004).
82. M. Mazzola, Assessment and management of soil microbial community structure for disease suppression, *Annu. Rev. Phytopathol.* 42, 35–59 (2004).
83. P. J. Hunter, G. M. Petch, A. A. Calvo-Bado, T. R. Pettitt, N. Parsons, J. A. W. Morgan, and J. M. Whipps, Microbial characteristics of peats associated with suppression of damping-off disease caused by *Pythium sylvaticum*, *Appl. Environ. Microbiol.* 72, 6452–6460 (2006).
84. A. Mendoza-Mendoza, M. J. Pozo, D. Grzegorski, P. Martínez, J. M. García, V. Olmedo-Monfil, C. Cortés, C. Kenerley, and A. Herrera-Estrella, Enhanced biocontrol activity of *Trichoderma* through inactivation of a mitogen-activated protein kinase, *Proc. Natl. Acad. Sci. USA* 100, 15965–15970 (2003).
85. J. M. Steyaert, H. J. Ridgway, Y. Elad, and A. Stewart, Genetic basis of mycoparasitism: a mechanism of biological control by species of *Trichoderma*, *N. Z. J. Crop Hort. Sci.* 31, 281–291 (2003).
86. J. M. Steyaert, A. Stewart, M. Jaspers, M. Carpenter, and H. J. Ridgway, Co-expression of two genes, a chitinase (chit42) and proteinase (prb1), implicated in mycoparasitism by *Trichoderma hamatum*, *Mycologia* 96, 1245–1252 (2004).
87. P. K. Mukherjee, J. Latha, R. Hadar, and B. A. Horwitz, Role of two G-protein alpha subunits, TgaA and TgaB, in the antagonism of plant pathogens by *Trichoderma virens*, *Appl. Environ. Microbiol.* 70, 542–549 (2004).
88. M. A. Carpenter, A. Stewart, and H. J. Ridgway, Identification of novel *Trichoderma hamatum* genes expressed during mycoparasitism using subtractive hybridization, *FEMS Microbiol. Lett.* 251, 105–112 (2005).
89. P. G. Liu and Q. Yang, Identification of genes with a biocontrol function in *Trichoderma harzianum* mycelium using the expressed sequence tag approach, *Res. Microbiol.* 156, 416–423 (2005).
90. Trichoest, available at www.trichoderma.org (2006).
91. Broad Institute, available at www.broad.mit.edu/annotation/fungi/sclerotinia_sclerotiorum (2006).
92. C. W. Rogers, M. P. Challen, J. R. Green, and J. M. Whipps, Use of REMI and *Agrobacterium*-mediated transformation to identify pathogenicity mutants of the biocontrol fungus, *Coniothyrium minitans*, *FEMS Microbiol. Lett.* 241, 207–214 (2004).

13. BIOLOGICAL CONTROLS AND THE POTENTIAL OF BIOTECHNOLOGICAL CONTROLS FOR VERTEBRATE PEST SPECIES

Peter Kerr*
CSIRO Entomology, GPO Box 1700, Canberra, ACT, 2601, Australia

Abstract. The introduction of myxoma virus into Australia to control the European rabbit is the classical example of biological control for a vertebrate pest species. The subsequent selection for reduction in virulence of myxoma virus strains and the increased resistance to myxoma virus of the new host is one of the paradigms for infectious disease biology. More recently, rabbit hemorrhagic disease virus has also been successfully introduced into Australia as a second biological control agent for rabbits and has been highly effective in the arid and semi-arid parts of the continent but less so in the higher rainfall zones. The use of biotechnology for vertebrate pest control has been explored through projects to develop virally vectored immunocontraceptives for rabbits, foxes, and mice, and although much progress has been made, it must be concluded that there is still a large gap between what biotechnology can deliver and what is needed for successful biological control of vertebrate pest species. The release of any biological control agent whether a naturally occurring virus or a genetically engineered organism requires very careful evaluation of the risks and benefits and two examples of this process are discussed.

Keywords: biological control, myxomatosis, RHDV, immunocontraception

13.1. Introduction

One of the best known examples of biological control was the introduction of the cactoblastis moth (*Cactoblastis cactorum*) to successfully control the introduced cactus prickly pear in Australia. Biological control can however go badly wrong as occurred when the cane toad (*Bufo marinus*) was introduced into Australia in 1935 as a control for cane beetles. Cane toads were not interested in cane beetles but have become a major pest in their own right and spread throughout northern Australia. The use of biological control for vertebrate species is much less common; there are only three successful examples and these all involve the use of viruses.

---

*To whom correspondence should be addressed, e-mail: peter.kerr@csiro.au

M. Vurro and J. Gressel (eds.), Novel Biotechnologies for Biocontrol Agent Enhancement and Management, 243–265.

The first well documented attempt at biological control of a vertebrate pest population was the use of the bacterium that causes chicken cholera (*Pasturella multocida*) by Louis Pasteur to control a population of European rabbits on the Pommeroy estate at Reims where the burrowing of the rabbits threatened to undermine the famous champagne cellars. Pasteur was interested in the problem of rabbit control and hoped to use the chicken cholera bacteria as a biological control for rabbits in Australia for which a large cash reward had been offered.[1,2] Despite the best attempts of Pasteur to promote the use of the chicken cholera agent as a biological control for rabbits it proved to not be transmitted from rabbit to rabbit and was not used in Australia, nor did Pasteur collect the prize. Although chicken cholera was not successful as a biological control for rabbits, it is with rabbits in Australia that the two most successful examples of biological control of vertebrates have been achieved.

Initially, biological control of rabbits in Australia was undertaken with the release of myxoma virus in 1950. This virus, which was native to South America, spread throughout Australia and was subsequently released in Europe and Britain. The release of myxoma virus and its subsequent co-evolution with the European rabbit provides a fascinating example of both the potential and the limitations of biological control.[1] A second and more recent use of a virus for biological control of rabbits was the release of rabbit hemorrhagic disease virus in Australia in 1995.[3] The only other successful use of biological control for a vertebrate pest population was the release of feline panleukopaenia virus into a feral cat population on Marion Island in the Indian Ocean. In this unusual case the cat population had been established from a small founder population that had not introduced this normally common feline virus.[4]

The advent of molecular biology has opened up a Pandora's box of potential for biological control using genetic engineering to produce recombinant viruses that could deliver antigens that decrease the fertility of the host by immunocontraception, or to interfere with development or even genetically engineer the pest species itself to be "daughterless" such that only males are produced. These novel possibilities of biotechnology also have the potential to create new problems and so the risks of biotechnology for vertebrate pest control need to be carefully assessed and managed. It is also possible that the potential for biotechnological solutions has been oversold for the present.

This paper will focus on the two highly successful examples of biological control for rabbits, myxomatosis and rabbit hemorrhagic disease in Australia and the limitations to biological control. It will then examine some of the biotechnological concepts for the control of vertebrate pests including virally-vectored immunocontraception, inhibition of cane toad (*Bufo marinus*) development and the possibilities of daughterless technology for control of carp (*Cyprinus carpio*). Finally, the potential risks and management of these risks will be briefly examined.

13.2. Myxomatosis as a Biological Control for the European Rabbit in Australia

13.2.1. THE EUROPEAN RABBIT IN AUSTRALIA

The European rabbit was introduced into Australia with European settlement in 1788. However, the spread of the rabbit and its history as Australia's pre-eminent vertebrate pest species dates from an introduction of wild rabbits in 1859.[2] From an initial focus in southern Australia the rabbit had by 1910 occupied virtually all of the non-tropical parts of the continent. Rabbits caused massive ecological disruption, competed with stock for grazing and provided a food resource for introduced predators such as the cat (*Felis catus*) and European red fox (*Vulpes vulpes*), which were responsible for the loss of small native marsupial species through predation.[5–7]

13.2.2. NATURAL HISTORY OF MYXOMA VIRUS

Myxoma virus is a poxvirus, a member of the genus *Leporipoxvirus*, and native to the South American tapeti or jungle rabbit (*Sylvilagus brasiliensis*) in which it causes an innocuous cutaneous fibroma at the inoculation site. The virus is spread on the mouthparts of biting arthropods such as mosquitoes or fleas when they probe through the skin seeking a blood meal, but it does not replicate in these vectors. A closely related virus is found in the brush rabbit (*Sylvilagus bachmani*) of California and the Baja peninsula. It is benign in its natural host, but introduced into European rabbits, myxoma virus causes the lethal fulminant disease called myxomatosis. This disease is characterized by a cutaneous primary lesion at the site of inoculation, with virus spread to the lymph nodes and then to other tissues such as the testes and mucocutaneous sites such as eyelids and nostrils and the skin of the ears, face and body where secondary swellings occur. The primary and secondary lesions have high titers of virus and this virus can be transferred to mosquitoes and fleas when they probe through the lesion. Death typically occurs within 10–14 days of inoculation with a small dose of virulent virus.[1]

13.2.3. MYXOMATOSIS IN AUSTRALIA

The potential of myxoma virus as a biological control for rabbits in Australia was recognized as early as 1919[8] but it was not until 1950 that the virus was successfully released into the Australian rabbit population, following its spread from an experimental site in northern Victoria driven by an irruption of predominantly *Culex annulirostris* mosquitoes, but also *Anopheles annulipes*, that were in plague numbers in the Murray–Darling river system that summer.

The virus was initially highly successful with some estimates suggesting that the rabbit population dropped by as much as 95%. The virus released killed over 99% of infected rabbits.[8] However, by the second summer epidemic in 1951/1952 rabbits were found in the field with serum antibodies to myxoma virus indicating that some rabbits had survived infection.

13.2.4. CO-EVOLUTION OF MYXOMA VIRUS AND RABBITS

Careful studies were undertaken to measure the virulence of field isolates of myxoma virus. Isolates were classified into five virulence grades based on the average survival time and proportion of rabbits killed when inoculated with a small dose of the virus. Grade 1 viruses killed 100% of rabbits with an average survival time of <13 days; Grade 2 viruses killed 95–99% with an average survival time of 13–16 days; Grade 3 viruses killed 70–95% with an average survival time of 17–28 days; Grade 4 viruses killed 50–70% with an average survival time of 29–50 days; Grade 5 viruses killed <50% of inoculated rabbits.[9] From the mid-1950's it was clear that the predominant viruses isolated from the field were of Grade 3 virulence (Figure 1). The percentage of myxoma virus isolates made from each virulence grade from 1 to 5 is indicated. The number of virus isolates tested in each period is shown above the bars and the virulence grade is indicated against each bar. The data in Figure 1 are from Fenner.[59]

These attenuated viruses were selected in the field because they were more efficiently transmitted by the mosquito vector than the highly virulent

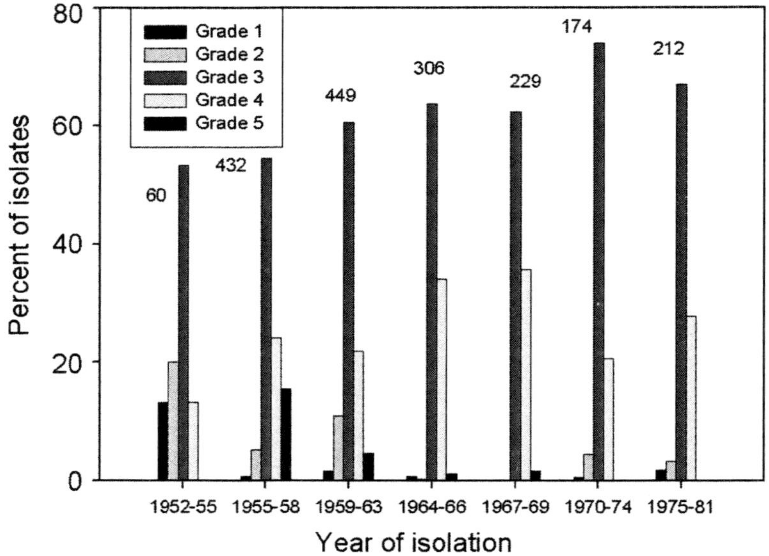

Figure 1. Virulence of myxoma virus isolates in Australia from 1952 to 1981

virus strains. Virus particles adhere to the mouthparts of the mosquito when it probes through a skin lesion. The virus titer in the lesion needs to be $\geq 10^7$ plaque forming units (pfu)/g of tissue for high infectivity. Highly lethal Grade 1 strains of virus reached this titer in the skin at the inoculation site 4 to 5 days after infection but the rabbits were dead within another 4 to 5 days leaving little time for transmission. The more attenuated strains reached similar titers of virus in the skin but allowed the rabbits to survive for much longer in an infectious state and thus had a higher probability of transmission. The relative rarity of highly attenuated Grade 5 strains of virus in the field was explained by their not reaching transmission threshold values in the skin.[10]

It is also likely that the very high summer temperatures in the inland regions of Australia contributed to rabbit survival particularly when the rabbits were infected with attenuated strains of virus.[11] Thus more rabbits were surviving myxomatosis but this was not just due to the attenuation of the virus. Rabbits were also being subjected to strong selection for resistance to myxomatosis.[12,13] This was clearly demonstrated at a study site at Lake Urana in New South Wales. Each year rabbit kittens born that year were trapped at this site prior to the annual epidemic of myxomatosis and reared in captivity. They were then challenged with a standard isolate of myxoma virus that initially killed 88% of challenged rabbits (Figure 2). Within 3 years the same virus was only lethal in approximately 50% of the challenged rabbits. Seven years after the introduction of myxoma virus to Australia the challenge virus killed only 26% of rabbits from Lake Urana. In addition to the high proportion of rabbits that survived infection, 30% of rabbits showed only mild clinical signs following infection compared to 0–2% following the initial epidemics.

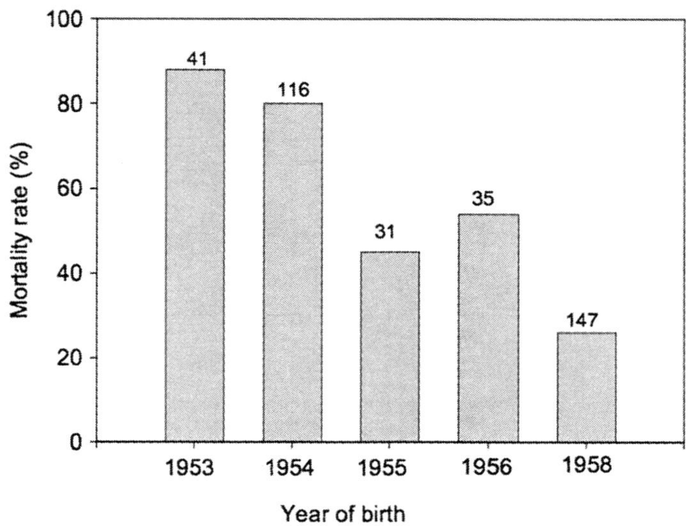

Figure 2. Evolution of genetic resistance to myxoma virus at Lake Urana, New South Wales

Overall this trial indicated that there had been very rapid selection for resistance to myxomatosis.

The mortality rates of rabbits from Lake Urana challenged with a grade 3 virulence strain of myxoma virus are plotted against the year of birth of the rabbits. This corresponds to 2, 3, 4, 5 or 7 epidemics of myxomatosis at Lake Urana. The number of rabbits tested from each year is shown above the histograms in Figure 2. Data are taken from Marshall and Fenner[12] and Marshall and Douglas[13].

The process of co-evolution of myxoma virus with the European rabbit in Australia is one of the best documented natural experiments in which a pathogen was introduced to a new host. Spread of the pathogen occurred on a continental scale but unlike in South America or California, there was no natural host of the virus to provide a source of re-introductions of virulent strains to the rabbit population. Highly virulent strains of the virus were out-competed by more attenuated strains that had a transmission advantage. At the same time the rabbit population underwent a massive selection for resistance to myxomatosis. As a result of this co-evolution, within 10 years of release myxoma virus was regarded as far less effective as a biological control agent.

13.2.5. INCREASING THE EFFECTIVENESS OF MYXOMATOSIS

Several approaches were undertaken to maximize the effectiveness of myxomatosis. Two new insect vectors were introduced into Australia. The first was the European rabbit flea (*Spilopsyllus cuniculi*) introduced in 1968. This provided a vector that was present in rabbit populations all year and thus could spread myxoma virus outside the spring/summer mosquito driven epidemics in temperate Australia. The second was the Spanish rabbit flea (*Xenopsylla cunicularis*) introduced in 1993 to provide a vector for myxoma virus that would be active in the arid parts of Australia. In addition, a highly virulent strain of myxoma virus known as the Lausanne strain (from its origin in the Lausanne culture collection) isolated in Brazil in 1949 was used for release on fleas or by direct inoculation of rabbits. Although this virus was more virulent than the originally introduced Standard Laboratory strain, it did not become established in the field in Australia presumably because it was out competed by better adapted field strains of the virus.[14] The Lausanne strain was released in France in 1952 and became established throughout Europe and Britain, but there it was not competing with previously established field strains of myxoma virus.

13.2.6. THE CURRENT SITUATION IN AUSTRALIA

Co-evolution has continued in the rabbit population in Australia and this has led to the re-emergence of highly virulent strains of myxoma virus in the field, highly virulent when assayed in unselected laboratory rabbits. When these strains were assayed in wild rabbits they had an attenuated phenotype with relatively long survival times of infected rabbits, as would be expected if selected for maximum transmissibility.[15]

Myxomatosis is now endemic in the Australian rabbit population with epidemics occurring in spring and summer associated with the entry of susceptible kittens into the populations. Deliberate release of highly virulent myxoma virus for rabbit control is rarely used, as the released virus tends to be out competed by successful field strains. It is difficult to estimate the proportion of rabbits killed by myxomatosis, which will undoubtedly vary from epidemic to epidemic. Estimates range from 30 to 60% indicating that the virus still provides a partial suppression of the rabbit population.

13.2.7. MYXOMA VIRUS IN EUROPE

Following its successful introduction into Australia, myxoma virus was released in France in 1952. The Lausanne strain released spread from its point release site across the range of the European rabbit and in the summer of 1953 was also illegally released into Britain. Myxoma virus is now endemic throughout Europe and Britain. Similarly to the situation in Australia, attenuated strains of virus emerged in Europe and Britain within a year or two of release although Grade 1 viruses were much more prevalent in the field for much longer than in Australia. Resistance to myxomatosis has also emerged in rabbits in Britain and Europe but again seems to have been slower to evolve than in Australia. This may reflect the fact that fewer studies were undertaken or possibly that the persistence of highly virulent strains of virus and cooler weather made survival from infection less likely and slowed selection for resistance.[16]

The European rabbit in Europe is not a pest species except in particular locations or industries such as forestry. It is a valued game animal for hunting as well as an essential component of many ecosystems. Thus the release of myxoma virus in Europe can be seen as an undesirable consequence of its release in Australia.

13.3. Rabbit Hemorrhagic Disease Virus as a Biological Control for the Rabbit in Australia

13.3.1. NATURAL HISTORY OF RABBIT HEMORRHAGIC DISEASE VIRUS

Rabbit hemorrhagic disease virus (RHDV) is a member of the *Caliciviridae* family (genus *Lagovirus*). It is a small un-enveloped virus approximately 30 nm in diameter with a single strand positive sense RNA genome of 7,437 nucleotides. Rabbit hemorrhagic disease first emerged in China in 1984 in domestic European rabbits recently imported from Germany.[3] A similar disease was recognized in Italy in 1986 and by 1988 the disease had spread over most of Europe and been introduced into many other countries. A related but distinct virus termed European brown hare syndrome virus (EBHSV) is found in hares where it causes a similar disease to rabbit hemorrhagic disease. More recently a non-pathogenic virus called rabbit calicivirus (RCV) has been identified in domestic rabbits in Italy.[17] This virus cross-protects rabbits from RHDV.

Serological and molecular evidence suggests that attenuated forms of RHDV have existed in European and British rabbit populations for over 50 years and that the virulent form of the virus probably emerged in Europe.[18,19] The epidemiology of the virus appears to be complicated by the occurrence of apparently avirulent forms of RHDV that are distinct from RCV[18,19] and also viruses that cross-react antigenically with RHDV but do not cross protect like RCV.[20–22] It has been suggested that this could be explained by an avirulent transmission model that creates long-term carriers.[23]

13.3.2. PATHOGENESIS OF RABBIT HEMORRHAGIC DISEASE

Rabbit hemorrhagic disease virus induces a rapidly lethal systemic disease with death occurring in 24–72 h after infection of adult rabbits depending on the route of infection. The key target organ is the liver but other tissues such as spleen, heart, lungs and kidneys may also be grossly affected with obvious hemorrhages. Interestingly, very young rabbits do not develop lethal disease following infection with RHDV and are immune to subsequent infection. This has important implications for the epidemiology of the virus and for its effectiveness for biological control. The virus is probably spread by ingestion, nasal inoculation, conjunctival contamination or potentially on the mouthparts of biting arthropods. Very high levels of virus are present in liver of infected rabbits and virus is also present in urine, faeces, saliva and other secretions.[3]

13.3.3. RHDV IN AUSTRALIA

The high mortality rates of rabbits infected with RHDV and the rapid spread of the virus stimulated interest in its potential for biological control. RHDV was imported into the Australian Animal Health Laboratory high security facility in 1991. Research was undertaken to confirm the lethality of the virus and that it did not infect animals other than rabbits. Following an extensive testing and consultation process the virus was then tested under semi-field conditions on Wardang Island, just off the coast of South Australia.[3] The virus spread from this facility probably on insect vectors to the mainland during late September/early October 1995 and despite intensive efforts to limit and control its spread it quickly became apparent that RHDV had moved hundreds of kilometers within a few weeks and that no control was possible. Plans to contain the spread of RHDV were formally abandoned in early November 1995. However, it was not until September 1996 that the virus was officially registered as a pest control agent and deliberate release permitted.[3] RHDV was illegally introduced into New Zealand in 1997, where it also became established in the wild rabbit population.

13.3.4. IMPACT OF RHDV IN AUSTRALIA

RHDV has caused a significant reduction in rabbit numbers in the arid and semi-arid parts of Australia. Populations in some areas are estimated to have dropped to less than 10% of pre-RHDV numbers and to have been maintained at a low level by annual epidemics.[24] This has been associated with regeneration of native vegetation and a large drop in the numbers of predators such as the European red fox.[25] However, in the more temperate and higher rainfall areas the impact has not been as great even though the virus has become established and epidemics have occurred.[3,24] The lower impact in high rainfall areas may be attributable to the presence of a pre-existing calicivirus of rabbits. Serological studies have indicated that in some areas there is considerable evidence for a virus that cross-reacts serologically with RHDV[20,21] and similar results have been obtained in New Zealand.[26,27] Unlike RCV, this cross-reacting virus does not provide complete protection from infection with RHDV, and until the virus or viruses are isolated and characterized the epidemiology of RHDV is likely to remain poorly understood.

A further complicating factor for RHDV use in biological control is that young rabbits less than around 6 weeks old recover from infection with RHDV but are subsequently immune to reinfection and have high titers of serum antibodies. These antibodies are passively transferred to the offspring of the females and can also provide partial or complete protection from RHDV for some weeks after birth.[28] Such partial protection can also protect against lethal infection and allow the infected animal to recover and be immune to subsequent infection.

Unlike myxoma virus there is no evidence that RHDV is becoming less virulent or that rabbits are becoming resistant to the virus. However, there have been no good studies to really examine whether such phenotypic changes in virus or rabbit are occurring. This suggests that not all biological control agents will necessarily co-evolve with their hosts in the same way as myxoma virus and the European rabbit. The relatively poorly understood epidemiology of RHDV-like viruses in Australia and Europe suggests that there is considerable capacity for RHDV to adapt to changing selection pressures. However, for the moment, RHDV has provided a second biological control for rabbits in

Australia and is having a major impact on rabbit populations and conservation biology, at least in the arid and semi-arid areas of Australia.

13.3.5. IMPACT OF RHDV ON RABBIT POPULATIONS IN EUROPE

RHDV has had a major impact on wild rabbit populations in parts of Europe, particularly France and the Iberian peninsula. In Spain, the fall in rabbit numbers has been associated with declines in resources for top level predators such as the Spanish lynx and imperial eagle. In Britain the impact has been more patchy, with loss of habitat maintenance by rabbit grazing a key feature.[19]

13.4. Biotechnological Approaches to Biological Control

13.4.1. IMMUNOCONTRACEPTION

The control of the fertility of a population provides a theoretical alternative to lethal control measures such as trapping, poisoning, shooting or potentially lethal measures such as habitat destruction. In particular fertility control is seen by many as being potentially more humane than conventional population controls.[29] There are three possible approaches to fertility control: surgical sterilization, which is only viable on small populations with high conservation or social values, for example, zoo animals or pets; hormonal manipulation that can be delivered by injections or implants or potentially by feeding and thus can have wider use than surgical sterilization and also offers the possibility that animals can return to fertility; and immunocontraception, which uses a vaccine to stimulate an autoimmune response that interferes with fertility.

Immunocontraception has been applied experimentally to a wide range of animals including deer, seals, elephants, horses, and primate models for human immunocontraception.[30] However, immunocontraception as it has been applied relies on injection of individual animals and so suffers from the same delivery limitations as other forms of fertility control. To be effective in widespread pest species such as the European rabbit, house mouse or European red fox, which are major pest species in Australia, some form of remote delivery that could operate on a continent-wide scale would be necessary. Thus was born the idea of virally-vectored immunocontraception.[31]

The concept of virally-vectored immunocontraception is quite simple. First, identify a protein antigen that will induce an immune response in the target species such that the immune response will block fertility. For example antibodies generated following immunization with oocyte or sperm proteins could block egg-sperm binding and hence prevent fertilization. The next step is to clone a cDNA copy of the gene that encodes the protein of interest and then insert that cDNA into a suitable virus vector such that the recombinant virus will express the foreign protein. The final step is to infect the target species with the virus so that the host develops an immune response both to the virus and to the immunocontraceptive antigen. The potential and the limitations of this use of biotechnology are probably best seen by examining the research that has been done on virally-vectored immunocontraception for mice, rabbits and foxes over the past 12 years. In each case three general questions must be answered:

1. What proportion of a wild population must be sterilized to reduce the impact of the pest?
2. Can an immunocontraceptive be developed that is species-specific, lasts for the lifetime of the animal, does not require boosting and is delivered by a recombinant virus?
3. Can such a recombinant virus successfully spread in the field in competition with field strains of virus and infect the required proportion of animals?[31]

These questions were addressed by a series of large scale ecological, epidemiological and laboratory experiments that brought together the disciplines of ecology, virology, immunology, reproductive biology, molecular biology and mathematical modeling.

13.4.1.1. *Virally-vectored Immunocontraception of Mice*

The first step in developing virally-vectored immunocontraception is to select an antigen. In the case of the mouse it was known that one of the proteins that forms the zona pellucida, a glycoprotein matrix surrounding the oocyte, ZP3, had potential as an immunocontraceptive. The cDNA for the ZP3 gene was readily available. The virus chosen initially was ectromelia virus which causes mousepox in mice. Ectromelia virus (ECTV) is a poxvirus (genus *Orthopoxvirus*); it has a double stranded DNA genome and techniques for constructing recombinant viruses were available. The ZP3 cDNA was inserted into the *thymidine kinase* (*tk*) gene of ECTV under the control of a strong late promoter. This insertion disrupted the *tk* gene and TK negative virus could be selected by addition of bromodeoxyuridine to the culture medium. This would be toxic in the presence of an active *tk* gene. In addition to the ZP3 gene, the *E. coli lacZ* gene was inserted to provide a color selection marker.[32] A control virus expressing only the *lacZ* gene was constructed in parallel. Ectromelia virus is normally lethal in Balb/c laboratory mice, however, the disruption of the *tk* gene also attenuated the virus allowing it to be tested in this strain of mice.

Female mice were inoculated with 10^6 plaque forming units (pfu) of either ECTV-ZP3 or ECTV-lac Z virus and had recovered from infection by two to three weeks after inoculation. The mice were then paired with males to determine their fertility compared to uninfected controls. The results were quite dramatic. Of the uninfected control mice 10/10 mice had litters with a mean litter size of 6.6 ± 0.8. This compared with 12/15 mice from the ECTV-lacZ control group which had a mean litter size of 7.3 ± 0.7 (mean of the mice that had litters) and 4/13 mice from the ECTV-ZP3 infected group which had a mean litter size of 1.8 ± 0.3. Mice immunized with the ZP3 expressing virus developed serum antibodies to ZP3 and these antibodies could be detected bound to the zona pellucida of developing oocytes in ovarian follicles in the ovaries of these mice. In addition, ovaries from 5/13 mice immunized with ECTV-ZP3 had abnormal morphology at autopsy and contained only small and medium follicles together with large clusters of "luteinized" cells.[32] In a subsequent long-term trial, infertility persisted for at least 3 months and as long as 37 weeks in one case. Interestingly, boosting the mice with 10^6 pfu of recombinant virus induced a delayed type hypersensitivity response at the inoculation site together with a rise in serum antibody levels to ZP3 and a return to infertility.[32]

This study provided proof of concept for virally-vectored immunocontraception. It demonstrated that a recombinant virus could infect the target species and induce an autoimmune response to a self-antigen that led to infertility in around 70% of infected animals and a reduction in fertility in the remainder. It also showed that infertility was not permanent and that as serum antibody levels dropped, mice returned to fertility. It did not demonstrate, nor was it intended to, that ectromelia virus would be a suitable vector for delivering immunocontraception to wild mice.

Subsequent studies on virally-vectored immunocontraception in mice demonstrated that murine cytomegalovirus (MCMV) could be used as a viral vector to deliver ZP3 to the mouse immune system.[33] From an epidemiological perspective, cytomegalovirus was seen as having advantages over ectromelia virus because it was widely distributed in wild mice and it was known that mice could be infected with multiple strains of cytomegalovirus and that the virus probably persisted for the life of the mouse.[34]

Female Balb/c mice were infected intraperitoneally with 2×10^4 pfu of recombinant MCMV expressing ZP3 (MCMV-ZP3). No litters were born to mice inoculated with this virus for the 250 days of the trial; the control females produced a total of 450 pups over the same period. Mice inoculated with the parental virus as a control produced similar numbers of pups to control mice. Serum antibodies to ZP3 were present in the immunized mice and significant ovarian pathology was described with no tertiary or secondary follicles present 150 days after inoculation.[33] Thus the use of a persistent virus appeared to have significant advantages over the use of the acute infection that occurs with ectromelia virus. In addition, ectromelia virus is exotic to Australian wild mice whereas MCMV was already widely distributed. However, it was noted that the recombinant MCMV-ZP3 was extremely attenuated in vivo compared to the wild-type parental virus. This may effect the transmission of the virus from mouse to mouse, as the virus did not reach high titers in salivary glands. Obviously any recombinant virus is likely to be less competitive in the field compared to naturally selected field strains of virus.

13.4.1.2. *Ecology of Virally-vectored Immunocontraception for Mice*

The recombinant virus will have to transmit between mice and potentially compete with wild-type field strains for virally-vectored immunocontraception to be successful in the field. Field studies on MCMV and mouse plagues, transmission studies in enclosures using wild type virus,[35] and modeling studies suggest that if this could be achieved, then recombinant MCMV has the potential to disrupt mouse plagues.[36] However, considerable effort may be needed to achieve a recombinant MCMV that can transmit effectively in the field.

13.4.1.3. *Virally-vectored Immunocontraception for Rabbits*

Myxoma virus was the obvious choice as a viral vector to deliver immunocontraception to rabbits. The virus is a large DNA virus with a genome size of 161 kb pairs that can easily accommodate additional genes; it is widespread in the field in Australia and is lagomorph-specific (rabbits and related species), only able to infect European rabbits and European hares in Australia. In addition, preliminary studies with recombinant myxoma virus expressing the haemagglutinin protein from influenza A virus showed that rabbits infected with a highly attenuated virus (MyxV-HA) developed high titers of serum antibodies to both the virus and the foreign antigen.[37]

Recombinant myxoma viruses were constructed by inserting the foreign DNA into an intergenic site between the M061R and M062R genes using homologous recombination within the flanking sequences to insert a construct consisting of a selectable marker gene under the control of a poxvirus early/late promoter, and the cDNA of interest under control of a separate promoter.[37,38] Some recombinants were also constructed using transient dominant selection in which the selectable marker was subsequently deleted from the recombinant virus.[39] The attenuated Uriarra (Ur) strain of myxoma virus,[40,41] which causes clinical myxomatosis but is only occasionally lethal in laboratory rabbits was used to construct recombinant viruses. This virus would not be a suitable vector in the wild as it is highly attenuated, but it was a convenient virus for testing in laboratory rabbits because it does not kill the rabbit. Most, but not all, of the recombinant viruses constructed were somewhat attenuated compared to the wild type Ur strain.

The zona pellucida glycoproteins were chosen as the antigens for the development of immunocontraception in rabbits, as was done with mice. There are at least three and probably four zona pellucida genes in rabbits, and cDNA sequences were available for *ZP1*, *ZP2* and *ZP3*. In addition, immunization of rabbits with whole porcine zona pellucida in Freund's adjuvant induced long term infertility;[42,43] immunization with rabbit zona pellucida did not induce infertility suggesting that it was seen as a self-antigen and did not induce an immune response.

Recombinant rabbit ZP1protein (also termed ZPB) was expressed from a rabbit cell line using a vaccinia virus expression system. This recombinant protein, emulsified in Freund's complete adjuvant, was used to immunize rabbits. Male rabbits developed a high titer serum antibody response to ZP1 following the first inoculation. By contrast female rabbits, in which ZP1 is a self-antigen were slower to develop a serum antibody response and this response did not reach high titers in all rabbits, even after two booster inoculations. Rabbits that had serum antibody titers of ≥12,800 (7/10) were infertile while those with lower titers were fertile.[37] Litter size in the immunized but fertile rabbits was similar to controls suggesting that infertility was an all or nothing phenomenon. Infertility was associated with some ovarian pathology in some rabbits, however, this was not consistent across the infertile rabbits.

Based on these results, rabbits were immunized with 1,000 pfu of a recombinant myxoma virus expressing rabbit ZP1. Control rabbits were matched full-siblings injected with the parental Ur strain of myxoma virus. Interestingly, when immunized with a recombinant virus both male and female rabbits had a rapid serum antibody response to ZP1. In other words, antigen presentation to the immune system in the context of a recombinant viral infection overcame self-tolerance in the female rabbits. However, antibody titers peaked at around day 15 after infection and quickly dropped thereafter in both males and females. Titers did not reach 12800 and only 25% of the rabbits were infertile, which was not significantly different to predicted fertility. All of the control rabbits became pregnant indicating that infection with wild-type myxoma virus did not induce infertility. Rabbits were solidly immune to subsequent infection with myxoma virus and it was not possible to boost with further virus inoculations. However, 80% of rabbits immunized with recombinant myxoma virus and then boosted with ZP1 protein, emulsified in Freund's incomplete adjuvant, developed high serum antibody titers (≥12800), delayed type hypersensitivity responses to the boosting , and were infertile.[38]

Following this trial, recombinant viruses expressing rabbit ZP2 and rabbit ZP3 glycoproteins were constructed and tested. Rabbits immunized with recombinant virus expressing rabbit ZP2 developed serum antibodies to ZP2 and these antibodies cross-reacted with zona pellucida in ovarian sections but the rabbits retained full fertility with 100% pregnant following mating at 30 days after immunization.[39] When rabbits were immunized with recombinant virus expressing rabbit ZP3, only 35% were pregnant at autopsy 10 days after mating. This was significantly less than predicted ($p < 0.001$).[39] However, serum antibody titers were relatively low and decreased quite rapidly after peaking at around 20 days after infection. Longer term fertility trials demonstrated that infertility was transient and that rabbits returned to normal fertility within two months of infection.[39]

Subsequent studies were done to optimize the presentation of ZP3 to the rabbit immune system. Expressing ZP3 under the control of a combined early/late promoter rather than the previously used late promoter substantially increased the proportion of rabbits that were infertile following immunization from 70% to 90–100%, and this infertility persisted in at least half of the rabbits on subsequent matings (Kerr, Perkins, and van Leeuwen, in preparation).

Recombinant myxoma virus expressing porcine ZP3 glycoprotein was used to test whether a heterologous antigen would be more effective at inducing an autoimmune response than a strict self-antigen. Immunization with this recombinant virus induced high and persistent serum antibody titers to porcine ZP3. There was significant infertility at the second and third mating after infection but not at the first mating. This suggested that co-immunization with recombinant viruses expressing rabbit ZP3 and porcine ZP3 might prove successful, however, the results of the co-immunization were similar to infection with virus expressing porcine ZP3 alone, indicating that the heterologous porcine antigen was immunodominant in the co-immunization experiments (Wijeratne, Kerr, Perkins, and van Leeuwen, unpublished data).

13.4.1.4. *Ecology of Immunocontraception for Rabbits*

Two large scale field trials were conducted in which surgical sterilization was applied to either 0, 40, 60 or 80% of the female wild rabbits in the population to determine whether fertility control could reduce rabbit impact under natural conditions.[44–46] Each trial ran for three breeding seasons and the surgical sterility was reimposed on the population at the end of each breeding season. These trials demonstrated that for fertility control to have an impact on rabbit populations it would need to be imposed at the 60–80% level for prolonged periods to overcome compensatory survival. Epidemiological studies were also conducted to determine whether it would be possible to introduce a recombinant myxoma virus into the field in competition with field strains. A large field trial demonstrated that a virus with a natural genetic deletion that enabled genetic typing could be released and spread in the field. This virus persisted at two of the four release sites but did not exclude field strains.[47]

13.4.1.5. *Future of Immunocontraception for Rabbits*

Combining the laboratory and the ecological studies on rabbit immunocontraception it is obvious that there is a large performance gap between what can currently be achieved in the laboratory with biotechnology and the very high levels of permanent infertility required to have an impact on rabbit populations in the wild.

13.4.1.6. *Virally–vectored Immunocontraception Research in Foxes*

The third target species for immunocontraception research was the European red fox. This animal was particularly challenging to work with because it is a seasonal breeder. In addition, only wild-caught foxes were available for trials in Australia and these bred poorly in captivity. Trials were conducted with either recombinant vaccinia virus or recombinant canine herpesvirus as a vector expressing either fox ZP3 or porcine ZP3. However, an immune response to the recombinant antigen was not achieved in the trials although immune responses to the viral vectors were induced.[48,49]

13.4.2. BIOTECHNOLOGY FOR CONTROL OF CANE TOADS

Despite extensive searches for a natural biocontrol agent for cane toads no suitable organism has been identified that could safely and effectively control the toads.[50] As a result of this failure, biotechnological approaches are being explored. The concept is to construct a recombinant virus that would interfere with metamorphosis from tadpole to toadlet or possibly with other life stages of the toad. This is based on the idea that proteins that are normally not expressed until after metamorphosis could be used to immunize tadpoles—where they would be seen as "foreign" by the immune system and that antibodies to these proteins would inhibit metamorphosis.[51,52]

13.4.3. BIOTECHNOLOGICAL APPROACHES FOR THE CONTROL OF CARP—THE DAUGHTERLESS CONCEPT

Carp (*Cyprinus carpio*) have emerged as a major vertebrate pest species of Australian waterways over the past 40 years. A novel biotechnological approach called "daughterless" is being explored to control this species. The aim is to create transgenic fish that express an inhibitor of an aromatase gene under the control of an ovary-specific promoter. Aromatase is required to convert androgen to estrogen. Embryos normally develop into males unless aromatase is active. If expression of the aromatase gene is inhibited then all embryos will develop along the default pathway into males. Modeling studies suggest that replacing 5% of wild type recruits each year with transgenic carp that carry the daughterless gene would lead to near extinction over a 30 year period.[53]

As a more conventional biological control the possibility of using koi herpesvirus a highly contagious lethal disease of carp is also being evaluated.

13.5. Risk Management of Biological Control and Biotechnology

The introduction of any organism for biological control requires extensive safety testing and public consultation. This becomes even more important when genetic manipulation is involved as public perceptions of recombinant organisms are often conditioned by highly emotive campaigns against genetic manipulation. As well as safety considerations such as species-specificity, the risk that the biological control agent will not have an impact on the pest species needs to be considered. If the biological control is unlikely to be effective then the precautionary principle would suggest that it should not be introduced. This section will briefly consider two examples of risk management. The first is the process that was used to determine whether RHDV should be used as a biological control agent in Australia and the second looks at the questions that would need to be addressed to introduce a putative recombinant myxoma virus as an immunocontraceptive.

13.5.1. BIOSAFETY ASSESSMENT OF RHDV

A strain of RHDV (Czech strain 351) was imported into Australia in 1991 under high security containment conditions for evaluation. It was already known that the virus had spread rapidly in other parts of the world, that it was rapidly lethal for European rabbits and appeared to only infect this species.[1,3] A series of trials were conducted to confirm that the virus was lethal in Australian wild rabbits and to further test species-specificity of the virus in a range of Australian and New Zealand native and introduced animals likely to be in contact with rabbits. Only one of the tested species showed any evidence of a response to RHDV. This was a kiwi that had been inoculated with 300,000 rabbit lethal doses of virus and developed a serum antibody titer to RHDV. In this case, the immune response was explicable by the very high dose of virus inoculated rather than due to virus replication.[1] Considerable attention was also paid to the animal welfare implications of introducing a lethal virus into the rabbit population.

A set of contained field trials were then conducted on Wardang Island to determine the impact of the virus on rabbit populations, how effectively the virus could spread in rabbits under natural conditions and persistence of the virus. Despite the high security and elaborate quarantine precautions taken it was during these trials that the virus escaped onto the mainland and spread throughout south-eastern Australia.[1] This escape forestalled the planned process of evaluation that was to have included public consultation and approval prior to a controlled release on the mainland and monitoring and assessment of the impact of the virus. However, following the escape of the virus an environmental impact assessment was conducted, which concluded that: "The disease is species-specific, highly effective as a control agent and likely to persist far into the future. It will greatly complement the existing rabbit control measures available."[54]

13.5.2. BIOSAFETY CONSIDERATIONS FOR INTRODUCTION OF AN IMMUNOCONTRACEPTIVE MYXOMA VIRUS

The potential release of a recombinant virus that would induce sterility would not be a simple task. Once released a virus cannot simply be recalled. Many of the socio-political and safety issues associated with such a release have been reviewed in detail[55–57] and will only be touched on here by way of example. Firstly, all studies with recombinant viruses were done under license from the Commonwealth of Australia Office of the Gene Technology Regulator (OGTR). This authority would have to approve and oversee any trials outside laboratory containment, which would require that both OGTR and the public were convinced of the safety and the effectiveness of the immunocontraceptive virus. The key scientific issues would be species-specificity of the virus and of the immunocontraceptive antigen and the risk that the virus could mutate to infect other species. At the international level, the risk of the recombinant virus being introduced to other countries where European rabbits have significant conservation or farming value would have to be considered.[57] In this context, it is worth remembering that myxoma virus was illegally introduced into France in 1952 and Britain in 1953 and RHDV was later illegally introduced into New Zealand. In addition to the European rabbit, myxoma virus originated in the South American species *Sylvilagus brasiliensis* and can also infect and potentially transmit from at least two North American lagomorph species (*Sylvilagus nuttalli* and *S. audubonni*)[58] and the hare species *Lepus europeus* and *Lepus timidus*.[8] It would probably be necessary to test any recombinant virus in each of these alternative hosts. If the virus sterilized one of these other species how would a decision be made on its use?

Whether myxoma virus could mutate to adapt to a novel host species is very difficult to predict. The most likely mechanism for this would be recombination with another poxvirus with a broader host-specificity such as vaccinia or cowpox viruses altering the host range of myxoma virus. As far as is known, these viruses do not occur in the field in Australia. Alternatively, that recombination transferred the immunocontraceptive gene to another poxvirus with a broad host-range. These scenarios seem unlikely as they would require that the two viruses were replicating in the same cell, in the same host and had sufficient nucleotide homology for recombination to occur. Even then the recombinant virus would still have to be transmitted from that host and persist.

The question of antigen specificity is interesting. Pig zona pellucida is an effective immunocontraceptive in many species.[30] Rabbit zona pellucida proteins are immunogenic in other species and could potentially be effective immunocontraceptives. Basically it is unlikely that antigen specificity could be obtained with the current technology. Therefore species-specificity would largely rely on the virus vector.

Finally, there is the question of whether the virus could be effective and have an impact on rabbit populations or more precisely on the damage that rabbits cause to the environment. The ecological trials conducted to measure the effect of fertility control on rabbit impact showed that the virus would need to be able to induce infertility in at least 60% and probably 80% of the female rabbits and that this infertility would probably need to persist for the life of the rabbit. This would set a very stringent test of effectiveness for any immunocontraceptive virus and one that cannot currently be met.

13.6. Conclusions

Effective biological control of vertebrate pest populations has been difficult to find and the two most successful have both been for the European rabbit. Whether this reflects the biology of the rabbit or simply the fact that two viruses essentially appeared without being looked for is not clear. The co-evolution of myxoma virus with the European rabbit demonstrates that pest-species are very adaptable and that the outcome of biological control may not be predictable. Similarly the introduction of RHDV into the rabbit population has revealed how little is actually known about the epidemiology of this and related viruses. Where searches have been done for biological control agents for other vertebrate pest species such as the cane toad there has been no success. Although koi herpesvirus may offer a possible biological control for carp, this virus was well known and not specifically sought out for biological control.

Biotechnology offers the opportunity to manipulate microorganisms and potentially also macroparasites to create novel biological control agents. This also has the potential of creating new and novel risks and requires careful management. However, experience so far suggests that developing biological control agents that can have widespread impact on a vertebrate pest species is going to be very difficult, albeit potentially worthwhile.

References

1. F. Fenner and B. Fantini, *Biological Control of Vertebrate Pests. The History of Myxomatosis—An Experiment in Evolution* (CAB International, New York, 1999).
2. E. C. Rolls, *They All Ran Wild* (Angus and Robertson, Melbourne, 1969).
3. B. D. Cooke and F. Fenner, Rabbit haemorrhagic disease and the biological control of wild rabbits, *Oryctolagus cuniculus*, in Australia and New Zealand, *Wildl. Res.* 29, 689–706 (2002).
4. P. J. J. Van Rensburg, J. D. Skinner, and R. J. Van Aarde, Effects of feline panleucopaenia on the population characteristics of feral cats on Marion Island, *J. App. Ecol.* 24, 63–73 (1987).
5. K. Myers, I. Parer, D. Wood, and B. D. Cooke, The rabbit in Australia, in *The European Rabbit. The History and Biology of a Successful Colonizer*, edited by H. V. Thompson and C. M. King (Oxford University Press, Oxford, 1999), pp. 108–157.
6. A. E. Newsome, I. A. Parer, and P. Catling, Prolonged prey suppression by carnivores: Predator-removal experiments, *Oecologia* 78, 458–467 (1989).
7. K. Williams, I. Parer, B. Coman, J. Burley, and M. Braysher, *Managing Vertebrate Pests: Rabbits* (Australian Government Publishing Services, Canberra, 1995).
8. F. Fenner and F. N. Ratcliffe, *Myxomatosis* (Cambridge University Press, Cambridge, 1965).
9. F. Fenner and I. D. Marshall, A comparison of the virulence for European rabbits (*Oryctolagus cuniculus*) of strains of myxoma virus recovered in the field in Australia, Europe and America, *J. Hyg. (Camb.)* 55, 149–151 (1957).
10. F. Fenner, M. F. Day, and G. M. Woodroofe, Epidemiological consequences of the mechanical transmission of myxomatosis by mosquitoes, *J. Hyg. (Camb.)* 54, 285–303 (1956).
11. I. D. Marshall, The influence of ambient temperature on the course of myxomatosis in rabbits, *J. Hyg. (Camb.)* 57, 484–497 (1959).
12. I. D. Marshall and F. Fenner, Studies in the epidemiology of infectious myxomatosis of rabbits, V: Changes in the innate resistance of wild rabbits between 1951 and 1959, *J. Hyg. (Camb.)* 56, 288–302 (1958).
13. I. D. Marshall and G. W. Douglas, Studies in the epidemiology of infectious myxomatosis of rabbits, VIII: Further observations on changes in the innate resistance of Australian wild rabbits exposed to myxomatosis, *J. Hyg. (Camb.)* 59, 117–122 (1961).
14. K. M. Saint, N. French, and P. Kerr, Genetic variation in Australian isolates of myxoma virus: an evolutionary and epidemiological study, *Arch. Virol.* 146, 1105–1123 (2001).
15. P. J. Kerr, J. M. Merchant, L. Silvers, G. Hood, and A. J. Robinson, Monitoring the spread of myxoma virus in rabbit populations in the southern tablelands of New South Wales, Australia, II: Selection of a virus strain that was transmissible and could be monitored by polymerase chain reaction, *Epidemiol. Infect.* 130, 123–133 (2003).
16. P. J. Kerr and S. M. Best, Myxoma virus in rabbits, *Rev. Sci. Technol. Off. Int. Epiz.* 17, 256–268 (1998).
17. L. Capucci, P. Fusi, A. Lavazza, M. L. Pacciarini, and C. Rossi, Detection and preliminary characterization of a new calicivirus related to rabbit haemorrhagic disease virus but non pathogenic, *J. Virol.* 70, 8614–8623 (1996).
18. S. R. Moss, S. L. Turner, R. C. Trout, P. J. White, P. J. Hudson, A. Desai, M. Armesto, N. L. Forrester, and E. A. Gould, Molecular epidemiology of rabbit haemorrhagic disease virus, *J. Gen. Virol.* 83, 2461–2467 (2002).

19. N. L. Forrester, R. C. Trout, S. L. Turner, D. Kelly, B. Boag, S. Moss, and E. A. Gould, Unraveling the paradox of rabbit haemorrhagic disease virus emergence, using phylogenetic analysis; possible implications for rabbit conservation strategies, *Biol. Conserv.* 131, 296–306 (2006).
20. B. D. Cooke, S. McPhee, A. J. Robinson, and L. Capucci, Rabbit haemorrhagic disease: Does a pre-existing RHDV-like virus reduce the effectiveness of RHD as a biological control in Australia, *Wildl. Res.* 29, 673–682 (2002).
21. A. J. Robinson, P. D. Kirkland, R. I. Forrester, L. Capucci, B. D. Cooke, and A. W. Philbey, Serological evidence for the presence of a calicivirus in Australian wild rabbits, *Oryctolagus cuniculus*, before the introduction of rabbit haemorrhagic disease virus (RHDV): Its potential influence on the specificity of a competitive ELISA for RHDV, *Wildl. Res.* 29, 655–652 (2002).
22. S. Marchandeau, G. Le Gall-Recule, S. Bertagnoli, J. Aubineau, G. Botti, and A. Lavazza, Serological evidence for a non-protective RHDV-like virus, *Vet. Res.* 36, 53–62 (2005).
23. P. J. White, R. A. Norman, and P. J. Hudson, Epidemiological consequences of a pathogen having both virulent and avirulent modes of transmission: the case of rabbit haemorrhagic disease virus, *Epidemiol. Infect.* 129, 665–677 (2002).
24. G. Mutze, P. I. Bird, J. Kovaliski, D. Peacock, S. Jennings, and B. D. Cooke, Emerging epidemiological patterns in rabbit haemorrhagic disease, its interactions with mxyomatosis, and their effects on rabbit populations in South Australia, *Wildl. Res.* 29, 577–590 (2002).
25. J. Read and Z. Bowen, Population dynamics, diet and aspects of the biology of feral cats and foxes in arid South Australia, *Wildl. Res.* 28, 195–203 (2001).
26. J. S. O'Keefe, J. E. Tempero, M. X. J. Motha, M. F. Hansen, and P. H. Atkinson, Serology of rabbit haemorrhagic disease virus in wild rabbits before and after the release of the virus in New Zealand, *Vet. Microbiol.* 66, 29–40 (1999).
27. J. P. Parkes, G. L. Norbury, R. P. Heyward, and G. Sullivan, Epidemiology of rabbit haemorrhagic disease (RHD) in the South Island, New Zealand, 1997–2001, *Wildl. Res.* 29, 543–555 (2002).
28. A. J. Robinson, P. T. M. So, W. J. Muller, B. D. Cooke, and L. Capucci, Statistical models for the effect of age and maternal antibodies on the development of rabbit haemorrhagic disease in Australian wild rabbits, *Wildl. Res.* 29, 663–671 (2002).
29. G. Oogjes, Ethical aspects and dilemmas of fertility control of unwanted wildlife: An animal welfarist's perspective, *Reprod. Fertil. Dev.* 9, 163–168 (1997).
30. S. K. Gupta, N. Srivastava, S. Choudhury, A. Rath, N. Sivapurapu, G. K. Gahlay, and D. Batra, Update on zona pellucida glycoproteins based contraceptive vaccine, *J. Reprod. Immunol.* 62, 79–89 (2004).
31. C. H. Tyndale-Biscoe, Virus-vectored immunocontraception of feral mammals, *Reprod. Fertil. Dev.* 6, 9–16 (1994).
32. R. J. Jackson, D. J. Maguire, L. A. Hinds, and I. A. Ramshaw, Infertility in mice induced by a recombinant ectromelia virus expressing mouse zona pellucida glycoprotein 3, *Biol. Reprod.* 58, 152–159 (1998).
33. M. L. Lloyd, G. R. Shellam, J. M. Papadimitriou, and M. A. Lawson, Immunocontraception is induced in BALB/c mice inoculated with murine cytomegalovirus expressing mouse zona pellucida 3, *Biol. Reprod.* 68, 2024–2032 (2003).
34. G. R. Shellam, The potential of murine cytomegalovirus as a viral vector for immunocontraception, *Reprod. Fertil. Dev.* 6, 129–137 (1994).

35. L. N. Farroway, G. R. Singleton, M. A. Lawson, and D. A. Jones, The impact of murine cytomegalovirus (MCMV) on enclosure populations of house mice (*Mus domesticus*), *Wildl. Res.* 29, 1–17 (2002).
36. A. D. Arthur, R. P. Pech, and G. R. Singleton, Predicting the effect of immunocontraceptive recombinant murine cytomegalovirus on population outbreaks of house mice (*Mus musculus domesticus*) in mallee wheatlands, *Wildl. Res.* 32, 631–637 (2005).
37. P. J. Kerr and R. J. Jackson, Myxoma virus as a vaccine vector for rabbits: antibody levels to influenza virus haemagglutinin presented by a recombinant myxoma virus, *Vaccine* 13, 1722–1726 (1995).
38. P. J. Kerr, R. J. Jackson, A. J. Robinson, J. Swan, L. Silvers, N. French, H. Clarke, D. F. Hall, and M. K. Holland, Infertility in female rabbits (*Oryctolagus cuniculus*) alloimmunized with the rabbit zona pellucida protein ZPB either as a purified recombinant protein or expressed by recombinant myxoma virus, *Biol. Reprod.* 61, 601–613 (1999).
39. S. M. Mackenzie, E. A. McLaughlin, H. D. Perkins, N. French, T. Sutherland, R. J. Jackson, B. Inglis, W. J. Muller, B. H. van Leeuwen, A. J. Robinson, and P. J. Kerr, The immunocontraceptive effects on female rabbits (*Oryctolagus cuniculus*) infected with recombinant myxoma virus expressing rabbit ZP2 and ZP3, *Biol. Reprod.* 74, 511–521 (2006).
40. R. J. Russell and S. J. Robbins, Cloning and molecular characterization of the myxoma virus genome, *Virology* 170, 147–159 (1989).
41. S. M. Best and P. J. Kerr, Coevolution of host and virus: The pathogenesis of virulent and attenuated strains of myxoma virus in resistant and susceptible European rabbits, *Virology* 267, 36–48 (2000).
42. D. M. Wood, C. Liu, and B. S. Dunbar, Effect of alloimmunization and heteroimmunization with zonae pellucidae on fertility in rabbits, *Biol. Reprod.* 25, 439–450 (1981).
43. S. M. Skinner, T. Mills, H. J. Kirchick, and B. S. Dunbar, Immunization with zona pellucida proteins results in abnormal ovarian follicular differentiation and inhibition of gonadotropin-induced steroid secretion, *Endocrinology* 115, 2418–2432 (1984).
44. L. E. Twigg and C. K. Williams, Fertility control of overabundant species; can it work for feral rabbits?, *Ecology Letters* 2, 281–285 (1999).
45. L. E. Twigg, T. J. Lowe, G. R. Martin, A. G. Wheeler, G. S. Gray, S. L. Griffin, C. M. O'Reilly, D. J. Robinson, and P. H. Hubach, Effects of surgically imposed sterility on free-ranging rabbit populations, *J. Appl. Ecol.* 37, 16–39 (2000).
46. C. K. Williams, C. C. Davey, R. J. Moore, L. A. Hinds, L. E. Silvers, P. J. Kerr, N. French, G. M. Hood, R. P. Pech, and C. J. Krebbs, Populations responses to sterility imposed on female European rabbits, *J. Appl. Ecol.*, in press.
47. J. M. Merchant, P. J. Kerr, N. Simms, G. M. Hood, R. Pech, and A. J. Robinson, Monitoring the spread of myxoma virus in rabbit (*Oryctolagus cuniculus*) populations on the southern tablelands of New South Wales, Australia, III: Release, persistence and rate of spread of an identifiable strain of myxoma virus, *Epidemiol. Infect.* 130, 135–147 (2003).
48. G. H. Reubel, S. Beaton, D. Venables, J. Pekin, J. Wright, N. French, and C. M. Hardy, Experimental inoculation of European red foxes with recombinant vaccinia viruses expressing zona pellucida C proteins, *Vaccine* 23, 4417–4426 (2005).
49. T. Strive, C. M. Hardy, N. French, J. D. Wright, N. Nagaraja, and G. H. Reubel, Development of canine herpesvirus based antifertility vaccines for foxes using bacterial artificial chromosomes, *Vaccine* 24, 980–988 (2006).
50. A. D. Hyatt, H. Parkes, and Z. Zupanovic, *Identification, Characterization and Assessment of Venezuelan Viruses for Potential Use As Biological Control Agents Against the Cane Toad (Bufo marinus) in Australia: A Report from the Australian Animal Health Laboratory* (CSIRO, Geelong, Australia, 1998).

51. G. M. Maniatis, L. A. Steiner, and V. M. Ingram, Tadpole antibodies against frog hemoglobin and their effect on development, *Science* 165, 67–69 (1969).
52. A. J. Robinson, A. D. Hyatt, J. Pallister, N. H. R. Hamilton, and D. C. T. Halliday. Biocontrol approaches to canetoad control, *Proceedings of Cane Toad control workshop, Kununurra* (in Press, 2006).
53. R. Thresher and N. Bax. The science of producing daughterless technology; possibilities for using daughterless technology; maximizing the impact of carp control in: *Proceedings of the National Carp Control Workshop*, Edited by K. L. Lapidge (Cooperative Research Centre for Pest Animal Control, Canberra, Australia, 2003).
54. B. Coman, *Environmental impact associated with the proposed use of Rabbit Calicivirus Disease for integrated rabbit control in Australia.* Prepared for the Australian and New Zealand Rabbit Calicivirus Program, (1996).
55. C. K. Williams, Development and use of virus-vectored immunocontraception, *Reprod. Fertil. Dev.* 9, 169–178 (1997).
56. P. Kerr, B. van Leeuwen, H. Perkins, M. Holland, W. Gu, R. Jackson, C. Williams, and A. Robinson, Development of fertility control for wild rabbits in Australia using a virally-vectored immunocontraceptive, in *Enhancing Biocontrol Agents and Handling Risks*, edited by M. Vurro and J. Gressel (IOS Press, Amsterdam, 2001), pp. 14–27.
57. E. Angulo and B. Cooke, First synthesize new viruses then regulate their release? The case of the wild rabbit, *Mol. Ecol.* 11, 2703–2709 (2002).
58. D. C. Regnery, The epidemic potential of Brazilian myxoma virus (Lausanne strain) for three species of North American cottontails, *Am. J. Epidemiol.* 94, 514–519 (1971).
59. F. Fenner, Biological control as exemplified by smallpox eradication and myxomatosis, *Proc. R. Soc. (Lond.)* B 218, 259–285 (1983).

14. GENETICALLY ENHANCING THE EFFICACY OF PLANT PATHOGENS FOR CONTROL OF WEEDS

Brian M. Thompson, Matthew M. Kirkpatrick, David C. Sands,*
and Alice L. Pilgeram
Plant Sciences and Plant Pathology, Montana State University, Bozeman, MT 59717, USA

Abstract. There are many plant pathogens that attack weeds, but only a few have proven virulent enough to control weed species and compete with chemical herbicides (R. E. McFayden, *Annu. Rev. Entomol.* 43, 369–393, 1998). One might surmise that there has been strong selection against highly virulent host-specific pathogens, as survival of the pathogen depends upon survival of the host. Total eradication of the host weed would not benefit the pathogen, an impasse that challenges researchers to develop innovative strategies using formulation, genetics, and synergy to enhance the effectiveness of biocontrol pathogens. Our research has capitalized on the inhibitory effects of certain amino acids on plant growth and development. Biocontrol pathogens that overproduce selected amino acids have increased virulence to the target weed and enhanced field performance. We report enhancement of virulence in three separate pathogen-host systems, two with *Fusarium* and one with *Pseudomonas*.

Keywords: *Cirsium arvense*, *Poa annua*, plant pathogen, amino acid, virulence

14.1. Introduction: Plant Disease Epidemics

Severe disease epidemics are rarely observed in native plant or dispersed weed populations.[1] Epidemics are more frequently observed in monocultures that lack genetic diversity and distance between susceptible plants.

Small changes in the fitness or susceptibility of a plant or small changes in the virulence of a pathogen can drastically alter the severity of a plant epidemic. Changes in crop plant resistance can occur rather rapidly due to breeding and more recently genetic engineering. In contrast, changes in pathogen

*To whom correspondence should be addressed, e-mail: dsands@montana.edu

M. Vurro and J. Gressel (eds.), Novel Biotechnologies for Biocontrol Agent Enhancement and Management, 267–275.

populations result from random mutations. Disease-resistant monocultures may enhance selection for pathogens with increased virulence driving further selection of disease resistance.

There is decreased selection pressure in native plant/pathogen populations. First, native plant populations have greater genetic diversity and tend to occur over much larger distances with variable density. Plant disease, regardless of severity, may be contained simply by distance between susceptible hosts.

Weed populations are intermediate in diversity between native plant populations and agricultural crops. Early in a weed infestation, plants are dispersed. However, many weed infestations (*Centaurea biebersteinii*, *Euphorbia esula*) rapidly progress to monoculture providing uniformly susceptible or resistant host populations. The invasiveness of a weed is often correlated to its adaptation to a new environment and the inability of pathogens and insect pests to match its rapid expansion in the new environment. Disease within a plant population should become more prevalent with decreasing diversity within the plant population.

Biocontrol researchers have exerted a tremendous effort to find naturally-occurring pathogens capable of controlling noxious weeds. There are pathogens that will attack weeds. However, there are very few pathogens that suppress weed expansion, much less actually eradicate a weed population. In pathogen-host interactions virulence is expensive and eradication of the host is suicidal. Therefore, parasitism becomes the more beneficial interaction for the pathogen, ensuring longer-term survival of the pathogen.

In our research we have found that every weed so far examined is inhibited by at least one amino acid. This observation leads to the conclusion that weeds have a weakness that can be readily exploited. Subsequent studies have found that the virulence and efficacy of bioherbicides can be greatly enhanced by selecting for variants of weed pathogens that overproduce and excrete amino acids that are inhibitory to a target plant.[2] The host range of the enhanced pathogen remains unaltered and very few plants within the population have been observed that are tolerant of such an amino acid imbalance. Alternatively, the fitness of a weed, and therefore its resistance to plant pests, can be reduced by direct application of inhibitory amino acids.

14.2. Enhancement of Bioherbicides

14.2.1. CRITERIA FOR SELECTION OF BIOCONTROL AGENTS

Classical biocontrol has proven successful in a few situations including biocontrol of rush skeletonweed with *Puccinia chondrilla* in Australia[3] and

Acacia saligna by the rust fungus *Uromycladium tepperianum* in South Africa.[4] These successes have utilized obligate pathogens that are highly host-specific, highly virulent, and capable of naturally spreading from a focal inoculation point. Such pathogens are few and far between. Since virulence of a pathogen can be increased, we can focus only on those pathogens with host specificity and a disseminative nature. Fortunately, there are still a few candidates available for most weeds. There are a number of genera of plant pathogenic fungi and bacteria where there are forma speciales or pathovars that display narrow host specificity including fungi (*Fusarium oxysporum*, several species of *Phomopsis* and *Colletotrichum*, and the rust fungi (*Basidiomycota, Uredinales*)) and bacteria (*Ralstonia*, *Pseudomonas syringae* and *Xanthomonas*). These agents offer the added advantage of being easily disseminated.

14.2.2. SELECTION OF BIOCONTROL AGENTS THAT EXCRETE TARGET AMINO ACIDS

The virulence and efficacy of bioherbicides is enhanced by selection of variants of the pathogen that overproduce and excrete amino acids that are inhibitory to the target plant.[2,5] This approach is modeled after "Frenching disease," a naturally occurring disease of tobacco.[6] Steinberg et al.[7] discovered that saprophytic bacteria growing on the roots of symptomatic plants overproduced a single amino acid, isoleucine. Isoleucine is synthesized in plants via the branched chain amino acid pathway. The end products of the pathway (valine, leucine, and isoleucine) allosterically regulate the activity of acetolactate synthase (ALS). The enzyme is differentially inhibited by these amino acids in different plant species. In "Frenching disease," overproduction of isoleucine by the saprophytic bacteria inhibited the activity of ALS in the tobacco, shutting down synthesis of valine and leucine, which in turn disrupted essential protein metabolism. Interestingly, several modern chemical herbicides mimic this strategy by inhibiting single biosynthetic enzymes in plants, rendering treated plants incapable of producing a metabolite essential for plant growth.[8]

The growth of *Cannabis sativa*, an illicit crop and a noxious weed, is inhibited by the amino acid valine. We isolated variants of *F. oxysporum* f. sp. *cannabis* that were resistant to valine analogs.[9] When analyzed these variants excreted 10–55 times more valine than their wild type parent (Table I).[2] Subsequently, valine-excreting strains of *F. oxysporum* f. sp. *cannabis* were more virulent to *C. sativa* than the wild type parent (Table I). The wild type strain resulted in 25% control of the target plant, while the valine mutants increased control to 70–90%. In addition, the development of wilt disease

TABLE 1. Valine excretion and virulence of wild type and valine overproducing variants of *F. oxysporum* f. sp. cannabis[9]

Strain	Description	Valine excretion* (mg/l)	Mortality rate†	%Kill
C95	Wild-type	0–0.18	6–8 weeks	25
4nv	Norvaline resistant‡	2.84	2–3 weeks	70
6pa	Penicillamine resistant‡	2.48	2–3 weeks	90
8pa	Penicillamine resistant‡	9.93	2 weeks	90

*Valine excretion was bioassayed by spectrophotometric analysis of growth of *Pediococcus cerevisiae* ATCC 8042 in culture supernatant.
†Mortality rate is the duration between inoculation and the first appearance of severe disease symptoms or death (greenhouse studies).
‡Spontaneous mutant strains were selected for their resistance to successively higher levels of valine analogs. Strain 4nv is resistant to norvaline and strains 6pa and 8pa are resistant to penicillamine.

was more rapid in the plants infested with the valine overproducers. Limited studies on fourteen other plant species did not reveal a change in host range.

Thus, overproduction of an essential amino acid provided a highly effective means of enhancing the virulence of a biocontrol agent and has been used to enhance the virulence of *Fusarium oxysporum* f. sp. *cannabis*,[9] *F. oxysporum* f. sp. papaveris,[2] *Pseudomonas syringae* pv. *tagetis* (N. Zidack, personal communication), *Fusarium oxysporum* for control of *Orobanche*[5] and *Xanthomonas campestris* pv. *poae* (A. Pilgeram, personal communication).

14.2.3. INHIBITION OF WEEDS BY AMINO ACIDS

Amino acids, when applied to plants or seeds have a definite effect on plant health.[5] In all cases where noxious weeds have been analyzed for amino acid sensitivity, an amino acid has been found that negatively affects the health of the plant. Inhibitory effects vary and include necrosis, wilting and stunting of growth. Certain amino acids actually enhance the growth and vigor of certain plants. Amino acids are applied to the soil at the base of the plant or drenched over the entire plant. In *Poa annua*, methionine stopped growth of the weed within days of application of the amino acid (Figure 1). Similarly when lysine is applied to *Cirsium arvense*, necrosis was observed on the leaves within days. Application of methionine plus lysine to *Cirsium arvense* resulted in yellowing on new leaf buds as well as necrosis. Other amino acids had little or no effect on the plants.

Figure 1. Growth of *Poa annua* 3 months after the application of 50 mM (left), 100 mM (middle), or 0 mM (right) methionine

14.3. General Methodology

14.3.1. DETERMINATION OF AMINO ACIDS OR COMBINATIONS OF AMINO ACIDS THAT ARE MOST INHIBITORY TO THE GROWTH AND DEVELOPMENT OF THE TARGET WEED

Surface sterilized seed are placed on plates of water agar (1.5% agar, 1 mM Tris, pH 6.8) that have been supplemented with a single amino acid (2–5 mM l-form). The inhibitory effects of amino acids in the branch chain pathway (valine, leucine, isoleucine), the aspartate pathway (lysine, threonine, and methionine) and the aromatic pathway (tyrosine, tryptophan, phenylalanine) can be evaluated as amino acid(s) that decrease seed germination, inhibit shoot growth or cause necrosis (Figure 2). Effects may be seen with single or combinations of amino acids depending on the plant involved.

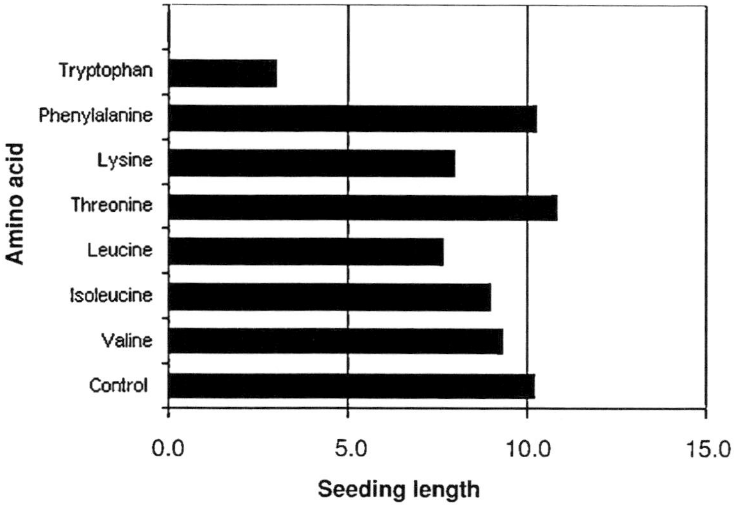

Figure 2. Inhibition of the growth of field bindweed seedlings by selected l-amino acids (Seeding growth was measured 14 days after placing the seed on a water agar plate supplemented with amino acid)

The lowest inhibitory concentrations of amino acids that are inhibitory to a target plant are determined by placing surface-sterilized seed on water agar that has been supplemented with increasing concentrations of the selected amino acid(s) (0 mM, 0.01 mM, 0.1 mM, 0.5 mM, 1 mM, 2 mM, 3 mM, 4 mM).

14.3.2. SELECTION OF VARIANTS OF THE BIOHERBICIDE RESISTANT TO ANALOGS OF THE SELECTED AMINO ACID

Amino acid overproducing strains of each fungus or bacterium can be selected by exposure to specific amino acid analogs.[10] For example, if the target weed is inhibited by lysine, then pathogens for control of that weed are exposed to lysine analogs to select mutants that overproduce lysine. Resistant colonies can be selected using a well zone–diffusion assay on CUTS minimal medium (Czapek-Dox Agar (Difco) (35 g/l) supplemented with ammonium sulfate (0.5 g/l), uracil (20 mg/l), thiamine (4 mg/l) and a vitamin mixture (100 mg of crushed Sesame Street Complete Vitamins). The zone diffusion plates are prepared by cutting a (blank mm) plug from the center of the CUTS plate with a sterile cork borer. The plates are then inoculated with 10^6–10^7 fungal spores, a suspension of $10^3 - 10^5$ mycelial fragments, or a suspension of 10^7–10^8 bacteria. A sterile solution of the amino acid analog (0.1 ml of a 100 mM solution) is then added to the well. The plates are incubated in a laminar flow hood for 4 h. An additional 0.1 ml of the analog solution is added to the well. The plates are incubated until the analog solution is absorbed into the agar and an additional 0.1 ml of analog solution is added to the well. The plates are then incubated at 28°C and monitored daily for the appearance of zones of inhibition and resistant colonies within the zone (Figure 3). Resistant colonies are isolated and analyzed for amino acid excretion. This selection may need to be repeated several times using increasing concentrations of analog and/or different analogs.

14.3.3. ASSAY FOR AMINO ACID EXCRETION

Amino acid excretion is measured assayed by growth of a bacterial auxotroph.[10] The auxotroph is seeded into media lacking the amino acid required for growth. Subsequent growth of the auxotroph in the media is dependent upon and proportional to the quantity of added amino acid. For example, in order to assay valine, a valine auxotroph of *E. coli* (strain CAG18431) is seeded into CUTS media. The auxotroph will not grow unless exogenous valine is added to the media. Colonies of the plant pathogenic fungi or bacteria that are resistant to a valine analog are sub-cultured onto the seeded media. The plates are incubated at 28°C for 2–3 days. If the resistant variants excrete

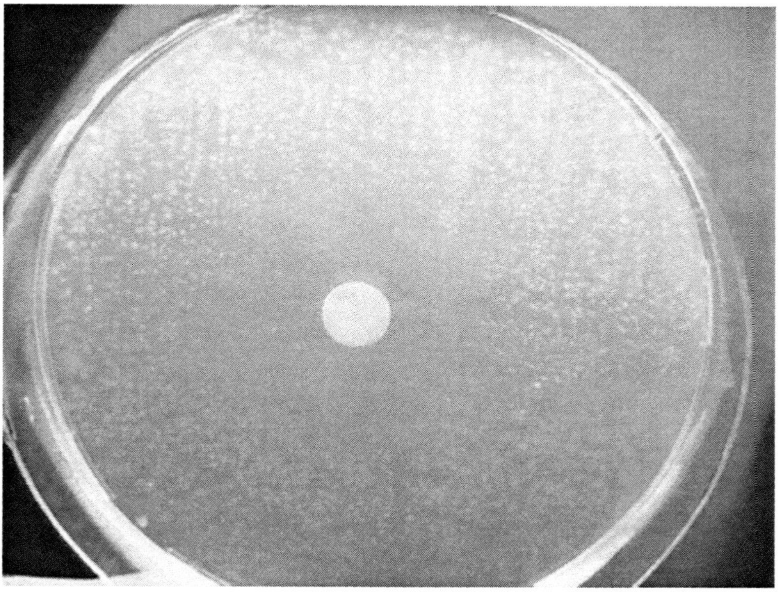

Figure 3. Zone diffusion assay for selection of variants resistant to an amino acid analog. The amino acid analog solution is placed on the disc in the center of the plate. Colonies that grow in the zone of inhibition are isolated and screened for amino acid excretion

valine, there will be a zone of auxotroph growth surrounding the sub-cultured colony. The size of the zone is an indication of the magnitude of valine excretion. A standard dose-response can be determined by placing discs containing various levels of amino acid onto the auxotroph seeded agar.

14.3.4. TESTING VIRULENCE AND HOST RANGE OF THE AMINO ACID OVERPRODUCING VARIANTS IN GROWTH CHAMBER STUDIES

The virulence (rate of kill and % mortality) of amino acid producing variants of each pathogen should be first evaluated in environmental growth chambers in order to eliminate as many external factors that may influence experimental results. In the initial studies, target weed plants are inoculated with each amino acid excreting variant and its respective wild type parent. Amino acid excreting variants that are more virulent than the parent are further evaluated in host range and scale-up experiments.

14.3.5. IMPROVING DISSEMINATION

A soil-applied pathogen will not be an efficacious mycoherbicide, even if it has specificity, sufficient lethality, and long-term soil survival, unless it can be delivered in a cost effective manner. Fungi grown in liquid or solid-phase fermentation are inherently expensive when applied to large acreages at 10^4 spores per gram of soil. Conventional formulation methods with spore suspensions and food-based formulations did not provide enough spores in the root zone of the target weed.[11–14] However, plant pathogenic fungi such as *Fusarium oxysporum* saprophytically colonizes the roots of many non-host plants and thus, *Fusarium oxysporum* mycoherbicides could be delivered to farmer's fields on non-host seed such as crops or grass, positioning the mycoherbicide directly in the rhizosphere of target weed.[2,13,15,16] The multiplication of fungal biomass in the rhizosphere of the carrier seedling allows for application of low levels of the mycoherbicide, greatly reducing the cost of inoculum production.

14.4. Conclusions

Over the last 30 years, numerous pathogens have been investigated as potential bioherbicides. Despite this intensive research effort, few pathogens have been successful as biocontrol agents. The inherent constraints associated with biological species are largely responsible for this failure, yet our preconceived ideas about these agents are also at fault. The authors believe that a paradigm shift must occur if bioherbicides are to enjoy wider success as a weed control method. In the past, researchers have focused on lethality and host specificity as requirements for a successful agent. However, many pathogens that do not meet these criteria could be enhanced by synergistic additions or genetic modification. Embracing new methodologies may allow many "unsuitable" pathogens to be developed into successful biocontrol agents. Likewise embracing collaborations with scientists with other approaches to biocontrol may provide the necessary synergy to implement a successful biocontrol project.

References

1. R. E. McFayden, Biological control of weeds, *Annu. Rev. Entomol.* 43, 369–393 (1998).
2. K. Tiourebaev, Virulence and dissemination enhancement of a mycoherbicide, Ph.D. Thesis (Montana State University, Bozeman, MT, 1999).
3. Hasan, S. and A. J. Wapshere, The biology of *Puccinia chondrillina* a potential biological control agent of skeleton weed, *Ann. Appl. Biol.* 74, 325–332 (1973).
4. M. J. Morris, Plant pathogens and biological control of weeds in South Africa: A review of projects and progress during the last decade, in *African Entomology Memoir No. 1,* edited by T. Olckers and M. P. Hill (Entomological Society of South Africa, Hatfield, 1999), pp. 125–128.
5. M. Vurro, Exogenous amino acids inhibit seed germination and tubercle formation by *Orobanche ramosa* (broomrape): Potential application for management of parasitic weeds, *Biol. Control* 36, 258–265 (2006).

6. R. A. Steinberg, A "Frenching" response of tobacco seedlings to isoleucine, *Science* 103, 329–330 (1946).
7. R. A. Steinberg, Accumulation of free amino acids as a chemical basis for morphological symptoms in tobacco manifesting frenching and mineral deficiency symptoms, *Plant Physiol.* 25, 279–288 (1950).
8. N. Amrhein, Specific inhibitors as probes into the biosynthesis and metabolism of aromatic amino acids, *Rec. Adv. Phytochem.* 20, 83–117 (1986).
9. K. S. Tiourebaev, Biological control of infestations of ditchweed (*Cannabis sativa*) with *Fusarium oxysporum* f. sp. *cannabis* in Kazakhstan, *Biocontrol Sci. Technol.* 11, 535–540 (2001).
10. D. C. Sands and L. Hankin, Selecting lysine-excreting mutants of lactobacilli for use in food and feed enrichment, *Appl. Microbiol.* 28, 523–524 (1974).
11. D. R. Fravel, Effect of temperature, soil type, and matrix potential on proliferation and survival of *Fusarium oxysporum* f. sp. *erythroxyli* from *Erythroxylum coca*, *Phytopathology* 86, 236–240 (1996).
12. B. A. Bailey, An alginate prill formulation of *Fusarium oxysporum* Schlechtend: Fr f. sp. *erythroxyli* for biocontrol of *Erythroxylum coca* var. *coca*, *Biocontrol Sci. Technol.* 7, 423–435 (1997).
13. M. Ciotola, Chlamydospore production, inoculation methods and pathogenicity of *Fusarium oxysporum* M12-4A, a biocontrol for *Striga hermonthica*, *Biocontrol Sci. Technol.* 10, 129–145 (2000).
14. W. J. Connick, Preparation of stable granular formulations containing *Fusarium oxysporum* pathogenic to narcotic plants, *Biol, Control* 13, 79–84 (1998).
15. L. W. Burgess, General ecology of the *Fusaria*, in *Fusarium: Diseases, Biology, and Taxonomy*, edited by P. E. Nelson, T. A. Toussoun, and R. J. Cook (Pennsylvania State University Press, University Park, PA, 1981), pp. 225–235.
16. A. Eparvier and C. Alabouvette, Use of ELISA and GUS-transformed strains to study competition between pathogenic and non-pathogenic *Fusarium oxysporum* for root colonization, *Biocontrol Sci. Technol.* 4, 35–47 (1994).

15. INTERACTIONS OF SYNTHETIC HERBICIDES WITH PLANT DISEASE AND MICROBIAL HERBICIDES

Stephen O. Duke,[1*] David E. Wedge,[1] Antonio L. Cerdeira,[2] and Marcus B. Matallo[3]

[1] *USDA, ARS, Natural Products Utilization Research Unit, P. O. Box 8048, University, MS 38677, USA*

[2] *Brazilian Dept. Agriculture, EMPRAPA/Environment, C.P. 69, Jaguariuna-SP-13820-00, Brazil*

[3] *Weed Science Laboratory, Instituto Biologico C.P. 70, Campinas, SP, 13001-970, Brazil*

Abstract. Synthetic herbicides have the potential to influence plant disease by several mechanisms. They can enhance disease or protect plants from pathogens due to direct effects on the microbe, to effects on the plant, or to effects on both organisms. The particular effect is a function of many factors including the herbicide class and its formulation, the disease species, the plant species, timing of herbicide application and infection, and environmental factors. These secondary effects of herbicides have not been sufficiently studied to fully understand their environmental toxicology implications or their potential for enhanced integrated pest management. Furthermore, understanding these interactions can sometimes be critical in the success of biocontrol of weeds with plant pathogens.

Keywords: glufosinate, glyphosate, herbicide, mycoherbicide, plant disease, plant pathogen, weed

15.1. Introduction

The effect of herbicides on plant disease is an important but generally overlooked aspect of integrated pest management. Nevertheless, understanding herbicide/plant disease interactions can be critical in designing effective and efficient integrated pest management programs. Those involved in biocontrol of weeds with plant pathogens have been keenly aware that these interactions can be crucial contributors to success or failure of this approach to

*To whom correspondence should be addressed, e-mail: sduke@olemiss.edu

M. Vurro and J. Gressel (eds.), Novel Biotechnologies for Biocontrol Agent Enhancement and Management, 277–296.

weed management. Indirectly, through their strong effects on plants, herbicides can influence almost any process or interaction of the plant, including its susceptibility to plant diseases. At sub lethal or non-toxic concentrations, herbicides can have often overlooked influences on plant disease. In some cases, herbicides also have direct effects on plant pathogens. In this short review, we will discuss both types of effect, and, where possible, provide the possible mechanism for the effect.

The topic of herbicide effects on plant diseases has been reviewed previously, either as a single topic[1–4] or as part of a more extensive review on secondary effects of pesticides (several of these are in the book by Altman[5]) or chemical effects on microbial agents for biocontrol of weeds.[6] There have been no recent reviews of this general topic other than a recent shorter version of this review.[7] Much of the older literature was descriptive, without much attempt to discuss the possible mechanisms of the interactions.

Recently, the influence of the herbicides glufosinate and glyphosate on diseases in the transgenic crops that are resistant to these herbicides has become a controversial topic. We discuss the meager literature on this topic in a separate section.

The topic of synthetic herbicide interactions with plant pathogens and plant disease is complicated by the complex interactions of herbicide dose, formulation, soil type and soil biota, environmental conditions, the plant pathogen, and the plant. Furthermore, the timing of the infection with the pathogen versus the herbicide treatment can have a profound influence on the interaction. Thus, the literature often appears to be conflicting, but the apparent conflicts may be due to differences in one or more of the factors involved.

15.2. Direct Effects on Plant Pathogens

Many herbicides are directly toxic to some plant pathogens at rates that are applied to crops or soil. Table I lists some of the herbicides that have this property. Obviously from these data, there are cases of incompatibility between some synthetic herbicides and certain microbial biocontrol agents. There generally seems to be no herbicide mode of action relationship to fungicidal activity, suggesting that the fungicidal activity may have a different mechanism than the herbicidal activity in some cases. However, some fungi and bacteria have the same molecular target sites as plants for herbicides. For example, herbicide enzymatic target sites in the shikimate pathway and branched chain amino acid synthesis pathways are found in both fungi and green plants.[8,9]

There are other papers that show little or no fungicidal activity of certain herbicides on particular plant pathogens, although results such as these are difficult to publish. For example, we have found that neither glyphosate nor its principle metabolic degradation product, AMPA (aminomethylphosphonic acid) to be fungitoxic to *Botrytis cinera, Colletotrichum acutatum,*

TABLE I. Direct inhibitory effects of some herbicides on plant pathogens in the absence of the host

Herbicide	Primary herbicide target	Plant pathogen	Ref.
Bentazon	Photosystem II	*Colletotrichum truncatum*	10
Bromoxynil	Photosystem II	*Rhizoctonia cerealis*	11
		Pseudocercosporella herpotrichoides	11
Clethodim	Acetyl-coenzyme A carboxylase (ACCase)	*Phomopsis amaranthicola*	12
Diclofop	ACCase	*Colletotrichum truncatum*	10
		Alternaria tenuia	10
Diquat	Photosystem I	*Cercospora rodmanii*	13
Diuron	Photosystem II	*Phomopsis amaranthicola*	12
Glufosinate	Glutamine synthetase	*Aspergillis flavus*	14
		Verticillium albo-atrum	15
		Rhizoctonia solani	16
Glyphosate	5-Enolpyruvyl-	*Puccinia lagenophora*	17
	shikimate-3-phosphate	*Dreschlera teres*	18
	synthase (EPSPS)	*Dactylaria higginsii*	19
		Calonectria crotalariae	20
		Phomopsis amaranthicola	12
		Pythium ultimum	21
		Fusarium solani	21
		F. nivale	22
		Rhizoctonia solani	23
Imazapyr	acetolactate synthase	*Phomopsis amaranthicola*	12
Linuron	Photosystem II	*Phomopsis amaranthicola*	12
MCPP	F-Box protein?	*Phomopsis amaranthicola*	12
Metalochlor	Very long chain fatty acid synthase	*Phomopsis amaranthicola*	12
Oxyfluorfen	Protoporphyrinogen oxidase	*Dactylaria higginsii*	19
Paraquat	Photosystem I	*Dreschlera teres*	18
Sethoxydim	ACCase	*Phomopsis amaranthicola*	12
		Dactylaria higginsii	19
Trifluralin	Tubulin	*Fusarium solani*	24
2,4-D	F-Box protein	*Puccina lagneiophorae*	17
		Cercospora rodmani	13

C. fragariae, C. gloeosporiodes, Fusarium oxysporum, Phomopsis obscurans, and *P. vitcola* at concentrations up to 1 mM in an in vitro microtiter plate bioassay (unpublished results). This could be due to different targets, lack of uptake, or degradation of the herbicide by the fungi. In some cases, such as with PSII inhibitors, the fungus does not have the herbicide target site. However, PSII inhibitors are known to also inhibit mitochondrial respiration, albeit at generally higher doses. Table I is not necessarily a good guide to what herbicides are not compatible with microbial biocontrol agents, as the existing literature indicates that interactions can change in the presence of the host plant and that direct effects of the herbicide can vary considerably between pathogens. Also, some of the studies listed in Table I used herbicide doses that might be unrealistically high.

The inhibition of fungal diseases can be through killing the spore or preventing its germination, as seen with herbicide effects on *Phomopsis amaranthicolca* spore viability (Figure 1). Some herbicide adjuvants can also adversely impact fungal spore germination.[12] Mycelial growth and sporulation can be greatly inhibited by some herbicides at field use rates (e.g., Figures 2 and 3). Yandoc et al.[19] also found glyphosate, oxyfluorfen, and sethoxydim to be strong inhibitors of the germination of *Dactylaria higginsii*. These authors observed that the effect on conidial germination was a function of the length of exposure to the herbicide.

It is not unusual for low rates of herbicides to stimulate in vitro pathogen growth[24] or sporulation.[25] Hormesis (the stimulatory effect of a sub toxic level of a toxin) is common with both fungicide effects on fungi and herbicide effects on plants.[26] Thus, dose rates are likely to be highly important in both direct and indirect effects of herbicides on plant disease.

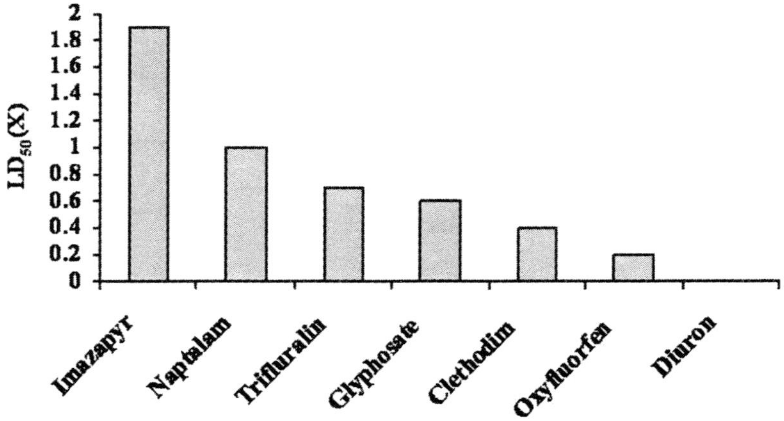

Figure 1. Effects of commercial formulations of several herbicides on conidial germination of *Phomopsis amaranthicola*. The LD_{50} (dose reducing germination by 50%) is given in terms of X, the highest labeled rate. Drawn from data of Wyss et al.[12]

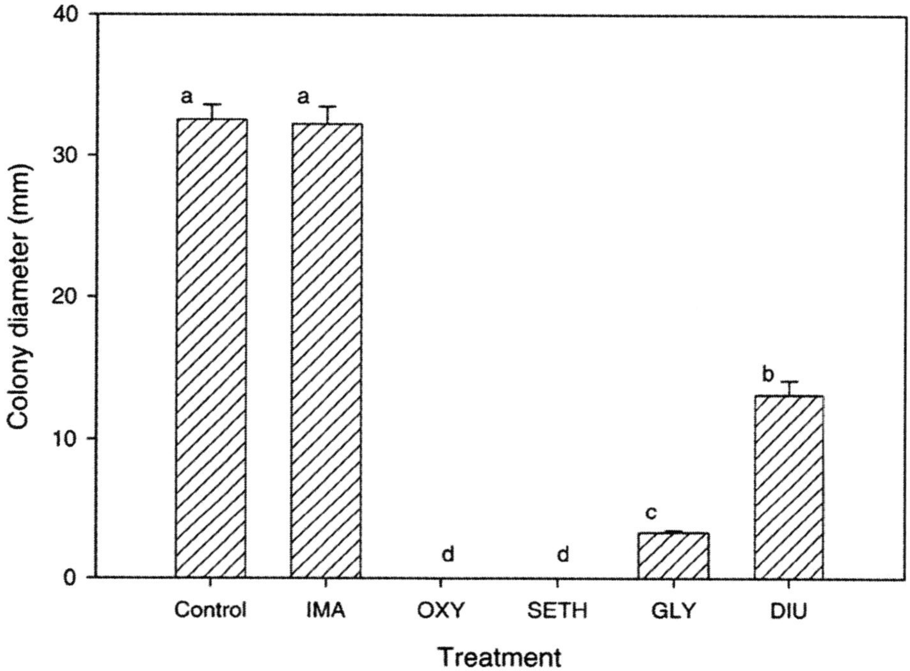

Figure 2. Effects of commercial formulations of herbicides at the recommended field rates on mycelial growth of *Dactylaria higginsii*. Bars with the same letter are not significantly different at $P = 0.05$. IMA = imazapyr, OXY = oxyfluorfen, SETH = sethyoxydim, GLY = glyphosate, DIU = diuron. Reproduced from Yandoc et al.[19] with the permission of the Weed Science Society of America

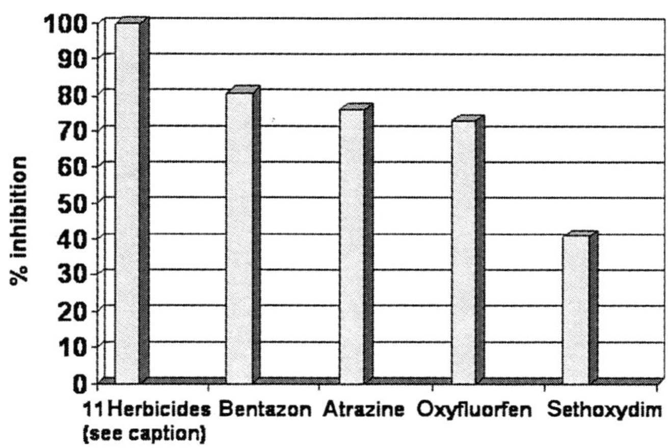

Figure 3. Effects of commercial formulations of herbicides at their highest labeled rate on percent reduction of sporulation of *Phomopsis amaranthicola*. Those causing 100% reduction are diuron, EPTC, glyphosate, imazypyr, linuron, metalachlor, naptalam, paraquat, pendimethylin, simazine, and trifluralin. Drawn from data in Wyss et al.[12]

Metabolic transformation of a herbicide by the plant to a more fungitoxic compound is possible. We are unaware of any examples of this, and we have not found studies looking for such a phenomenon. Such a mechanism might be more probable with herbicides for which crops are naturally resistant due to rapid metabolic degradation of the herbicide.

15.3. Indirect Effects

15.3.1. ENHANCED DISEASE RESISTANCE

Sub toxic levels of herbicides can increase resistance to plant diseases via indirect effects on the crop. Sub lethal oxidative stress can induce synthesis of phytoalexins,[27] and inhibitors of protoporphyrinogen oxidase (Protox) cause oxidative stress via accumulation of the photosensitizing pigment protoporphyrin IX.[28] Protox inhibitors cause enough oxidative stress at sub lethal levels to induce production of phytoalexins in several plant species (Figure 4).[27,29] These authors also found levels of medicarpin and wyerone to be increased by sub lethal aciflurofen treatment. High levels of glyceollin are induced by lactofen in soybeans, resulting in some protection from white mold (Sclerotinia stem rot) (Figure 5).[30]

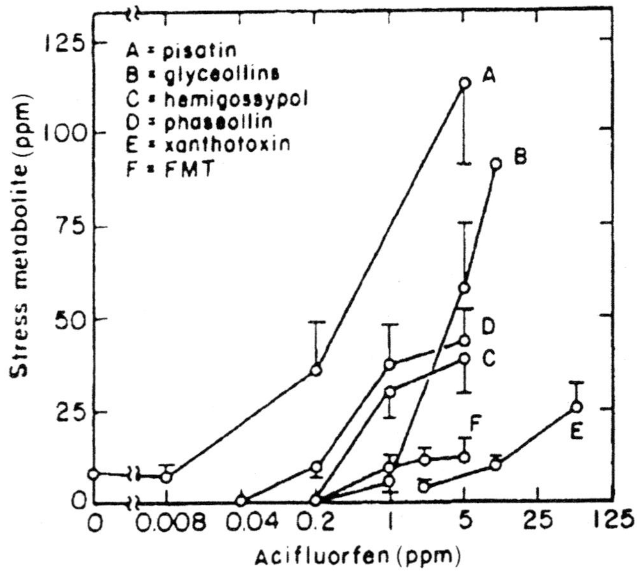

Figure 4. Effects of acifluofen on phytoalexin and secondary product levels in A: peas, B: soybean, C: cotton, D: bean, E: celery and F: spinach. FMT = N-feruloyl-3-methoxytyramine Reprinted with permission form Kömives and Casida[29] (Copyright 1983, American Chemical Society)

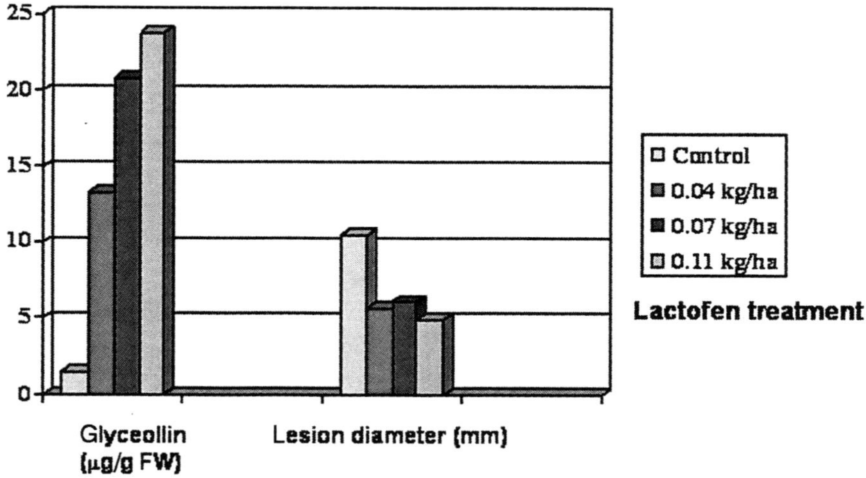

Figure 5. Lactofen effects on glyceollin and lesion diameter in field-grown soybeans infected by *Sclerotinia schlerotiorum*. Drawn from data in Dann et al.[30]

Most other Protox-inhibiting herbicides also induce synthesis of glyceollin in soybeans (Figure 6). The use label for lactofen in the USA indicates that it can be used for white mold management in soybeans (Figure 7). The information on effects of this class of herbicides on plant disease resistance mechanisms suggests that an unintended effect of sub lethal doses of this class of herbicides to off target vegetation or to crop and weed species that are naturally resistant is induction of resistance to plant pathogens, although the cause of reduced resistance to plant pathogens with this class of herbicides can also be a direct effect (see Section 15.2).

Herbicides with other mechanisms of action can also stimulate production of phytoalexins and thereby influence plant disease resistance. For example, pretilachlor and butachlor trigger accumulation of the phytoalexins momilactone A and sakurantetin in rice leaves.[31] A study by Grinstein et al.[32] found that trifluralin potentiated cotton and tomato to produce fungitoxic compounds when treated with vascular wilt-causing fungi. A later study showed that a related herbicide, pendimethalin, induces the synthesis of the phytoalexin tomatine in tomato.[33]

Although the herbicides discussed in this section might be useful for reducing some diseases in some crops, the literature would suggest that they would antagonize efficacy of microbial biocontrol agents. However, some literature[34] indicates that interactions of some of these herbicides with some plant diseases can be synergistic, rather than antagonistic, under some application conditions.

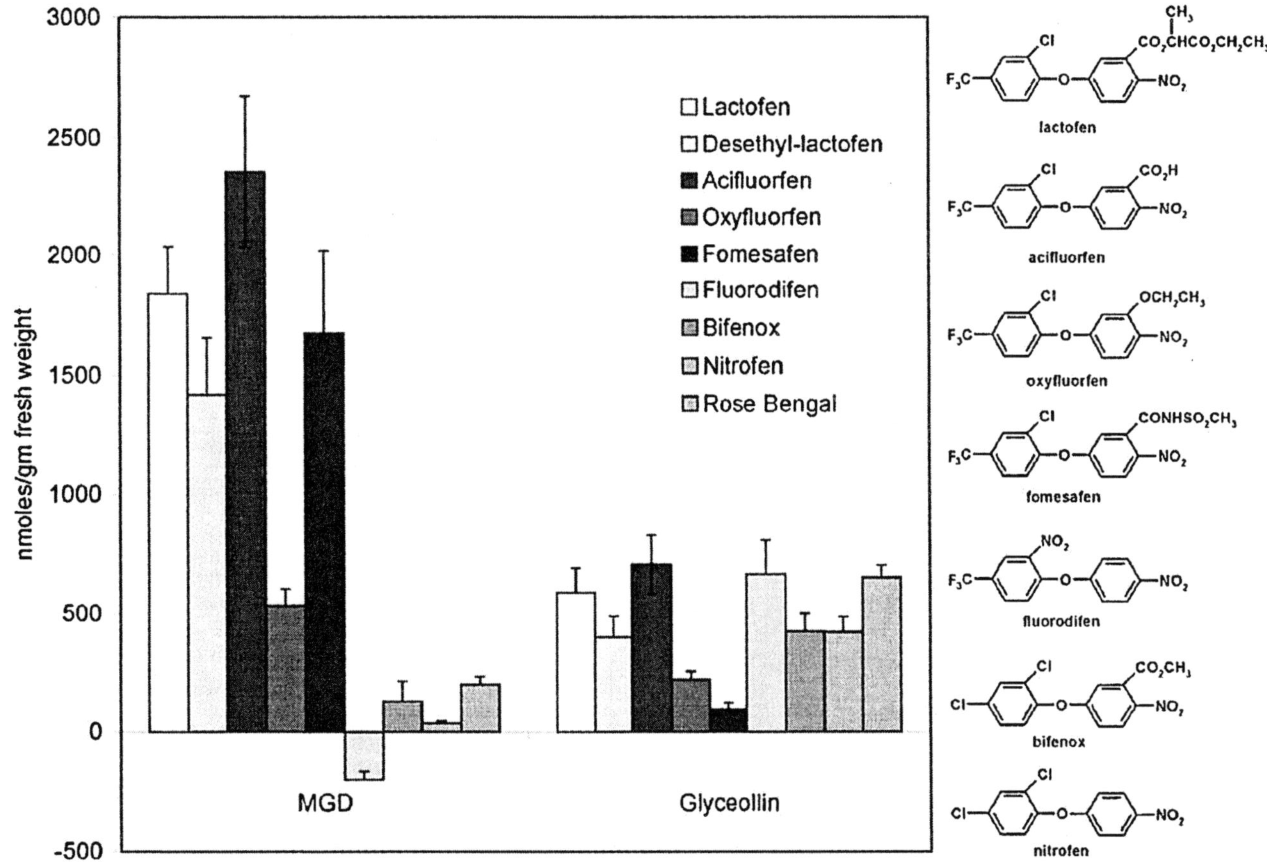

Figure 6. Effects of several Protox inhibitor herbicides and rose bengal, a photosensitizing dye, on accumulation of glyceollin and the 7-*O*-glucosyl-6″-*O*-malonate conjugate of the isoflavone daidzein (MGD) in soybean leaves. Reproduced from Landini et al.[35] with permission from Elsevier

Table 6S.

WHITE MOLD* SUPPRESSION BY *COBRA*
Soybean Growth Stage - 6 to 8 fl. oz./Acre**

Common Name	Soybean Growth Stage	Adjuvant Recommendation
White Mold (*Sclerotinia* stem rot) Suppression	Applications of *Cobra* for white mold suppression in soybeans must be made at or just before 1st bloom. Generally this occurs after the 4th trifoliate is fully expanded.	Crop Oil Concentrate or Methylated Seed Oil 0.5 to 1.0 pt./A Non-ionic Surfactant 0.25% v/v

* The soybean disease white mold is caused by *Sclerotinia sclerotiorum*.

** Rates higher than 8 oz./A will not increase the level of white mold suppression. In cases where rates higher than 8 oz./A are desired for weed control, please refer to Table 8S or Table 2S.

Figure 7. Commercial label information for the herbidide Cobra® with the active ingredient lactofen, for white mold suppression in soybeans in the USA

15.3.2. REDUCED DISEASE RESISTANCE

When a herbicide is toxic enough to a plant to cause significant harm, it may debilitate the defense mechanisms of that plant to pathogens. This is probably at least one of the mechanisms of synergism between several herbicide classes with several mycoherbicides reported by Christy et al. (Table II).[34] Of the 26 combinations, there was no synergy in only six cases. Other herbicide doses

TABLE II. Interactions of herbicides and plant pathogens on their hosts. + = synergy, − = no synergy, NT = not tested. From Christy et al.[34]

	Mycoherbicide species		
Herbicide	*Alternaria cassiae*	*Colletotrichum coccodes*	*C. truncatum*
Acifluorfen	+	+	+
Bentazon	+	+	+
Chlorimuron	−	+	+
Diclofop	+	NT	−
Fluazifop	+	NT	+
Imazaquin	+	NT	+
Metribuzin	+	NT	−
Mefluidide	+	NT	−
Oryzalin	+	NT	+
Sethoxydim	+	NT	−
Thidiazuron	−	+	+

might have been more effective. It was noted that the herbicides did not alter host spectrum for the plant pathogens.

Christy et al.[34] and Hoagland[6] reviewed the literature of other cases of herbicides enhancing virulence of plant pathogens. Since then, other papers have appeared, such as that by Vurro et al.,[36] on the enhancement of efficacy of the mycoherbicide species *Ascochyta caulina* on the weed *Chenopodium album* by reduced rates of the herbicides metribuzin and rimsulfuron. But, the mechanisms of these interactions were unknown or unclear.

Glyphosate is so effective at lowering resistance to plant diseases that it was tested extensively as a synergist for microbial weed biocontrol products (Figure 8).[10] The sulfonium salt of glyphosate synergized the efficacy of an

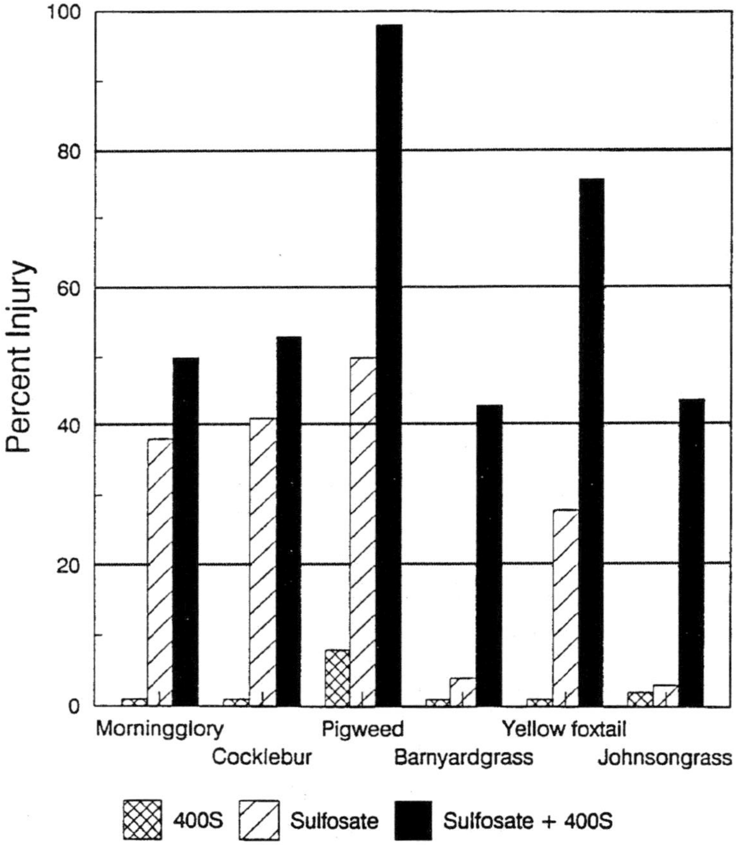

Figure 8. Synergy between the sulfonium salt of glyphosate (sulfosate) at 0.067 kg ai/ha and a pathogenic bacterial preparation (400S) on several weed species. morningglory = *Ipomoea* sp., cocklebur = *Xanthium strumarium*, pigweed = *Amaranthus* sp., barnyardgrass = *Echinochloa crus-galli*, yellow foxtail = *Setaria glauca*, johnsongrass = *Sorghum halepense*. Reproduced with permission from Christy et al.[34] (copyright 1993, American Chemical Society)

TABLE III. Studies in which a correlation of effects of glyphosate on reduced phytoalexin levels and increased susceptibility to a plant pathogen were found

Plant	Phytoalexin	Pathogen	Reference
*Cassia obtusifolia**	Chromenes	*Alternaria cassiae*	38
Glycine max	Glyceollin	*Phytophthora megasperma*	39, 40
		Pseudomonas syringae	41
Medicago sativa	Medicarpin	*Verticillium albo atrum*	42
Phaseolus vulgaris	Phaseolin	*Colletotrichum*	43
		lindemuthianum, Pythium spp.	44

*Renamed *Senna obtusifolia*.

undefined bacterial plant pathogen preparation, but the consistency of the effect in the field was not good. These findings suggest that for some plant pathogens, the indirect effects of glyphosate on the plant outweigh the direct effects of on the microbe (see Section 15.2).

The mode of action of glyphosate is inhibition of the shikimic acid pathway by inhibition of EPSPS.[37] The shikimic pathway produces aromatic amino acids, from which are derived secondary plant products involved in responses of plants to plant pathogens. Soon after the mode of action of glyphosate was discovered, several laboratories showed that sub-lethal treatments of glyphosate caused lowered phytoalexin levels and increased susceptibility to plant pathogens (Table III). Glyphosate reduces synthesis of the shikimate pathway-derived phytoalexin camilexin in *Arabidopsis thaliana*, although this paper did not evaluate the effect of this reduction on plant disease.[27]

Synthesis of the phytoalexin (2-(*p*-hydroxyphenyloxy)-5,7-dihydroxy-chromene) of the weed *Senna obtusifolia* that is induced by inoculation with the mycoherbicide *Alternaria cassiae* is greatly reduced by sublethal doses of glyphosate (Figure 9). Damage to the weed was remarkably enhanced by applying conidia in a solution of 50 μM glyphosate (Figure 10). The effect was observed at a large range of inoculum levels (Figure 10(a)). At a low inoculum dose that caused only scattered necrotic spots alone, a glyphosate concentration that caused no phytotoxicity (50 μM) improved the efficacy of the mycoherbicide to cause death (Figure 10(b)).

Effects on phytoalexin synthesis may only be part of the cause for increased virulence of plant pathogens to glyphosate-treated plants. Liu et al.[44] found reduced lignification and alterations in root exudates caused by sub-lethal exposure to glyphosate also contributed to susceptibility to *Pythium* spp. (Figure 11). Similarly, the growth-stimulation caused by glyphosate on plants at very low doses has been proposed to be due to reduced lignin synthesis.[26] Neumann et al.[45] proposed that glyphosate could reduce resistance to pathogens by limiting micronutrient availability, primarily Mn, but

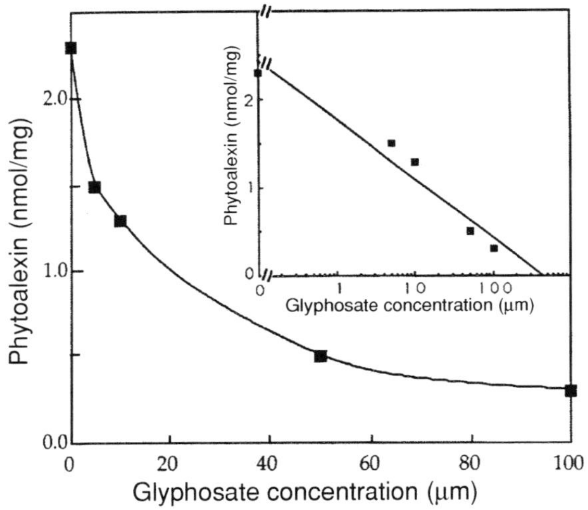

Figure 9. Concentration dependence of accumulation of *Senna obtusifolia* phytoalexin in leaf discs from detached leaves of *S. obtusifolia* inoculated with *A. cassiae* conidia. The I_{50} was 15 μM. Reproduced from Sharon et al.[38]

also Zn, Fe, and B. The mechanism was speculated to be adverse effects on Mn-reducing microbes, but glyphosate chelates 100% of Zn^{2+} at pH $\geq$ 7 and chelates 50% and close to 100% of Mn^{2+} at pH values of 7 and 9, respectively.[46]

Despite the effects glyphosate on plant disease, very little has been done to develop related information that can be used in ecotoxicology assessments, biocontrol of weeds, and/or integrated pest management. These questions have

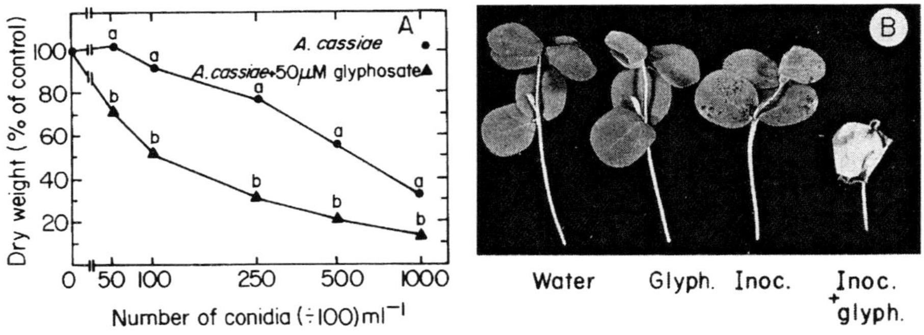

Figure 10. A. Damage to *Senna obtusifolia* caused by *A. cassiae* as measured by seedling dry weight 10 days after treatment, when one true leaf seedlings were spayed to runoff with different innoculum levels with and without 50 μM glyphosate. B. Shoots of *S. obtusifolia* 7 days after treatment with *A. cassiae* conidia . Glph. = glyphosate, Inoc. = 10^4 conidia/ml. Reproduced from Sharon et al.[38] with permission of ASPB

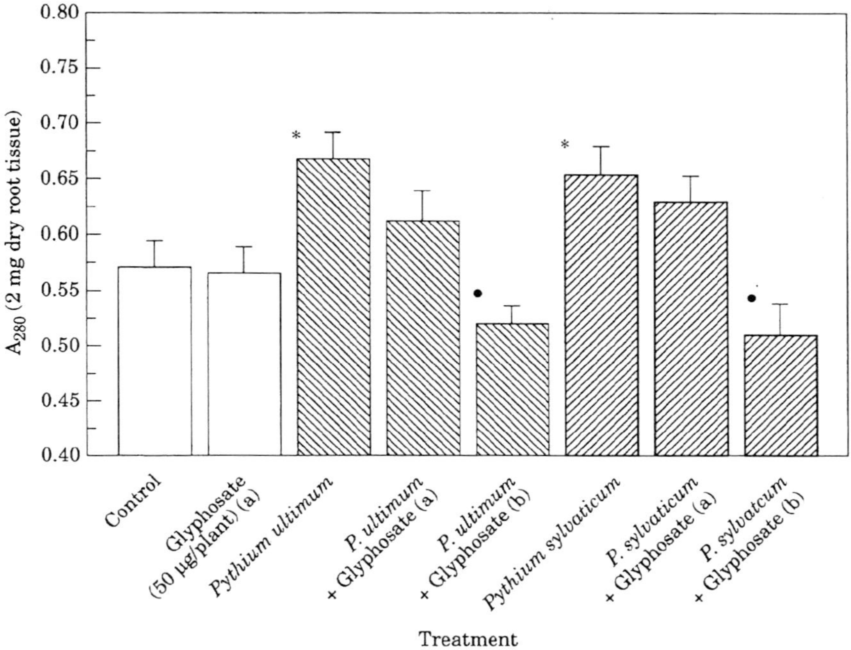

Figure 11. Glyphosate may enhance the virulence of *Pythium* spp. on bean seedling by its reduction of the lignin content induced by *Pythium* spp. in roots of bean seedlings grown in a hydroponic system. Seedlings were inoculated with *Pythium* spp. 2 days after transfer to the hydroponic system. Glyphosate was applied 2 days (a) or immediately (b) after transfer. Lignin content was measured 3 days after inoculation. Asterisks represent significant differences from treatments without *Pythium* spp. and solid dots represent significant difference from *Pythium* spp. inoculation alone. Reproduced from Liu et al.[44] with permission from Elsevier

become especially important, in that glyphosate has become by far the dominant herbicide throughout the world, due primarily to the advent of transgenic, glyphosate-resistant crops.

15.4. Herbicides and Plant Disease in Herbicide-Resistant Crops

Theoretically, there is unlikely to be a significant phytoalexin-mediated effect of glyphosate on disease resistance in glyphosate-resistant crops, as the shikimic pathway is not blocked by the herbicide in these transgenic crops, even though the native EPSPS is inhibited. However, as discussed in detail by Gressel,[47] incomplete expression of the gene for the glyphosate-resistant EPSPS in certain tissues or under some environmental conditions could reduce shikimic acid pathway-mediated disease resistance mechanisms (e.g., lignification, phytoalexins), especially since in non-transgenic crops, very low

TABLE IV. Reports of glyphosate interactions and lack of interactions with plant disease in glyphosate-resistant crops

Crop	Disease	Effect	Reference
Soybean	*Phakopsora pachyrhizi*	Reduces	9, 48
	Fusarium spp.	Increases	49
	S. sclerotiorum	No effect	50
		Increases	51
	F. solani	Increases	52, 53
Cotton	*Rhizoctonia solani*	Reduces	54
Wheat	*Puccinia triticina*	Reduces	9

doses of glyphosate have profound effects on production of lignin and phytoalexins. On the other hand, the sometimes fungicidal activity of glyphosate (Table I) might prove beneficial to glyphosate-resistant crops. However, reports of both enhanced and reduced disease severity have been reported in glyphosate-resistant crops (Table IV).

Recently, glyphosate was reported to have both preventative and curative properties on rust diseases in both glyphosate-resistant wheat and soybean (Figures 12 and 13).[9,48] The effects are apparently through direct effects on the fungi, as there were no effects of glyphosate on systemic acquired resistance proteins. Feng et al.[9] proposed that glyphosate may be fungicidal through effects on fungal EPSPS, based on their analysis of amino acid sequences

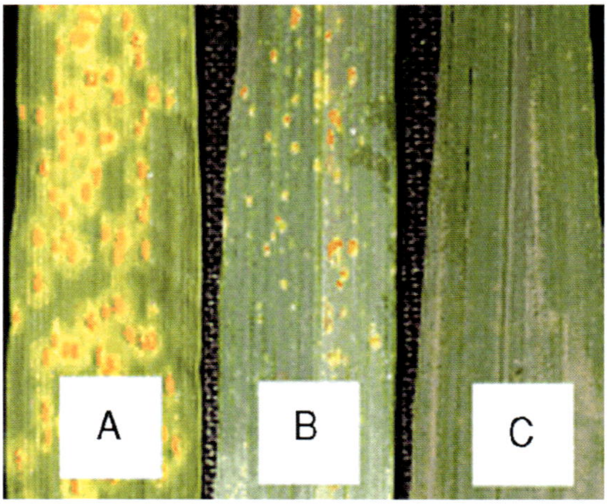

Figure 12. Glyphosate control of wheat rust (*P. triticina*) on transgenic, glyphosate-resistant wheat 13 days after infection. A: No treatment B: Treated 13 days before infection with 0.84 kg glyphosate/ha C: Treated 1 day before infection with 0.84 kg glphosate/ha. Reprinted from Feng et al.[9] (copyright 2005, National Academy of Sciences, USA)

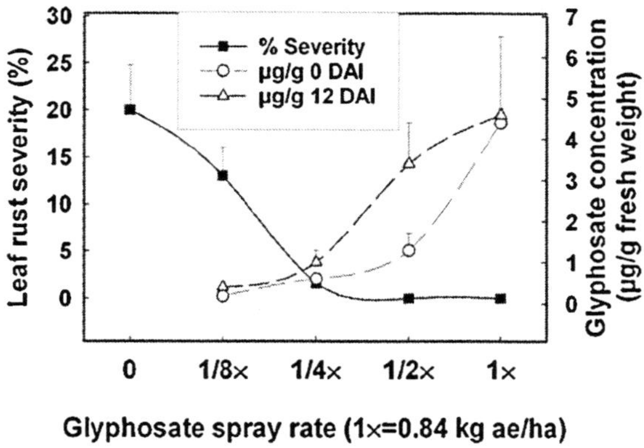

Figure 13. Relationships between glyphosate dose, severity of leaf rust (*P. triticana*), and systemic concentration of glyphosate in inoculated leaf or glyphosate-resistant wheat. Plants were inoculated with the rust 1 day after herbicide treatment, and disease severity was evaluated days after inoculation. Reprinted from Feng et al.[9] (copyright 2005, National Academy of Sciences, USA)

of known fungal versions of the enzyme. Glyphosate is a highly systemic herbicide and is thought to degrade slowly, if at all, in most plant cells.[37] The apparent correlation of systemic glyphosate and disease resistance in the study of Feng et al.[9] suggests that it acts as a systemic fungicide in the case of this disease. There is evidence in transgenic soybean with only site of action resistance that much of the applied glyphosate is eventually metabolically degraded to AMPA.[55] Results of Feng et al.[9] indicate that degradation was insufficient to reduce glyphosate levels sufficiently to impact the effect of the herbicide on soybean rust during the time period of their experiments. Such a systemic activity might be eliminated by glyphosate resistance genes that encode glyphosate-detoxifying enzymes, such as the glyphosate oxidase that is used in glyphosate-resistant oilseed rape and the glyphosate acyltransferase[56] that is being developed for a new generation of glyphosate-resistant crops by DuPont/Pioneer.

The mechanism of enhancement of certain plant diseases by glyphosate in glyphosate-resistant crops is unclear from the literature. Theories to explain these phenomena include: reduced production of shikimate-based defensive compounds, altered root exudates, altered mineral nutrition, adjuvant effects, effects of glyphosate on beneficial microbes, and indirect effects from dying weeds associated with the crop. Definitive research in this area is needed.

The effects of glufosinate in reducing plant disease in glufosinate-resistant crops may be due primarily to direct fungitoxic effects (Tables I and V). This hypothesis is supported out by the fact that glufosinate reduces plant disease in all transgenic, glufosinate-resistant crops in which an effect on disease has been reported, protecting crops from both bacterial and fungal diseases.

TABLE V. Reports of glufosinate interactions with plant disease in glufosinate-resistant crops

Crop	Disease	Effect	Reference
Bentgrass	*Rhizoctonia solani*	Reduces	16, 57
	Sclerotinia homeocarpa	Reduces	16, 57
Rice	*Rhizoctonia solani*	Reduces	58
Soybean	*Pseudomonas syringae*	Reduces	59

Considering the huge areas being planted in herbicide-resistant crops, especially those resistant to glyphosate, and the considerable opposition to these transgenic crops in Europe, one would think that there would be more interest in discovering indirect effects of glyphosate and glufosinate use on plant disease, whether one supports or opposes this technology. Adoption of these crops is increasing, and more herbicide-resistant crops, both additional crop species and resistance to additional herbicides, are being developed.[60] Therefore, the importance of this aspect of the interactions of herbicides and plant disease will grow.

15.5. Conclusions

There have been no organized efforts to analyze the data that exist on herbicide-plant disease interactions in order to understand the conditions, the herbicides and their doses, the species of plants, and the species of pathogens involved to produce principles or generalizations that might be used to predict these interactions. Understanding the mechanisms of the interactions, such as has been accomplished in most cases with glyphosate and protoporphyrinogen oxidase-inhibiting herbicides, should aid in such an effort. Clearly, herbicides have the potential to affect plant disease by many mechanisms. In some cases the direct effects of a herbicide on the pathogen and the indirect effects on the host plant may be in opposition (e.g., glyphosate). Herbicide/plant disease interactions have not been sufficiently studied to fully understand their environmental toxicology implications or for an adequate knowledge of them to enhance integrated pest management. Availability of this information is especially important in the context of biocontrol of weeds with plant pathogens. Much more research will be required to fill our knowledge gaps in this area.

References

1. J. Altman, Herbicide-pathogen interactions in plant disease, *Pestic. Outlook* 2(1), 17–21 (1991).

2. J. Altman and C. L. Campbell, Herbicides and environment: A review on stimulating and inhibiting interactions with plant diseases, *Z. Pflanzenkrankheiten Pflanzenschutz* 86, 290–302 (1979).
3. J. Katan and Y. Eshel, Interactions between herbicides and plant pathogens, *Res. Rev.* 45, 145–177 (1973).
4. C. A. Levesque and J. E. Rahe, Herbicide interactions with fungal root pathogens, with special reference to glyphosate, *Annu. Rev. Phytopathol.* 30, 579–602 (1992).
5. J. Altman (editor), *Pesticide Interactions in Crop Production: Beneficial and Deleterious Effects* (CRC Press, Boca Raton, FL, 1993), 579 p.
6. R. E. Hoagland, Chemical interactions with bioherbicides to improve efficacy, *Weed Technol.* 10, 651–674 (1996).
7. S. O. Duke, D. E. Wedge, A. L. Cerdeira, and M. B. Matallo, Herbicide effects on plant disease, *Outlooks Pest Manag.* 17: in press (2007).
8. S. C. Falco and K. S. Dumas, Genetic analysis of mutants of *Saccharomyces cerevisiae* resistant to the herbicide sulfometuron methyl, *Genetics* 109, 21–35 (1985).
9. P. C. C. Feng, G. J. Baley, W. P. Clinton, G. J. Bunkers, M. F. Alibhai, T. C. Paulitz, and K. K. Kidwell, Glyphosate inhibits rust diseases in glyphosate-resistant wheat and soybean, *Proc. Natl. Acad. Sci. USA* 102, 17290–17295 (2005).
10. J. D. Caulder and L. Stowell, Synergistic herbicidal compositions comprising *Alternaria cassiae* and chemical herbicides, US Patent 4,776,873 (1988).
11. H. R. Kataria and U. Gisi, Interactions of fungicide-herbicide combinations against plant pathogens and weeds, *Crop Protect.* 9, 403–409 (1990).
12. G. S. Wyss, R. Charudattan, E. N. Rosskopf, and R. C. Littell, Effects of selected pesticides and adjuvants on germination and vegetative growth of *Phomopsis amaranthicola*, a biocontrol agent for *Amaranthus* spp, *Weed Res.* 44, 469–482 (2004).
13. R. Charudattan, The role of pesticides in altering biocontrol efficacy, in *Pesticide Interactions in Crop Production: Beneficial and Deleterious Effects,* edited by J. Altman (CRC Press, Boca Raton, FL, 1993), pp. 421–432.
14. K. M. Tubajika and K. E. Damann, Glufosinate-ammonium reduces growth and aflatoxin B1 production by *Aspergillus flavus, J. Food Prod.* 65, 1483–1487 (2002).
15. I. Ahmad and D. Malloch, Interaction of soil microflora with the bioherbicide phosphinothricin, *Agric. Ecosystems Environ.* 54, 165–174 (1995).
16. C.-An Liu, H. Zhong, J. Vargas, D. Penner, and M. Sticklen, Prevention of fungal diseases in transgenic, bialaphos- and glufosinate-resistant creeping bentgrass (*Agrostis palustris*), *Weed Sci.* 46, 139–146 (1998).
17. G. S. Wyss and H. Müller-Schärer, Effects of selected herbicides on the germination and infection process of *Puccinia lagenophora*, a biocontrol pathogen of *Senecio vulgaris, Biol. Control* 20, 160–166 (2001).
18. H. Toubia-Rahme, D.-E. Ali-Haimoud, G. Barrett, and L. Albertini, Inhibition of *Dreschlera teres* sclerotioid formation in barley straw by application of glyphosate or paraquat, *Plant Dis.* 79, 595–598 (1995).
19. C. B. Yandoc, E. N. Rosskopf, R. L. C. M. Pitelli, and R. Charudattan, Effects of selected pesticides on conidial germination and mycelial growth of *Dactylaria higginsii*, a potential bioherbicide for purple nutsedge (*Cyperus rotundis*), *Weed Technol.* 20, 255–260 (2006).
20. D. K. Berner, G. T. Berggren, and J. P. Snow, Effects of glyphosate on *Calonectria crotalariae* and red crown rot of soybean, *Plant Dis.* 75, 809–813 (1991).
21. S. C. Kawate, S. Kawate, A. G. Ogg, and J. M. Kraft, Response of *Fusarium solani* f. sp. *pisi* and *Pythium ultimum* to glyphosate, *Weed Sci.* 40, 497–502 (1992).

22. E. Grossbard, Effects of glyphosate on the microflora: with reference to the decomposition of treated vegetation and interaction with some plant pathogens, in *The Herbicide Glyphosate,* edited by E. Grossbard and D. Atkinson (Butterworths, London, 1985), pp. 159–185.
23. B. D. Black, J. S. Russin, and J. S. Griffin, Herbicide effects on *Rhizoctonia solani* in vitro and Rhizoctonia foliar blight of soybean (*Glycine max*), *Weed Sci.* 44, 711–716 (1996).
24. S.-M.Yu, G. E. Templeton, and D. C. Wolf, Trifluralin concentration and the growth of *Fusarium solani* f. sp. *cucurvitae* in liquid medium and soil, *Soil Biol. Biochem.* 20, 607–612 (1988).
25. R. Charudattan, Integrated control of waterhyacinth (*Eichornia crassipes*) with a pathogen, insects, and herbicides, *Weed Sci.* 34(Suppl. 1), 26–30 (1986).
26. S. O. Duke, N. Cedergreen, E. D. Velini, and R. G. Belz, Hormesis: Is it an important factor in herbicide use and allelopathy, *Outlooks Pest Manag.* 17, 29–33 (2006).
27. J. Zhao, C.C. Williams, R. L. Lasta, Induction of Arabidopsis tryptophan pathway enzymes and camalexin by amino acid starvation, oxidative stress, and an abiotic elicitor, *Plant Cell* 10, 359–370 (1998).
28. F. E. Dayan, J. G. Romagni, and S. O. Duke, Protoporphyrinogen oxidase inhibitors, in *Handbook of Pesticide Toxicology, Vol. 2.: Agents*, 2nd edition, edited by R. I. Kriegr, J. Doull, D. Ecobichon, D. Gammon, E. Hodgson, L. Reiter, and J. Ross (Academic Press, San Diego, CA, 2001), pp. 1529–1542.
29. T. Kömives and J. E. Casida, Acifluorfen increases the leaf content of phytoalexins and stress metabolites in several crops, *J. Agric. Food Chem.* 31, 751–755 (1983).
30. E. K. Dann, B. W. Diers, and R. Hammerschmidt, Suppression of Sclerotinia stem rot of soybean by lactofen herbicide treatment, *Phytopathology* 89, 598–602 (1999).
31. S. Tamogami, O. Kodama, K. Hirose, and T. Akatsuka, Pretilachlor[2-chloro-*N*-(2,6-diethylphenyl)-*N*-(2-propoxyethyhl) acetamide]- and butachlor[*N*-(butoxymethyl)-2-chloro-*N*-(2,6-diethlphenyl)acetamide]-induced acumulation of phytoalexin in rice (*Oryza sativa*) plants, *J. Agric. Food Chem.* 43, 1695–1697 (1995).
32. A. Grinstein, N. Lisker, J. Katan, and Y. Eshel, Herbicide-induced resistance to wilt diseases, *Physiol. Plant Pathol.* 24, 347–356 (1984).
33. A. El-Shanshoury, R. El-Raheem, S. M. Abu El-Sououd, O. A. Awadalla, and N. B. El-Bandy, Formation of tomatine in tomato plants infected with *Streptomyces* species and treated with herbicides, correlated with reduction of *Pseudomonas solanacearum* and *Fusarium oxysporum* f. sp. *lycopersici, Acta Microbiol. Polonica* 44, 255–266 (1995).
34. A. L. Christy, K. A. Herbst, S. J. Kostka, J. P. Mullen, and P. S. Carlson, Synergizing weed biocontrol agents with chemical herbicides, *Am. Chem. Soc. Symp. Ser.* 524, 87–100 (1993).
35. S. Landini, M. Y. Graham, and T. L. Graham, Lactofen induces isoflavone accumulation and glyceollin elicitation competency in soybean, *Phytochemistry* 62, 865–874 (2003).
36. M. Vurro, M. C. Zonno, A. Evidente, A. Andolfi, and P. Montemrro, Enhancement of efficacy of *Ascochyta caulina* to control *Chenopodium album* by use of phytotoxins and reduced rates of herbicides, *Biol. Control* 21, 182–190 (2001).
37. S. O. Duke, S. R. Baerson, and A. M. Rimando, Herbicides: Glyphosate, in *Encyclopedia of Agrochemicals*, edited by J. R. Plimmer, D. W. Gammon, and N. N. Ragsdale (Wiley, New York, 2003), available at: http://www.mrw.interscience.wiley.com/eoa/articles/agr119/frame.html
38. A. Sharon, Z. Amsellem, and J. Gressel, Glyphosate suppression of an elicited response. Increased susceptibility of *Cassia obtusifolia* to a mycoherbicide, *Plant Physiol.* 98, 654–659 (1992).

39. N. T. Keen, M. J. Holliday, and M. Yoshikawa, Effects of glyphosate on glyceollin production and the expression of resistance to *Phytophthora megasperma* f. sp. *glycinea* in soybean. *Phytopathology* 72, 1467–1470 (1982).
40. E. W. B. Ward, Suppression of metalaxyl activity by glyphosate. Evidence that host defence mechanisms contribute to metalaxyl inhibition of *Phytopthora megasperma* f. sp. *glycinea* to soybeans, *Physiol. Plant Pathol.* 25, 381–386 (1984).
41. M. J. Holliday and N. T. Keen, The role of phytoalexins in the resistance of soybean leaves to bacteria: Effect of glyphosate on glyceollin accumulation, *Phytopathology* 72, 1470–1474 (1982).
42. A. O. Latunde-Dada and J. A. Lucas, Involvement of the phytoalexin medicarpin in the differential response of callus lines of lucerne (*Medicago sativa*) to infection by *Verticillium albo atrum, Physiol. Plant Pathol.* 26, 31–42 (1985).
43. G. S. Johal and J. E. Rahe, Glyphosate, hypersensitivity and phytoalexin accumulation in the incompatible bean anthracnose host-parasite interaction, *Physiol. Mol. Plant Pathol.* 32, 267–281 (1988).
44. L. Liu, Z. K. Punja, and J. E. Rahe, Altered root exudation and suppression of induced lignification as mechanisms of predisposition by glyphosate of bean roots (*Phaseolus vulgaris* L.) to colonization by *Pythium* spp., *Physiol. Mol. Plant Pathol.* 51, 111–127 (1997).
45. G. Neumann, S. Kohls, E. Landsberg, K. Stock-Oliveira Souza, T. Yamada, and V. Römheld, Relevance of glyphosate transfer to non-target plants via the rhizosphere, *J. Plant Dis. Protect.* (special issue) 20, 963–969 (2006).
46. H. E. L. Madsen, H. H. Christensen, and C. Gottlieb-Peterson, Stability constants of copper(II), zinc, manganese(II), calcium, and magnesium complexes of *N*-phosphonomethyl)glycine (glyphosate), *Acta Chem. Scand., Ser. A: Phys. Inorg. Chem.* A 32, 79–83 (1978).
47. J. Gressel, *Molecular Biology of Weed Control* (Taylor and Francis, London, 2002), 504 p.
48. J. A. Anderson and J. A. Kolmer, Rust control in glyphosate tolerant wheat following application of the herbicide glyphosate, *Plant Dis.* 89, 1136–1142 (2005).
49. R. J. Kremer, P. A. Donald, A. J. Keaster, and H. C. Minor, Herbicide impact of *Fusarium* spp. and soybean cyst nematode in glyphosate-tolerant soybean. P. 104 in 2001 Agronomy abstracts (ASA, Madison, WI, 2001).
50. C. D. Lee, D. Penner, and R. Hammerschmidt, Influence of formulated glyphosate and activator adjuvants on *Sclerotinia shlerotiorum* in glyphosate-resistant and susceptible *Glycine max, Weed Sci.* 48, 710–715 (2003).
51. K. A. Nelson, K. A. Renner, and R. Hammerschmidt, Cultivar and herbicide selection affects soybean development and the incidence of Sclerotinia stem rot, *Agron. J.* 94, 1270–1281 (2002).
52. S. Sanogo, X. B. Yang, and P. Lundeen, Field response of glyphosate-tolerant soybean to herbicides and sudden death syndrome, *Plant Dis.* 85, 773–779 (2001).
53. V. N. Nijiti, O. Myers, D. Schroeder, and D. A. Lightfoot, Roundup ready soybean: Glyphosate effects on *Fusarium solani* root colonization and sudden death syndrome, *Agron. J.* 95, 1140–1145 (2003).
54. J. H. Pankey, J. L. Griffin, P. D. Colyer, R. W. Schneider, and D. K. Miller, Preemergence herbicide and glyphosate effects on seedling disease in glyphosate-resistant cotton, *Weed Technol.* 19, 312–318 (2005).
55. S. O. Duke, A. M. Rimando, P. F. Pace, K. N. Reddy, and R. J. Smeda, Isoflavone, glyphosate, and aminomethylphosphonic acid levels in seeds of glyphosate-treated soybean, *J. Agric. Food Chem.* 51, 340–344 (2003).

56. D. L. Siehl, L. A. Castle, R. Gorton, Y. H. Chen, S. Bertain, H.-J. Cho, R. Keenan, D. Liu, and M. W. Lassner, Evolution of a microbial acetyltransferase for modification of glyphosate: A novel tolerance strategy, *Pest Manag. Sci.* 61, 235–240 (2005).
57. Y. Wang, M. Browning, B. A. Ruemmele, A. Bridger, J. M. Chandlee, A. P. Kausch, and N. Jackson, Glufosinate reduces fungal diseases in transgenic glufosinate-resistant bentgrasses (*Agrostis* spp.), *Weed Sci.* 51, 130–137 (2003).
58. H. Uchimiya, M. Iwata, C. Nojiri, P. K. Samarajeewa, S. Takamatsu, S. Ooba, H. Anzai, A. H. Christensen, P. H. Quail, and S. Toki, Bialaphos treatment of transgenic rice plants expressing the *bar* gene prevents infection by the sheath leaf blight pathogen (*Rhizoctonia solani*), *Bio/Biotechnol.* 11, 835–836 (1993).
59. W. A. Pline, G. H. Lacy, V. Stromberg, and K. K. Hatzios, Antibacterial activity of the herbicide glufosinate on *Pseudomonas syringae* pathovar glycina, *Pestic. Biochem. Physiol.* 71, 48–55 (2001).
60. S. O. Duke and A. L. Cerdeira, Potential environmental impacts of herbicide-resistant crops, in *Collection of Biosafety Reviews,* (International Centre for Genetic Engineering and Biotechnology, Trieste, Italy, 2005), Vol. 2, pp. 66–143. Available at: http://www.icgeb.org/~bsafesrv/resources/dukecerdeira.pdf

16. APPROACHES TO AND SUCCESSES IN DEVELOPING TRANSGENICALLY ENHANCED MYCOHERBICIDES

Jonathan Gressel,* Sagit Meir, Yoav Herschkovitz, Hani Al-Ahmad,** Inbar Greenspoon,† Olubukola Babalola,‡ and Ziva Amsellem
Plant Sciences, Weizmann Institute of Science, Rehovot, 76100, Israel

Abstract. Inundative mycoherbicides have not been successful in weed control in row crops, probably due to evolutionary barriers, and adding virulence factors was considered essential. Exogenous addition of the products of various genes was used to ascertain synergy as a prelude to adding them transgenically. Transgenically over-expressing single "soft" genes (host lytic enzymes such as pectinase, cellulase and expansins, or natural hormones such as IAA), or "hard" genes encoding toxins such as NEP1 and CP1, has enhanced virulence, but not enough. Gene stacking to obtain synergies among the various genes is considered a top priority, both to achieve sufficient virulence and to delay the evolution of weed resistance to the fungal pathogens.

Keywords: carbohydrases, mycoherbicides, *NEP*1, phytotoxins, transgenic enhancement

16.1. The Need for Enhancement—Exogenous Synergists versus Endogenous Transgenes

Inundative mycoherbicides have rarely been commercialized in row crop agriculture, where they must compete with conventional herbicides. That is not to say there is no need for them; there are many row crop situations where no conventional herbicide can selectively distinguish between crop and related weed. The barrier is often evolutionary: if the specific pathogen had the extreme virulence needed in row crops, it would kill all host plants, and both might become extinct. Thus the need to enhance the potential of mycoherbicides with virulence factors from other sources.

---

*To whom correspondence should be addressed, e-mail: Jonathan.Gressel@weizmann.ac.il
**Present address: Department of Biology & Biotechnology, An-Najah National University, Nablus, Palestinian Authority.
†Present address: Faculty of Agriculture, Hebrew University of Jerusalem, Rehovot, Israel.
‡Present address: Department of Microbiology, Olabisi Onabanjo University, Ago-iwoye, Ogun state, Nigeria.

M. Vurro and J. Gressel (eds.), Novel Biotechnologies for Biocontrol Agent Enhancement and Management, 297–305.

16.1.1. SYNERGISTS AS GENE MODELS

As discussed in Chapter 15, synergists that help overcome host defenses can assist in enhancing virulence of a biocontrol agent. This has a cost of the synergists, and they cannot always be used except in a laboratory situation; e.g., when a biocontrol agent is to be soil applied, the likelihood of its persistence long enough to be effective is minimal. Thus, it is suggested that the synergist be made by the biocontrol agent, genetically engineering the appropriate genes. Exogenously added synergists have a biosafety advantage over engineered synergists insofar as the biocontrol agent is only hypervirulent when the synergist is present (as discussed in Chapter 19).

When a synergist does provide hypervirulence, it provides an inkling about what genes might be transformed to provide hypervirulence. One way to choose putative synergists for testing is to scan the literature on characterized mutants that lost virulence and ask whether adding the missing gene products to the wild type enhances its virulence. We have thus seen that fungal mutants losing the ability to produce auxins, various cell wall and middle lamellae hydrolases, oxalate biosynthesis, callose biosynthesis, as well as phytoalexin biosynthesis have less virulence. This led to demonstration that an antimetabolite preventing phytoalexin biosynthesis,[1] and agents that complex a key co-factor in callose biosynthesis[2] could serve as synergists. This led to engineering one such co-factor (Section 16.2.5). Adding pectinase or cellulase to fungal inocula could synergize virulence (Figure 1), leading to using genes for overproducing cell wall/middle lamellae degrading enzymes (sections 16.2.2–16.2.4).

There are times where there can be an apparent failure from engineering genes for overproduction, based on synergies. Adding genes encoding IAA overexpression to a *Colletotrichum coccodes* specific to *Abutilon theophrasti* did not increase virulence, (Amsellem and Gressel, unpublished data), even though the same gene enhanced *Fusarium* spp. on *Orobanche.*[3] As the requisite enzymes were expressed, we hypothesized that they had insufficient endogenous substrate, and added tryptophan, which greatly enhanced the activity of the transgenic fungi, but not the wild type (Amsellem and Gressel, unpublished data). Thus, you can even synergize a transgenic biocontrol agent. One could also genetically (Chapter 14), or transgenically enhance tryptophan production.

16.1.2. CONCEPT OF "SOFT" GENES VERSUS "HARD" GENES

We divide the genes that are being engineered into mycoherbicidal agents as "soft" and "hard," based on their modes of action, prevalence, and virulence. Those genes whose products are already present in the human food supply and would have "Generally Regarded as Safe" (GRAS) toxicological

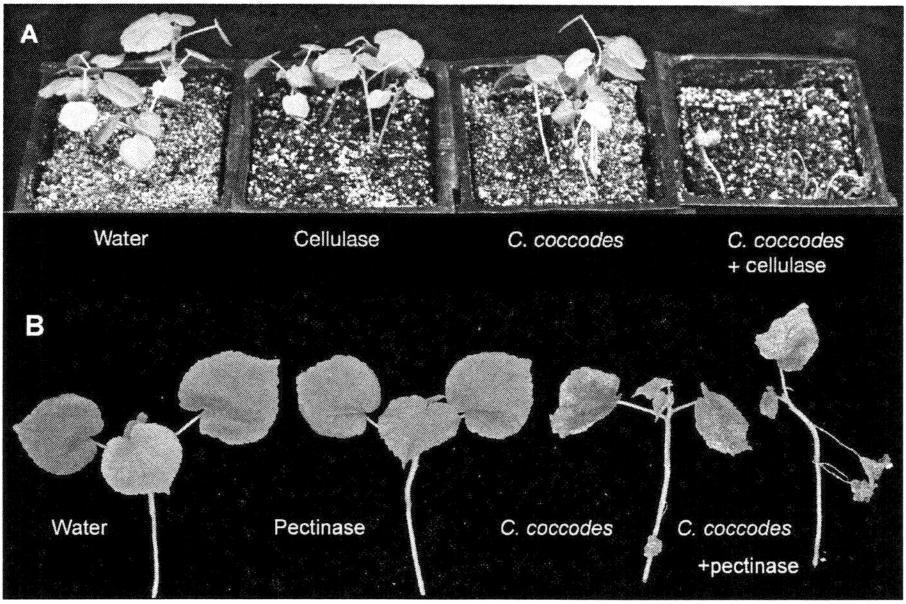

Figure 1. Exogenous application of cellulase and of pectinase increases the virulence of *Colletotrichum coccodes*. Seedlings of *Abutilon theophrasti* were sprayed with chopped mycelia amended with: (A) 10 units/ml of Cellulysin (Calbiochem-Behring) added to the mycelial suspension (2.2×10^6 propagules/ml) or; (B) 1.4 units/ml pectinase (Sigma) added to the mycelial suspension (4×10^6 propagules/ml). The results represent averages of 20 replicates ± SE. The photographs were taken 6 days after spraying

status, would be considered "soft," e.g., carbohydrases, auxin and oxalate. Their affects are also not expected to be as dramatic as "hard" genes such as those encoding phytotoxins. Organisms with hard genes may be harder to get through regulatory channels, but their greater efficacy requires that they be considered.

Nature rarely uses a single solution for a problem, unlike too many of the single "stand alone" solutions used for pest control. It is advisable to learn from nature, and combine genes for hypervirulence. This should give synergistic interactions (or at least additive ones) such that one can get closer to cost effective weed control. A multitude of genes will also make it harder for weeds to evolve resistance to the transgene products in the hypervirulent biocontrol agent.

16.1.3. CONSTRUCTION OF A UNIVERSAL CASSETTE

All the genes we wished to test had already been isolated and cloned. It was necessary to prepare a universal cassette with many cloning sites so that the genes graciously made available to us by colleagues could be easily inserted into a vector that would have the same high expression trpC promoter that we have successfully used on previous occasions.[4] Such a cassette was constructed (Al-Ahmad et al., unpublished) along with a second cassette with a different high expression toxA promoter.[5] The protoplast transformation system that we routinely use allows us to co-transform many genes simultaneously. We have both hygromycin and bleomycin selectable markers so that we can transform strains that have previously been transformed with other genes, and the other selectable marker.

16.2. "Soft" Genes

The various genes that we have obtained from a variety of sources are being transformed into three "real life" model systems being used in our laboratories: (1) Two local isolates of *Fusarium* spp. that specifically attack parasitic *Orobanche* spp., which are higher plants that attack crops; (2) *Alternaria cassae* that attacks the weed *Senna obtusifolia;* and *Colletotrichum coccodes* attacking the weed *Abutilon theophrasti.*

16.2.1. AUXINS

The two genes responsible for bacterial biosynthesis of auxin from tryptophan, *IAA*H, and *IAA*M were transformed into the *Fusarium* spp. When the fungus was pre-cultured on tryptophan prior to preparing inocula, the level of virulence was doubled.[3] While this was statistically significant, it was far from the orders of magnitude increased virulence that was necessary.

When the same genes were transformed into *Colletotricum coccodes*, they had no effect on *Abutilon*, except when tryptophan was added to the mycelial inoculum, as described above. Only then did the epinasty and death typical of auxin herbicides occur.

16.2.2. PECTINASE

Pectinases (polygalacturonidases) are typically used by fungi to separate the cells during penetration, and adding pectinases enhanced virulence. Pectinase genes originating from higher plants have no sequence homology to those of fungi. Thus, we inserted an apple pectinase gene[6] into our universal cassettes, with a feeling of surety that there would be no co-suppression of the gene due to homology with the fungal gene. We found a moderate increase of the fungi virulence (Figure 2). We then co-transformed the pectinase gene with the cerato-platinin gene to further increase of the fungal virulence. Preliminary results of the co-transformants of *Colletotrichum coccodes* show promising results (data not shown).

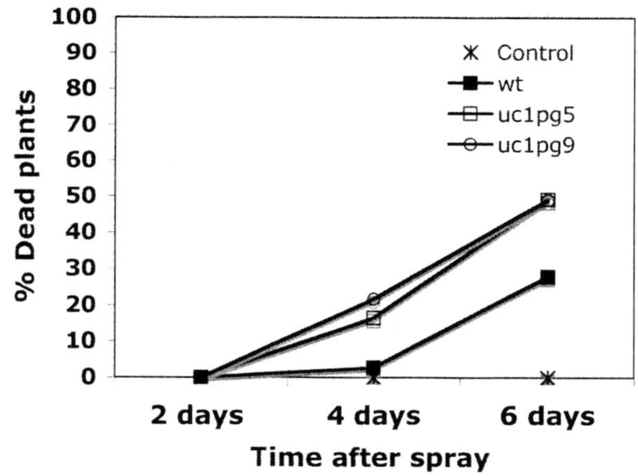

Figure 2. Transformation of pectinase (PG) gene into *Colletotrichum coccodes* enhances the death of *Abutilon theophrasti* seedlings. The pectinase gene[6] was transformed into the fungus under the control of the trpC promoter. The seedlings were sprayed with chopped mycelia. Each treatment is an average of about 20 seedlings. The experiment was repeated three times

16.2.3. EXPANSINS

Expansins are similar to pectinases insofar as they separate cell walls. They are secreted by nematodes upon penetration into plant tissue, allowing them to slither between the cells. We inserted the nematode expansin *Gr-Exp1* gene[7] into our universal cassettes and transformed them to our model fungal systems. The virulence of the *Fusarium* spp transformants increased significantly towards its hosts (Figure 3). We co-transformed the soft gene expansin with the hard gene encoding cerato-platanin to *Colletotrichum coccodes* to further enhance the virulence, which was increased.

16.2.4. CELLULASES

Cellulases are routinely secreted by fungi to assist in dissolving cell walls, releasing free sugars and allowing fungal penetration into cells. Bacterial cellulase genes have little sequence homology to the fungal genes, and thus the bacterial cellulases *celY* and *celZ*[8] will be cloned into the universal cassettes, with the hope that there would be no co-suppression of the fungal gene upon transformation.

16.2.5. OXALATE SYNTHESIS

Oxalate is naturally used by fungi in invasions of plants. Oxalate irreversibly complexes calcium, a macro-element in plants used as a signal and co-factor in many defense responses against fungi, especially callose biosynthesis. We

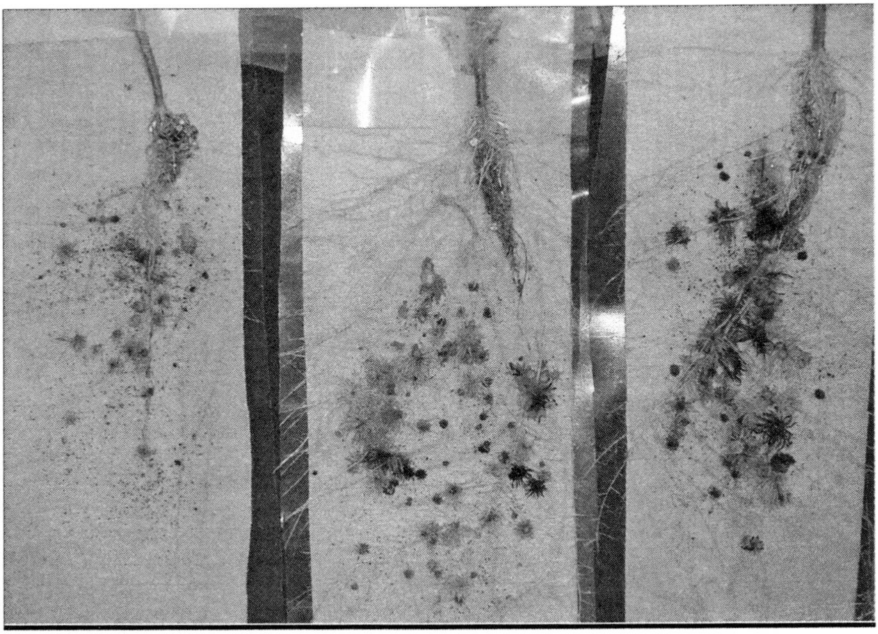

Figure 3. Transformation of expansin *GR-Exp1* (*Exp*) into *Fusarium oxysporum* (FOXY) causes rapid death of *Orobanche aegyptiaca* tubercles attached to the roots and parasitizing tomato. The nematode *GR-Exp1* gene[7] was transformed into *Fusarium oxysporum* (FOXY). The *Orobanche* tubercles attached to and parasitizing tomato roots, were sprayed with chopped mycelia. Photograph taken 5 days after infection. Each treatment is an average of five plants with about 180 tubercles (total). The experiment was repeated three times. Note that tomato itself was unaffected by the transformed fungus

have inserted the *Botrytis cinerea OahA* gene into the universal cassettes. We transformed the *OahA* gene alone and co-transformed it with the cerato-platanin gene to *Colletotrichum coccodes* that is known to induce callose synthesis.[2] Transforming with *oahA* greatly enhanced virulence, which was enhanced even more by co-transformation with *CP*.

16.3. Hard Genes

16.3.1. *NEP1*

NEP1 is a *Fusarium oxysporum* gene encoding a "necrosis enhancing protein," which was once considered to be a potential natural herbicide.[9] It was rapidly realized that it could hardly be made to penetrate plants when used as a stand-alone. We utilized this gene with the high-expression cassette provided by Bailey[9] and found it to be exceedingly potent in enhancing virulence of *Colletotrichum* on *Abutilon,*[4] of *Alternaria* on *Senna* (Safran et al., unpublished data) and *Fusarium* sp. CNCM I-1621 on *Orobanche* (Meir et al., unpublished data). It did not enhance the virulence of the forma specialis of *Fusarium oxysporum* that attacks *Orobanche.* We rapidly discovered that all forma speciales of *F. oxysporum* that we checked bear the gene, but express it at very low levels. For this reason we are excising the native gene, and are reinserting the high expression gene, hoping to obtain hypervirulence with this weed/fungus pair.

The *NEP1* gene in *Colletotrichum* expanded the host range beyond its high specificity to *Abutilon* and it was pathogenic to species such as tomato and tobacco.[4] This is probably because the fungus caused minor injury that allowed the phytotoxin to enter leaves, causing a necrotic lesion that allowed the fungus to attack as a heterotrophic organism—not a true pathogen. When *Fusarium* sp. CNCM I-1621 overexpressing *NEP1* colonized tomato roots "waiting" for *Orobanche* to attack the tomato, it had no deleterious effects on the tomato plants. Thus, when the fungus does not scar the plant, the NEP1 protein does not affect it.

The *Fusarium* sp. CNCM I-1621 with *NEP*1 is still insufficiently virulent for commercial use and will need to be stacked with other virulence genes.

16.3.2. CERATO-PLATANIN

The phytotoxic protein cerato-platanin is produced by the plant pathogenic fungus *Ceratocystis fimbriata f. platani.*[10] This fungus attacks *Plantanus* species (London plane, oriental plane and American sycamore) and causes a canker stain disease. The disease is characterized by foliar wilting and spreading lesions that involve phloem, cambium and extensive regions of sapwood. Cerato-platanin shares some structural and functional characteristics with other fungal hydrophobins.

We inserted the cerato-platanin gene into our universal cassettes and transformed it into our model fungal systems. The cerato-platanin transformants showed virulence enhancement in *Colletotricum coccodes* and *Fusarium oxysporum* (Figure 4). Overexpression of cerato-platanin alone did not enhance the virulence in *Fusarium* sp. CNCM I-1621, thus we co-transformed the cerato-platanin gene with *NEP*1 gene to obtain hypervirulence strain.

16.4. Preserving Enhanced Virulence

It is typical of pathogenic fungi that they lose virulence when continually cultivated on rich media. Such instability of virulence also seems to be the case

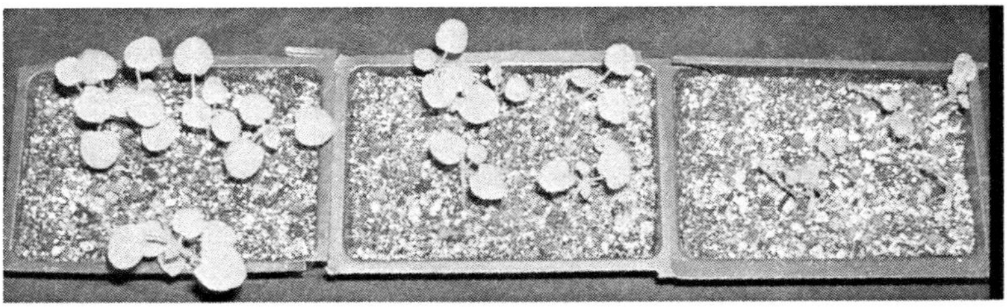

Figure 4. Cerato-platanin (CP) gene enhanced the virulence of *Colletotrichum coccodes* (Coll) on the weed *Abutilon theophrasti*. The CP gene[10] under the control of the trpC promoter was transformed and seedlings were sprayed with chopped mycelia (10^5 propagules/ml). Each treatment was tested on 25 seedlings. The experiment was repeated 4 times. The representative photograph was taken 5 days after spraying

with transformed fungi; they can lose their hypervirulence if not continually passed through host plants, with initial isolates maintained as glycerol stocks at –80°C. In the case of transgenics, Alan Watson (personal communication) found that one of our lines lost virulence, although the transgene was still present. Thus, there is a form of gene silencing that must remain a major concern.

Acknowledgments

The technical assistance of Adi Maoz and Mayan Shaviv at various stages of this project is acknowledged. Bryan Bailey and Mary Strem, USDA, kindly supplied the *NEP*1 gene construct and the polyclonal antibody against the gene product. Linda Ciufetti, David Straney, Amir Sharon, Luigia Pazzagli, Hans Helder, Ross Atkinson, Peter Schaap, and Lonnie Ingram kindly provided the genes used in our research, and Alan Watson and Doug Boyette provided the highly specific isolates of *Colletotrichum coccodes* and *Alternaria casseae.* This research was supported as part of the EU 6th Framework Priority 5—Food Quality and Safety Project: Enhancement and Exploitation of Soil Biocontrol Agents for Bio-Constraint Management in Crops (contract no. FOOD-CT-2003-001687). The information/opinions provided in the paper do not necessarily represent the official position/opinion of the European Commission.

References

1. A. Sharon, Z. Amsellem, and J. Gressel, Glyphosate suppression of an elicited defense response, *Plant Physiol.* 98, 654–659 (1992).

2. J. Gressel, D. Michaeli, V. Kampel, Z. Amsellem, and A. Warshawsky, Ultralow calcium requirements of fungi facilitate use of calcium regulating agents to suppress host calcium-dependent defenses, synergizing infection by a mycoherbicide, *J. Agric. Food Chem.* 50, 6353–6360 (2002).
3. B. Cohen, Z. Amsellem, R. Maor, A. Sharon, and J. Gressel, Transgenically-enhanced expression of indole-3- acetic acid (IAA) confers hypervirulence to plant pathogens, *Phytopathology* 92, 590–596 (2002).
4. Z. Amsellem, B. A. Cohen, and J. Gressel, Transgenically conferring sufficient hypervirulence on an inundative mycoherbicidal fungus for efficient weed control, *Nat. Biotechnol.* 20, 1035–1039 (2002).
5. L. M. Ciuffetti, R. P. Tuori, and J. M. Gaventa, A single gene encodes a selective toxin causal to the development of tan spot of wheat, *Plant Cell* 9, 135–144 (1997).
6. R. G. Atkinson, A cDNA clone for endopolygalacturonase from apple, *Plant Physiol.* 105, 1437–1438 (1994).
7. L. Qin, U. Kudla, E. H. A. Roze, A. Goverse, H. Popeijus, J. Nieuwland, H. Overmars, J. T. Jones, A. Schots, G. Smant, J. Bakker, and J. Helder, Plant degradation: A nematode expansin acting on plants, *Nature* 427, 30 (2004).
8. S. G. Zhou and L. O. Ingram, Synergistic hydrolysis of carboxymethyl cellulose and acid-swollen cellulose by two endoglucanases (CelZ and CelY) from *Erwinia chrysanthemi*, *J. Bacteriol.* 182, 5676–5682 (2000).
9. B. Bailey, R. Collins, and J. Anderson, Factors influencing the herbicidal activity of Nep1, a fungal protein that induces the hypersensitive response in *Centaurea maculosa*, *Weed Sci.* 48, 776–785 (2000).
10. L. Pazzagli, G. Cappugi, G. Manao, G. Camici, A. Santini, and A. Scala, Purification, characterization, and amino acid sequence of cerato-platanin, a new phytotoxic protein from *Ceratocystis fimbriata* f. sp. *platani*, *J. Biol. Chem.* 274, 24959–24964 (1999).

17. FUNCTIONAL GENOMICS: FUNCTIONAL RECONSTITUTION OF PORTIONS OF THE PROTEOME IN INSECT CELL-LINES

Protein Production and Functional Genomics in Cell-lines

Thomas A. Grigliatti* and Tom A. Pfeifer
Department of Zoology, Life Sciences Institute, University of British Columbia, 2350 Health Sciences Mall, Vancouver, BC, Canada V6T 1Z3

Abstract. This chapter describes the assembly and use of a gene expression system that allows a wide variety of proteins to be cloned and expressed in insect cell-lines grown in culture. The system faithfully produces substantial amounts of the gene product and it performs those post-translational modifications that are appropriate for the given protein, and the protein is trafficked to the appropriate sub-cellular compartment. In cases where the researcher wants to have the product exported out of the cell, this can be accomplished using vectors that contain secretion signals, and the product can be recovered from the tissue culture medium using removable protein recovery tags. The system has various applications that include producing large amounts of protein for research or veterinary or medicinal purposes, and assembling testing and modifying constructs to be used in bio-control of genetically engineered pests. Several different genes can be inserted into a single cell type; to date, up to 10 different genes have been placed into a single cell, but this is not the upper limit. Since the individual proteins function as they do in situ, the expressed proteins will interact as they do in situ. Thus, it is possible to reconstitute any portion of the proteome, including biochemical pathways, protein complexes, and combinations thereof. To date we have reconstituted and examined over 20 different pathways and gene complexes. The dynamic and kinetic properties of the individual components and the assembled pathways are indistinguishable from their native counterparts. Thus, the system can be used to determine what is necessary and sufficient for virtually any physiological process. The system can also be used to determine how mutations, either those that occur naturally in the population or genetically engineered, alter the physiological process. Finally, since the cell-lines are permanent and stable in the absence of any selection, they serve as platforms for both the discovery

*To whom correspondence should be addressed, e-mail: grigliat@zoology.ubc.ca

M. Vurro and J. Gressel (eds.), Novel Biotechnologies for Biocontrol Agent Enhancement and Management, 307–325.

and development of bio-active compounds that can be used in bio-control or, in the case of vertebrates, drugs for humans or veterinary purposes.

Keywords: protein expression, insect cell lines, functional proteomics

17.1. Introduction

The following chapter (see Chapter 18) discusses the concept of using genetic engineering and the release of transgenic organisms into a population as a pest management system. The transgenic organisms would introduce a genetically engineered expression vector, which conferred the phenotype "*susceptibility to management,*" to members of the target pest population, with susceptibility to management resulting from the potential to express an incapacitating gene that is driven by an inducible promoter. The engineered construct, conferring "*susceptibility to management,*" would spread to all members of the population within a few generations of its release. Thus, the target population would be pre-conditioned for direct management, and activation of the management process would be elective. There are many possibilities for incapacitation or management strategies. The expression construct would be activated only when the pest became a serious economic or health treat. Population trials conducted in very defined laboratory conditions suggest this concept is feasible. Whether it should attempted in limited field trials, or not, needs significant consideration and debate.

Constructing transgenic pest organisms requires an enormous amount of time and effort. It usually entails a tedious trial and error approach, based on modifications of transgenic protocols and systems that may have been established for very distantly related organisms. Furthermore, to have a hope of success in higher metazoans, genetic transformation demands detailed knowledge of the early embryogenesis and the precise timing and position of establishment of the germ-line tissue in each of the pest species that are targeted for transformation. Even when germ-line transformation is achieved, the success rate, at least initially, is often less than 1%. Hence, germ-line transformation is, at best, a very inefficient method of constructing and testing the expression constructs that might be used in both establishing the transformation protocol and the pest management system. How does one assemble, test and modify genetic constructs to be used in genetically engineered expression constructs, which can be used for pest management strategies and for germ-line transformation?

The most efficient method of creating, testing, and optimizing the efficacy of gene expression constructs is to use cell-lines grown in culture. There are a variety of insect cell-lines that can be used for these purposes. Although these cell-lines are not always derived from the particular pest species, cell-lines from related species or at least genera, often exist. This is certainly the case for insects. Indeed, several hundred different insect cell-lines exist. Since these cell-lines have been derived from many different genera and from different tissue types and stages of development, they provide a large library for testing a wide variety of expression vectors for use in pest management and transformation.

17.2. Background to Gene Expression System in Insect Cell Lines

Some years ago, we faced the problem of being unable to make a useful antibody for an insect protein that had been produced from bacteria, which is the traditional source for proteins or peptides for antibody production. The antibody made against the bacterially expressed protein simply did not detect the protein in its natural state. This forced us to produce the protein in insect cells grown in tissue culture. This work led us to the realization that the codon bias used in insects very closely resembles the codon usage in humans. Basically, we share the same dialect for reading the genetic code. Thus, it became apparent that we could quite likely produce human proteins in insect cells. While this chapter is not about producing human proteins or human therapeutics, that is what provided the research support to create the InsectSelect™expression system, and provided a number of examples of the different types of proteins that can be made in insect cell-lines. Over the years other similar insect-based expression systems have also been created.[1] If a wide variety of human proteins can be expressed in insect cell-lines grown in cell culture conditions, then the same system should easily express insect genes and make insect proteins. By extension, and a slight modification of this logic, one should be able to make expression constructs for virtually any target organism. This includes plants, since many plants have a codon bias that is similar to humans and thus to insects and insect cell lines.

To be useful, an expression vector for insects or insect cell-lines grown in cell culture must have the following properties:

1. It must be able to replicate and survive in bacteria for cloning and maintaining the cloned constructs.
2. It must have a selectable marker for expression and recovery in bacteria.
3. It must be able to "shuttle" between bacteria and the insect or insect cell-line, and thus it must have a selectable marker and be able to survive and function in the eukaryotic host cell-line.
4. Some vectors can enter the cell and function for a short period of time, but are then lost. We require the vector to integrate into the genome of the insect (host) cell-line where it replicates as part of the host genome and thus becomes a permanent part of the genome.
5. It must have an insect promoter that will drive the expression of any gene, regardless of its source.

6. It should have a transcription start site upstream of a multiple cloning site so that any gene can be inserted and expressed, i.e., the nucleotide position at which the gene to be expressed is inserted into the expression cassette is flexible.
7. The vector must be small enough, the smaller the better, to accommodate one or more large genes and still be easy to handle and employ.

17.3. Gene Expression and Protein Production in Insect Cell Lines

We have created several cloning and expression vectors that have all of the properties listed above. We have used several different selectable markers, each of which functions both in bacteria and in insects. Using a single, multifunctional selectable marker reduces the size of the vector. Having several vectors, each with different selectable markers, allows several different genes, or different sets of genes, to be inserted into the insect or insect cell-line sequentially, if necessary. All of the expression cassettes have an insect promoter,[1] a transcription start site and a multiple cloning site that allows any gene to be very easily inserted into, and function within, the expression cassette portion of the vector. These expression vectors and the expression system are very easy to use and produce substantial amounts of product from both transiently transfected or genetically transformed, i.e., permanently altered, cell-lines. Hence, they incorporate all of the features listed above. These cloning and expression vectors proved to be very popular and thus, for the purposes of easy distribution, the system was licensed to Invitrogen, which markets them under the name InsectSelect™. For simplicity, and for ease of reference to the various constructs that are available, we will refer to the system by its trade name, InsectSelect™. An example of one of the vectors is shown in Figure 1. It is very small, about 2.7 kb in size, and functions as a shuttle vector, that is, it can be grown in both bacteria and insect cell-lines. All of the basic vectors are approximately 3.0 kb in size, and thus one or more genes can be added to the vector and it will remain ease to handle. The vector shown in Figure 1 utilizes Zeocin, a broad spectrum antibiotic, as the selectable marker. Zeocin has two advantages: (a) it functions in both prokaryotes and eukaryotes, which helps to keep the size of the vector small, and (b) it provides very rapid selection, usually within 2 to 3 cell-cycles. The vectors are available with several other selectable markers used in place of the Zeocin. In all cases, the vectors contain selectable markers that function in both prokaryotic and eukaryotic cells. This particular vector contains a recovery tag (6×His), which allows rapid purification of the protein product. A variety of different vectors, with and without specific recovery tags, and or secretion signals and with different selection markers, are available.

Once a vector has been constructed, tested, and its function optimized, the two main problems for the production of functional proteins *in vivo* are

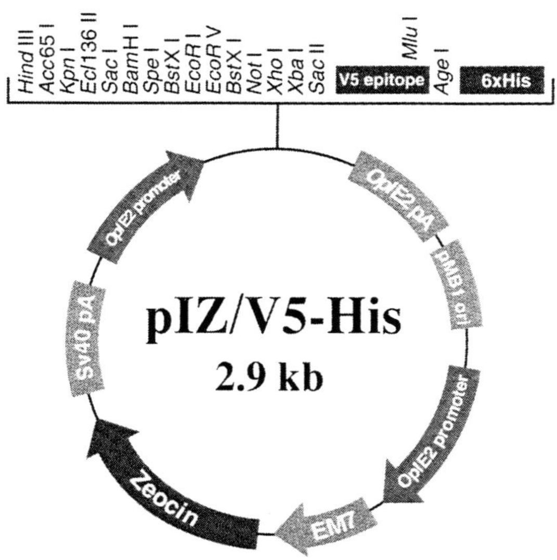

Figure 1. An example of one of the InsectSelect™ vectors. All vectors contain a multiple cloning site, in this case the HindIII–SacII region shown in the line at the top of the vector, and some vectors contain a recovery tag, in this case the 6×His tag. OpIE2 is an immediate early gene promoter derived from a baculovirus: Zeocin is an antibiotic resistance gene

the accretion of the proper post-translational modifications, and trafficking the protein to the correct compartment of the cell to allow proper function. Post-translational modification and protein trafficking to the appropriate intra-cellular compartments require the appropriate intracellular enzymes and cellular machinery, respectively. The absence or malfunction of either process is often fatal for protein production, and the production of an inappropriately modified protein is sometimes fatal to the cell. There are about 160 different types of post-translational modifications that occur in the various cells and tissues of higher organisms. Clearly, cell and tissue-types differ in their post-translational capacities, since not all cells require all of these modifications. Likewise, cell types differ in their protein trafficking capabilities. Hence, no single cell type is sufficient for testing the function of all gene expression cassettes, and the production and function of every "protein of interest," This provides a significant challenge to constructing and testing the function of various expression vectors and "disabling" genes or systems.

The solution is to create a small library of cell-lines derived from different tissue types, different stages of development, and/or different insect genera, and test the function of the expression cassette and production of the protein of interest in all cells in this library. Fortunately, this library need not be extensive. We've found that five to seven different cell-types are generally sufficient to express and produce functional protein from virtually any gene. To date, we've expressed over 50 different genes, producing a variety of types of protein product, and have had 100% success. No protein was produced efficiently in all cell-lines, but every protein was produced in goodly amounts in at least one, and often several, different cell-lines. Clearly, these data indicate that a particular protein will be expressed well in a subset of tissues in the transformed organism.

17.3.1. CREATING TRANSFORMED CELL-LINES

17.3.1.1. *Testing Protein Expression*

Genomic DNA (introns will be removed by the insect cell-lines) or cDNA of the gene to be expressed (protein to be produced) is cloned into the expression vector by inserting it into the multiple cloning site (a multiple cloning site is a collection of restriction endonuclease target sites—often 6–12—that are clustered together and present at only this one site in the vector, see Figure 1). The expression vector is then introduced into the insect cell lines via liposomes, electroporation, ballistic transformation, or calcium phosphate based infusion processes. We generally use liposomes, since the protocol is simple and efficient.[2,3] The construct will enter the nucleus within an hour and its protein product is usually detectable within 6 h. Protein production usually peaks about 48 h post-transfection, and the protein continues to be produced for up to 7 days. This gene expression and protein production is transient, since the expression construct, while resident within the nucleus, typically has not inserted into the genome and is eventually destroyed. However, it is quite easy to transfect five or six different types of cell-lines at the same time, and, within a day or two, to determine which cell-lines are producing the product and which are not. In fact, one can collect several milligrams of the product within a week from these transiently transfected cell-lines, if the process is scaled up appropriately, and this is often more than enough for antibody production, and for many other functional assays and uses. However, permanently transformed cell-lines can be established with little effort.

17.3.1.2. *Establishing Permanently Transformed (Stable) Cell-lines*

Permanently transformed (stable) cell-lines expressing the heterologous gene can be made quite simply using the following protocol. After determining which cell-lines are producing functioning product, which takes about 24–48 h (see above), the appropriate compound for the selectable marker is added to the medium in which the transiently transfected cell-lines are growing.[3] We often use Zeocin, since, as stated above, it selects against unprotected cells within a few cell cycles, and it works in both bacterial and eukaryotic cells, allowing us to use one selection system and thus keep the vectors small. The cells are grown under selection for several cell cycles. We usually establish polyclonal cell lines with 2–4 weeks of selection. Only those cells in which the expression vector has integrated, intact, into the genome of the cell will survive the selection process. The Zeocin selection kills all of the cells in which the vector did not integrate into the genome. Hence, polyclonal, permanently transformed cell-lines can be established in 2–4 weeks.[2] Clonal cell-lines, i.e., cell-lines in which all cells are derived from a single cell and thus are genetically identical, can be established in 6–8 weeks using cloning rings or dilution-enrichment protocols.[3]

17.3.2. STABILITY OF TRANSFORMED INSECT CELL LINES

Once the expression construct has inserted into the genome, the selection agent, e.g., Zeocin, can be removed and the expression construct remains integrated within the genome and continues to function, i.e., to produce its protein product. This is not necessarily the case for transformed mammalian cell-lines. In mammalian cell-lines, if antibiotic selection is removed, as it must be for the production of proteins or peptides that are destined to be used as therapeutics, the inserted gene is often silenced after several weeks, and, in some cases, the expression construct is eventually purged from the genome and lost. So, it was surprising that insect cells continued to produce large amounts of product over many months when grown in the absence of any antibiotic selection.

The stability of the transformed insect cell-lines was examined in two different types of experiments. First, we simply removed selection and then grew the cell lines for 24 months in the absence of any antibiotic selection pressure. Since the cell-cycle time is about 30 ± 5 h, depending on the cell type and its origin, this represents about 500 cell cycles. At least once each month, over the 2-year period, we measured the amount of protein produced and performed a Southern blot analysis on an aliquot of the cells. Protein production remained constant over the 2-year period, that is, the cells made about the same amount of protein per cell at each sample time. The Southern blot analyses showed several bands, suggesting that each cell contained several copies of the expression construct, and these constructs were integrated into different sites within the genome. Moreover, the pattern of the Southern blots did not change over time, suggesting that the number and position of the integrated expression cassettes did not change over time. In the second test, the transformed cells were removed from selection after four weeks, and the cells were allowed to grow in the absence of selection for 6 months. Then, the culture was divided into six aliquots of equal size and subjected to selection using Zeocin at 0 (no selection), 50, 100, 250, 500, 1000 μg/ml. Selection at these six concentrations was applied for 1 month (about 24 cell cycles) and then the number and position of the inserted expression cassettes was determined by Southern blot analyses. Again, the number and size of the bands on the

Southern blots were indistinguishable, suggesting that neither the number nor the position of the integrated expression constructs had changed, regardless of the intensity of the selection. Three conclusions can be drawn from these data. First, multiple copies of the expression cassette are integrated into the genome of each cell. Second, they integrate at different sites within the genome. Third, once the genetically engineered expression cassettes have integrated into the genome, they do not appear to move, either when taken off selection, or in response to renewed selection. These attributes are the likely foundation for the stability of the transformed insect cell lines and there is no reason to believe that they would differ in germ-line transformation. Indeed, in the insects that have been transformed to date, germ-line transformation generally has had the same result. That is, the expression construct can integrate at many different sites within the euchromatic portion of the genome, and when more than one copy of the construct integrates, each copy is usually found at a different site within the genome. Multiple integration provides genetic redundancy, and it assures that relatively large amounts of functional protein is made from the inserted gene or genes, in those cell-types that have the appropriate post-translational modification and trafficking capabilities.

17.3.3. TYPES OF GENES EXPRESSED AND PROTEINS PRODUCED

Over the past decade, more than 50 different proteins have been expressed using the InsectSelect™ system. Often the proteins produced were those that were problematic in other systems.[4] We have been able to express all of these genes and produce their protein products in ample quantities. We've produced proteins that function as intracellular proteins, for example Mek, Erk, Ro52, and ß-glucocerebrosidase, those that function as membrane associated proteins such as G-protein coupled receptors, receptor tyrosine kinases, glutamate transporters, melanotransferrin, and ion transporters, and those that are secreted out of the cell, such as Factor X, and Interleukin-6. In cases, where the researcher wanted to purify the expressed protein from the medium in which the cells grow, rather than harvest the cells, we were able to engineer the construct so that the protein product was secreted out of the cell into the medium where it could be easily purified using a recovery tag.[5] In all cases the recovery tag is engineered so that it can be removed quite easily from the protein after purification. Some examples follow.

17.3.3.1. *Human Melanotransferrin*

Human melanotransferrin is a 97 Kd protein with a very complex structure, including 14 cysteines and 7 disulfide bridges, and is heavily modified after translation.[6] In human cells, the protein is exported out of the cell and then bound to the outside surface of the cell membrane of by a glypiated anchor. The protein was not functional when produced in bacteria or yeast expression systems. The human melanotransferrin was produced in both a *Drosophila* SL2 cell-line and a *Spodoptera frugiperda* (fall armyworm) Sf-9 cell-line, but was not produced in *Lymantria dispar* (gypsy moth) Ld652Ycell-line or in the *Drosophila* Kc cell-line, in fact expressing the gene caused the cells to die.[7] This provides a clear demonstration that cell-lines, from different origins, differ in their protein production, post-translational modification, and/or trafficking capabilities. In the lepidopteran Sf9 cell lines the melanotransferrin was not only produced, it was also exported and attached to the exterior of the cell membrane just as it is in mammalian cells. Hence, this series of experiments demonstrates that different cell-types, even those derived from the same species, can differ in their post-translational modifications and/or protein trafficking capabilities. Moreover, it demonstrates that mis-expression, that is expression in cell-lines that lack one of more of these post-translational or protein trafficking capabilities such as the *Drosophila* Kc line, can lead to cellular dysfunction and eventually cell death. Hence, mis-expression of a normal cellular function, that is expression in an inappropriate cell or tissue type, may be sufficient to incapacitate an organism.

17.3.3.2. *Schistocerca gregaria Ion Transport Protein*

For the purposes of insect pest control we expressed a gene that encodes an ion transport peptide from the pest insect, *Schistocerca gregaria* (desert locust). The Ion Transport Peptide (ITP) is a neuropeptide produced in the nervous corpora cardiaca (NCC). It stimulates an electrogenic Cl^- pump in the apical membrane of the ileum and rectum.[8] ITP stimulates the uptake of Cl^-, K^+, Na^+, and fluid resorption in the locust mid- and hind-gut.[9] Consequently, it is an antidiuretic factor. Transcription the ITP gene and translation of its mRNA produces an inactive preproprotein that is 130 amino acids long. To be activated, the preproprotein must be cleaved at two internal locations to produce a 73 amino acid peptide, which is then amidated, and subsequently secreted into the circulatory system. The gene was expressed and the ITP protein was properly cleaved and post-translationally modified in Drosophila Kc1, *Tricoplusia ni* Hi-5, and *Spodoptera frugiperda* Sf9 cell-lines.[10] The ITP product from all 3 cell-lines was active, but the Kc cell-lines were able to produce the most product per cell. The mature, functional ITP was exported from the cell-lines into the cell culture medium, just like it is from the NCC cells in situ. The secreted product was very bioactive; so active, that we only needed to spin the cells down from the culture medium and then apply 5 μl of the spent culture medium to the ileum to get a dramatic response in a living bioassay.[10] Unfortunately, at this time, there is no protocol for transformation of *Schistocerca gregaria*. Hence, the system cannot be tested in the pest itself, other than by bioassays on the tissue. Nonetheless, the results of tissue assays suggest that either over-expressing the gene or expressing an antagonist would effectively devastate the organism. An antagonist of the ITP could easily be constructed by modifying the gene. The antagonist need only bind to the receptor with an affinity equal to or higher than the 73 amino acid protein, and thus block activation of the pump. This requires some simple genetic engineering. And again, the action and competitiveness, binding affinity, of the antagonist is easily tested using products produced in the cell-line system.

17.3.4. INDUCIBLE PROMOTERS

One of the tenets of the TAC–TICS system is the ability to induce the expression of the "disabling gene" or genes only in response to a very specific, and otherwise benign, compound or trigger. To do this, the expression of the "disabling gene" must be driven by an inducible promoter, a promoter that activates transcription only in response to a very specific activating agent. Several examples of inducible promoters exist. In bacteria, the expression of the lactose operon is induced by the presence of lactose sugar or the compound IPTG (isopropyl-beta-D-thiogalactopyrnaoside) in the medium, and is epistatically repressed by the presence of glucose, the preferred energy and carbon source, in the medium. In eukaryotes, the promoters of the heat shock genes are the most studied and best characterized inducible promoters. The heat shock promoters respond to a rapid change in temperature and activate the expression of a number of genes whose products effectively protect the cell from the temperature shock. In fact, the heat shock promoters respond to a number of different stress conditions, in addition to heat shock. Nonetheless, a wide variety of other types of inducible systems exist in eukaryotes. These include ecdysone, juvenile hormone, and other hormone inducible systems, the ITP system we described above, the metallothionein promoters, and a variety of others. In addition, several synthetic inducible systems have been created.

To examine the utility of inducible promoters we used the metallothionein promoter, which responds to low levels of certain metals, to drive the expression of the human melanotransferrin gene. As stated above, when expressed in *Drosophila* Kc1 cell-lines, the human melanotransferrin protein was toxic if constitutively expressed. We created an expression cassette where the *Drosophila* metallothionein promoter drives the heterologous "gene of interest." The gene, in this example the human melanotransferrin gene, is not expressed in the absence of the inducer. However, when low concentrations of copper or zinc are added to the medium, the gene is transcribed and the protein is produced. Using this inducible promoter, the expression of the human melanotransferrin gene could be induced for short periods of time, about 3–4 h of synthesis per cell cycle, and the Kc1 cells would produce and accumulate the human melanotransferrin protein within the cell. We have done similar experiments using the heat shock promoter in transformed cell-lines and in genetically transformed insects, and the gene expression is activated, and product made, only in response to induction. Hence, inducible promoters can be used to control the induction of gene expression and the duration of protein production, and thus to some extent they can be used to control the amount of product made.

17.3.5. APPLICATIONS OF PROTEIN PRODUCTION IN CELL CULTURE

While the numerous experiments and tests that were required to develop the final series of InsectSelect^{TM}expression vectors has not been described herein, it should be clear that without using the insect cultured cell-lines this process would have been a daunting task. These gene expression cassettes were then placed into the Drosophila transformation vector, containing several hundred base pairs of the P transposable element ends. This vector was then used to transform Drosophila pre-blastoderm embryos, using a helper plasmid. The helper plasmid supplies the functioning transposase, but it lacks the P transposable ends, and thus it cannot insert into the genome. The transformation/expression vector inserts into the nuclei of the syncitial blastoderm, including those nuclei that become the nuclei of the germ-line cells. Thus, the offspring of the transformed insect contain the expression cassette and genetically engineered insect strains or lines are established. The expression cassette functioned in the insects just as it did in the cell-lines grown in tissue culture. Of course, since we used a "helper" construct to provide the transposase in trans, these transformed lines lack the transposase (the "helper" construct is unable to integrate into the genome and is lost), and thus the expression cassette is not dispersed throughout the population. In the TAC–TICS system, we use a transformation construct that contains an intact transposase gene as well as the expression cassette. While slightly larger in size, the TAC construct is still easily handled and manipulated.

In addition to testing the function of TAC–TICS-type expression cassettes in cell-lines prior to using them for germ-line transformation of the target organism, there are many useful applications of the InsectSelect^{TM}, or similar expression systems. The applications include, but are not limited to, the production of protein for use as reagents, the production of therapeutic proteins and peptides for medicinal or veterinary uses, the production of large amounts of protein for 3D crystallographic studies, the production of large amounts of peptides for pest control use, and the production of large amounts of antibodies.

17.4. Functional Genomics: Reconstituting Physiological Pathways

The expression system allows us to produce proteins whose functions are virtually indistinguishable from their in situ produced counterparts. Since this is the case, we wondered if we could transform a single cell with several different expression constructs and reconstitute virtually any portion of the proteome and have it function just as it does in situ. The simple answer is yes; we've been able to put up to 10 different genes into a single cell and have them function. Ten is not the upper limit, we just have not tried to place more than ten into a single cell.

What value does this have? We believe that virtually any metabolic pathway, any protein complex, or combinations thereof, for any physiological or developmental event can be reconstituted in this system. Clearly there will be some exceptions, but so far they have not materialized. The ability to reconstitute a portion of the proteome and have it function properly, allows one to define what is necessary and sufficient for the function of that physiological pathway or response. It allows one to study the effect of mutations in any of the genes/proteins in the pathway, both naturally occurring and synthetically created variants, and define their impact on the physiological system. In addition, as the cell lines are stable, they can be used as platforms to screen for peptides or bio-active compounds that disrupt the pathway and thus alter the physiological process and its outcome. When applied to a pest management scenarios, one could screen for compounds that block a pathway and thus disrupt the normal function, or one could screen for compounds that activate the pathway in a particular cell-type or at a developmental interval where or when it should not occur. This includes identifying agents that can disrupt feeding, mating, mobility, reproduction, or development of a pest species. Disrupting, or markedly destabilizing, any of these functions would dramatically impede the pest, and thus severely alter the dynamics of the pest population growth. We will present two examples in which portions of a proteome can be reconstituted in insect cell lines for potential screening of bio-control related agents. The two examples are: (a) G-protein coupled receptors, and (b) intracellular signaling cascades.

17.4.1. WHAT ARE G-PROTEIN COUPLED RECEPTORS?

G-protein coupled receptors (GPCRs) are the largest family of cell membrane associated proteins and indeed of all protein families. They play a critical role in mediating cellular responses via signal transduction pathways. Dysfunction of GPCRs has been found in a growing number of human diseases. GPCRs have seven membrane spanning domains; these receptors have an extracellular N-terminus and three extracellular loops, and three intracellular loops, with a cytoplasmic C-terminal tail. Thus the receptor sits within the membrane and is effectively divided into extracellular, transmembrane, and cytoplasmic domains. The role of GPCRs is to transduce signals that originate either from the external environment or from within the organism itself (other cell or tissue types) and induce the appropriate cellular, and thus tissue, response. GPCRs respond to specific ligands, the chemical nature and spectrum of which are diverse, and include biogenic amines, peptides, glycoproteins, nucleotides, and proteases.

We chose to reconstitute G-protein coupled receptors (GPCRs) and the heterotrimeric G-proteins to which they couple for several reasons. First, they are responsible for coordinating about 50% of the total inter-cellular communication that occurs within an organism. So, they provide a large set of control points that regulate a wide variety of physiological events within the organism. Second, they represent the single largest family of proteins in the genome. While the various GPCRs differ in their gene sequence, they all function in a very similar manner. Hence, if you can reconstitute one, or a few of these systems, you can probably reconstitute most of them. Third, they reside within and transit the cell membrane, with part of the receptor lying in the extracellular milieu and part within the cytosol of the cell. Therefore, the agonist or antagonist need not enter the cell; it simply has to be able to contact the GPCR. This means that it can be an environmental cue such as light (visual system), an odorant or pheromone (olfactory or mating system), or enter the tracheal or circulatory system; of course, the agonist can be also be a cellular product, i.e., a peptide, biogenic amine, glycoprotein, and so forth that is released into the circulatory system. In humans, about 50% of the drugs that are currently on the market target G-protein coupled receptors and these drugs were discovered and designed at a time when only about 15% of the total GPCRs were known.[12–14] This fact, underscores the central role that GPCRs play in regulating physiological responses and the utility of targeting them to disrupt, or modify in the case of drugs, specific physiological responses.

How do GPCRs orchestrate physiological responses to environmental and physiological cues? How do they, activate and silence specific genes to provide the appropriate response to an altered environmental or internal physiological cue?

When a ligand binds to the GPCR, it triggers an allosteric change in the shape of the receptor. This stimulates a G-protein complex that resides in the cytosol, in the region just below the cell membrane. The G-protein is a complex comprised of three gene products called G-proteins, Gα, Gß, and Gγ, respectively. When the G-protein complex is activated, it splits into two components and these stimulate one or more intracellular signaling cascades. The final protein component of the intracellular signaling pathway transits the nuclear membrane and stimulates and/or represses specific target genes and thus elicits the appropriate cell and tissue response(s). The genome of most metazoans is comprised of several hundred, and sometimes over a thousand different GPCRs. Accordingly, GPCRs are among the main control points that collectively coordinate the cell, tissue and therefore organism response to both environmental and internal cues.

We have now reconstituted more than 20 different GPCR signaling pathways.[15,16] These include a wide variety of different GPCR sub-types, for example, several different members of the adrenergic, cholinergic, dopaminergic, glutaminergic, histaminergic, serotonergic, all five pain receptors, and others. The success rate, so far, is 100%, that is, all of the human GPCR signaling pathways function when reconstituted, by expressing their human genes, in insect cell-lines. We have focused on human GPCRs for research funding reasons. However, it is reasonable to say that if human GPCRs and their signaling pathways can be reconstituted and function in insect cell-lines, then their counterparts from other organisms should also function in the system.

17.4.2. INTRACELLULAR SIGNALING CASCADES

Activation of the GPCR initiates an intracellular cascade of kinase reactions. The phosphorylation of the last protein in this pathway causes it to enter the nucleus and stimulate and/or repress the expression of the appropriate target genes. Thus the cell responds appropriately to the external (environment) or internal (physiological) cue.

There is another large family of membrane bound receptors called Receptor Tyrosine Kinases (RTK). Like GPCRs these receptors transit the cell membrane. However, they do not have the seven-transmembrane helical domains that are characteristic of GPCRs. Also unlike GPCRs, they do not function via a G-protein. Instead, RTKs have their own kinase domain and autophosphorylate in response to ligand binding. However, they also activate an intracellular kinase signaling cascade in which the last component transits into the nucleus and initiates the appropriate gene response(s).

We have reconstituted several of these intracellular signaling cascades. The Ras-Raf-Mek-Erk pathway is one example. In this pathway, Ras activates Raf, which activates the Mek kinase, and Mek phosphorylates, and thus activates, Erk, which is the final step in the pathway. The activated Erk protein transits the nucleus to activate the expression of specific target genes. For the purposes of this brief discussion we'll focus on the Mek-Erk part of the pathway. Since Mek is the upstream kinase in this cascade, the cell-lines were engineered to express considerably less human Mek than its target, human Erk. The idea was to limit the phosphorylation to the appropriate target. This was accomplished by simultaneously transfecting the cell with two independent expression cassettes at a ratio of about 1:10, one expression cassette contained the human Mek gene and the other contained the human Erk gene. Since each construct inserts independently and into different sites within the genome, theoretically the two constructs should be inserted into the genome at a ratio of about 1:10 and the various inserts should be in different regions of the genome. Therefore, the cell should produce about 10 times more Erk protein, the downstream target in this pathway, than Mek. This appeared to be the case; the cell-lines produced far more of the target than the upstream activator. More importantly, the Mek activated the Erk and the activated Erk induced the appropriate genetic response.

17.4.3. SCREENING FOR AGONISTS OR ANTAGONISTS OF A RECEPTOR OR AN ENZYMATIC PATHWAY

Since these cell-lines are permanently transformed and their gene products are produced over a long period of time, they can be used to screen for compounds that either inappropriately stimulate or block the receptor or one of the steps in metabolic cascade. Thus, these compounds are potential bio-control agents or lead compounds for the development of rather specific bio-control agents. Using these engineered cell-lines, a small lab can routinely screen libraries for new "lead" compounds at a level that is considered moderate throughput, about 5000 compounds per week. Using robotics and the same engineered cell-lines, bio-pharmaceutical or bio-agriculture companies can easily screen 1 million or more compounds in a 3 month period.

Receptors for pheromone, odorant detection, or gustatory signaling are excellent targets for the identification and development of new and novel bio-control agents. Most of these responses are mediated through GPCR signaling. These receptors and their cognate G-proteins can be reconstituted in engineered cell-lines, and the engineered cell-lines can be used as a stable platform in screens for new pest control compounds. These new pest control compounds should have very high specificity, since they target a specific receptor or a specific protein in a signal transduction cascade, and thus they should have very low environmental impact. Since different compounds would target the receptor vs. the intracellular signal transduction cascade, several different compounds, all of which target the same pest species, could be isolated and used to manage the pest population in alternative years, or situations.

Of course, the GPCRs and intracellular signaling components of the pest insect must be identified. To this end, the seven-transmembrane signature of the GPCRs coupled with genome projects allows the identification of GPCRs in those insects that have been, or are scheduled to be, sequenced. For example, a complete set of the GPCRs, including all of the olfactory receptors, have been identified in several Drosophila species as the complete genomes have become available[17]; eight other species are being sequenced at this time, and there is a proposal to sequence a total of 40 to 50 Drosophilid species that diverged over a period of 3 to about 55 million years, and which occupy a variety of different ecological niches. This should give the scientific community a very powerful data set for the analyses of biological evolution of gene structure, gene regulatory elements, and the complexity of gene expression patterns in managing different environmental challenges. The genome of the *Anopheles gambia* mosquito, which is the vector for malaria, has been completed[18] and annotation Release 1.0 of the genome of the *Aedes aegypti* mosquito, which is the vector for yellow fever and dengue fever was made public in June 2006.[19] The genome of the silk moth, *Bombyx mori*, is complete.[20–22] An international genomics effort is underway for the honeybee,[23] arguably the best studied and most economically important member of the Hymenoptera. The International Lepidopteran Genome Project has been charged with applying new technologies to compare the genomes of a growing list of agriculturally important moths and butterflies. The crop-feeding heliothine moths are among these. Working groups have formed to undertake the genomics of insects belonging to other orders, such as Coleoptera (beetles) and Homoptera (true bugs, those with piercing/sucking mouthparts). We are part of a small working group that has submitted a proposal to examine the blackfly genome.

Even in the absence of full genome projects, the membrane-transmembrane domain architecture and reasonable conservation, enables the identification and cloning of the genes that encode GPCRs and their cognate G-proteins.[24] Thus, the widespread use of GPCRs to coordinate physiological responses both within the body and between the organism and its external environment, coupled with its gene structure, makes them very useful targets for a wide variety of pest management systems. This includes the isolation of peptides, biogenic amines, glycoproteins, nucleotides, proteases, other natural chemical compounds, including plant compounds. or synthetic mimics of these plant compounds, that can be used to disrupt very specific pathways in the olfactory, gustatory, and mating responses, locomotion, and various developmental pathways including gametogenesis. Likewise, a series of studies similar to the pharmcogenetic studies described by Harvey et al.,[25] can be used to determine whether naturally occurring mutations in the pest population would influence the action of a pest control agent or system long before it is deployed in a field setting.

It also includes the use of these agents in a TAC–TICS type system. Since sensory cells, such as olfactory receptors, usually express only one or two different types of GPCRs in a given cell type, it is possible to express genes, driven by inducible promoters, that would interfere with the signaling from these specific receptors and thus drastically disrupt the behavioral response of the pest. The olfactory and gustatory receptors are obvious targets, but there are many others, such as the ITP and diuresis system described in Section 17.3.3.2. We are limited only by our knowledge of the genetic control of various physiological responses and developmental events.

References

1. T. A. Pfeifer, Expression of heterologous proteins in stable insect cell culture, *Curr. Opin. Biotech* 9, 518–521 (1998).
2. T. A. Pfeifer, D. D. Hegedus, T. A. Grigliatti, and D. A. Theilmann, Baculovirus immediate-early promoter-mediated expression of the Zeocin^{TM} resistance gene for use as a dominant selectable marker in Dipteran and Lepidopteran insect cell lines, *Gene* 188, 183–190 (1997).
3. D. D. Hegedus, T. A. Pfeifer, J. Hendry, D. A. Theilmann, and T. A. Grigliatti, A series of broad host range shuttle vectors for constituitive and inducible expression and secretion of heterologous proteins in insect cell lines, *Gene* 207, 241–249 (1998).
4. G. Sinclair, T. A. Pfeifer, T. A., Grigliatti, and F. Y. M. Choy, Secretion of human glucocerebrosidase from stable transfected insect cells using native signal sequences, *Biochem. Cell Biol.* 84, 148–156 (2006).
5. T. A. Pfeifer, M. M. Guarna, E. M. Kwan, G. Lesnicki, D. A. Theilmann, T. A. Grigliatti, and D. G. Kilburn, Expression analysis of a modified FactorX in stable insect cell lines, *Prot. Exp. Purif.* 23, 233–241 (2001).
6. T. M. Rose, G. D. Plowman, D. B. Teplow, W. J. Dreyer, K. E. Hellstrom, and J. P. Brown, Primary structure of the human melanoma-associated antigen p97 (melanotransferrin) deduced from the mRNA sequence, *Proc. Natl. Acad. Sci, USA* 83(5), 1261–1265 (1986).
7. D. D. Hegedus, T. A. Pfeifer, M. L. Kennard, W. A. Gabathuler, D. A. Jeffries, D. A. Theilmann, and T. A. Grigliatti, Intergenic differences in the expression of human melanotransferrin in insect cell lines, *Prot. Exp. Purif.* 15, 296–307 (1999).
8. J. E. Phillips, C. Wiens, N. Audsley, L. Jeffs, T. Bilgen, and J. Meridith, Nature and control of chloride transport in insect absorptive epithelia, *J. Exp. Zool.* 275, 294–299 (1996).
9. N. Audsley, C. Mcintosh, and J. E. Phillips, Actions of ion-transport peptide from locust Corpus Cardiacum on several hindgut transport processes, *J. Exp. Biol.* 173 (1). 275–288 (1992).
10. T. A. Pfeifer, D. Hegedus, Y-J. Wang, Y. Zhao, J. Meredith, J. W. Brock, , J. E. Phillips, T. A. Grigliatti, and D. A . Theilmann, Analysis of an insect neuropeptide, *Schistocerca gregaria*, ion transport protein (ITP), expressed in insect cell systems. *Arch. Insect Biochem. Physiol.* 42, 245–252 (1999).
11. V. M. Gorenflo, T. A. Pfeifer, G. Lesnicki, E. M. Kwan, T. A., Grigliatti, D. G. Kilburn, and J. M. Piret, Production of a self-activating CMB-Factor X fusion protein in a stable transformed Sf9 insect cell line using high cell density perfusion culture, *Cytotechnology* 44, 93–102 (2004).
12. T. Gudermann, B. Nurnberg, and G. Schultz, Receptors and G proteins as primary components of transmembrane signal transduction, Part 1: G-protein-coupled receptors: Structure and function. *J. Mol. Med.* 73(2), 51–63 (1995).
13. J. Drews, Genomic sciences and the medicine of tomorrow, *Nat. Biotechnol.* 14(11), 1516–1518 (1996).
14. G. Muller, Towards 3D structures of G protein-coupled receptors: A multidisciplinary approach, *Curr. Med. Chem.* 7(9), 861–888 (2000).
15. P. J. K. Knight, T. A. Pfeifer, and T. A. Grigliatti, A functional assay for G-Protein coupled receptors using stably transformed insect tissue culture cell lines, *Anal. Biochem.* 320, 88–103 (2003).
16. P. Knight and T. A. Grigliatti, Chimeric G proteins extend the range of insect cell-based functional assays for human G protein-coupled receptors, *J. Recept. Sig. Transd.* 24, 241–256 (2004).

17. M. D. Adams, S. E. Celniker, R. A. Holt, C. A. Evans, J. D. Gocayne, P. G. Amanatides, S. E. Scherer, P. W. Li, R. A. Hoskins, R. F. Galle, R. A. George, S. E. Lewis, S. Richards, M. Ashburner, S. N. Henderson, G. G. Sutton, J. R. Wortman, M. K D. Yandell, Q. Zhang, Lin X. Chen, R. C. Brandon, Yu-Hui, C. Rogers, R. G. Blazej, M. Champe, B. D. Pfeiffer, K. H. Wan, C. Doyle, E. G. Baxter, G. Helt, C. R. Nelson, G. L. G. Miklos, J. F. Abril, A. Agbayani, H-J An, C. Andrews-Pfannkoch, D. Baldwin, R. M. Ballew, A. Basu, J. Baxendale, L. Bayraktaroglu, E. M. Beasley, K. Y. Beeson, P. V. Benos, B. P. Berman, D. Bhandari, S. Bolshakov, D. Borkova, M. R. Botchan, J. Bouck, P. Brokstein, P. Brottier, K. C. Burtis, D. A. Busam, H. R. Butler, E. Cadieu, A. Center, I. Chandra, J. M. Cherry, S. Cawley, C. Dahlke, L. B. Davenport, P. Davies, B. de Pablos, A. Delcher, Z. Deng, Anne D. S. Mays, I. Dew, S. M. Dietz, K. Dodson, L. E. Doup, M. L. Downes, S. Dugan-Rocha, B. C. Dunkov, P. Dunn, K. J. Durbin, C. C. Evangelista, C. Ferraz, S. Ferriera, W. Fleischmann, C. Fosler, A. E. Gabrielian, N. S. Garg, W. M. Gelbart, K. Glasser, A. Glodek, F. Gong, J. H. Gorrell, Z. Gu, P. Guan, M. Harris, N. L. Harris, D. Harvey, T. J. Heiman, J. R. Hernandez, J. Houck, D. Hostin, K. A. Houston, T. J. Howland, M.-H. Wei, C. Ibegwam, M. Jalali, F. Kalush, G. H. Karpen, Z. Ke, J. A. Kennison, K. A. Ketchum, B. E. Kimmel, C. D. Kodira, C. Kraft, S. Kravitz, D. Kulp, Z. Lai, P. Lasko, Y. Lei, A. A. Levitsky, J. Li, Z. Li, Y. Liang, X. Lin, X. Liu, B. Mattei, T. C. McIntosh, M. P. McLeod, D. McPherson, G. Merkulov, N. V. Milshina, C. Mobarry, J. Morris, A. Moshrefi, S. M. Mount, M. Moy, B. Murphy, L. Murphy, D. M. Muzny, D. L. Nelson, D. R. Nelson, K. A. Nelson, K. Nixon, D. R. Nusskern, J. M. Pacleb, M. Palazzolo, G. S. Pittman, S. Pan, J. Pollard, V. Puri, M. G. Reese, K. Reinert, K. Remington, R. D. C. Saunders, F. Scheeler, H. Shen, B. C. Shue, I. Sidén-Kiamos, M. Simpson, M. P. Skupski, T. Smith, E. Spier, A. C. Spradling, M. Stapleton, R. Strong, E. Sun, R. Svirskas, C. Tector, R. Turner, E. Venter, A. H. Wang, Xin Wang, Z.-Y. Wang, D. A. Wassarman, G. M. Weinstock, J. Weissenbach, S. M. Williams, T. Woodage, K. C. Worley, D. Wu, S. Yang, Q. A. Yao, J. Ye, R.-F. Yeh, J. S. Zaveri, M. Zhan, G. Zhang, Q. Zhao, L. Zheng, X. H. Zheng, F. N. Zhong, W. Zhong, X. Zhou, S. Zhu, X. Zhu, H. O. Smith, R. A. Gibbs, E. W. Myers, G. M. Rubin, and J. Craig Venter. The genome sequence of *Drosophila melanogaster*. *Science* 287, 2185–2195 (2000).
18. R. A. Holt, G. M. Subramanian, A. Halpern, G. G. Sutton, R. Charlab, D. R. Nusskern, P. Wincker, A. G. Clark, J. M. C. Ribeiro, R. Wides, S. L. Salzberg, B. Loftus, M. Yandell, W. H. Majoros, D. B. Rusch, Z. Lai, C. L. Kraft, J. F. Abril, V. Anthouard, P. Arensburger, P. W. Atkinson, H. Baden, V. de Berardinis, D. Baldwin, V. Benes, J. Biedler, C. Blass, R. Bolanos, D. Boscus, M. Barnstead, S. Cai, A. Center, K. Chatuverdi, G. K. Christophides, M. A. Chrystal, M. Clamp, A. Cravchik, V. Curwen, A. Dana, A. Delcher, I. Dew, C. A. Evans, M. Flanigan, A. Grundschober-Freimoser, L. Friedli, Z. Gu, P. Guan, R. C. Guigo, M. E. Hillenmeyer, S. L. Hladun, J. R. Hogan, Y. S. Hong, J. Hoover, O. Jaillon, Z. Ke, C. Kodira, E. Kokoza, A. Koutsos, I. Letunic, A. Levitsky, Y. Liang, J-J. Lin, N. F. Lobo, J. R. Lopez, J. A. Malek, T. C. McIntosh, S. Meister, J. Miller, C. Mobarry, E. Mongin, S. D. Murphy, D. A. O'Brochta, C. Pfannkoch, R. Qi, M. A. Regier, K. Remington, H. Shao, M. V. Sharakhova, C. D. Sitter, J. Shetty, T. J. Smith, R. Strong, J. Sun, D. Thomasova, L. Q. Ton, P. Topalis, Z. Tu, M. F. Unger, B. Walenz, A. Wang, J. Wang, M. Wang, X. Wang, K. J. Woodford, J. R. Wortman, M. Wu, A. Yao, E. M. Zdobnov, H. Zhang, Q. Zhao, S. Zhao, S. C. Zhu, I. Zhimulev, M. Coluzzi, A. della Torre, C. W. Roth, C. Louis, F. Kalush, R. J. Mural, E. W. Myers, M. D. Adams, H. O. Smith, S. Broder, M. J. Gardner, C. M. Fraser, E. Birney, P. Bork, P. T. Brey, J. C. Venter, J. Weissenbach, F. C. Kafatos, F. H. Collins, S. L. Hoffman. The genome sequence of the malaria mosquito *Anopheles gambiae*. *Science* 298(5591), 129–149 (2002).

19. TIGR (The Institute for Genome Research) *Aedes aegypti* annotation Release 1.0. This annotation was produced jointly by The Institute for Genomic Research and VectorBase with support from The Broad Institute of Harvard/MIT. Available at: msc.tigr.org/aedes/release.shtml.
20. K. Mita, M. Kasahara, S. Sasaki, Y. Nagayasu, T. Yamada, H. Kanamori, N. Namiki, M. Kitagawa, H. Yamashita, Y. Yasukochi, K. Kadono-Okuda, K. Yamamoto, M. Ajimura, G. Ravikumar, M. Shimomura, Y. Nagamura, T. Shin-I, H. Abe, T. Shimada, S. Morishita, and T. Sasaki, The genome sequence of silkworm, *Bombyx mori, DNA Res.* 11(1), 27–35 (2004).
21. Q. Xia, Z. Zhou, C. Lu, D. Cheng, F. Dai, B. Li, P. Zhao, X. Zha, T. Cheng , C. Chai, G. Pan, J. Xu, C. Liu, Y. Lin, J. Qian, Y. Hou, Z. Wu, G. Li, M. Pan, C. Li, Y. Shen, X. Lan, L. Yuan, T. Li, H. Xu, G. Yang, Y. Wan, Y. Zhu, M. Yu, W. Shen, D. Wu, Z. Xiang, J. Yu, J. Wang, R. Li, J. Shi, H. Li, G. Li, J. Su, X. Wang, G. Li, Z. Zhang, Q. Wu, J. Li, Q. Zhang, N. Wei, J. Xu, H. Sun, L. Dong, D. Liu, S. Zhao, X. Zhao, Q. Meng, F. Lan, X. Huang, Y. Li, L. Fang, C. Li, D. Li, Y. Sun, Z. Zhang, Z. Yang, Y. Huang, Y. Xi, Q. Qi, D. He, H. Huang, X. Zhang, Z. Wang, W. Li, Y. Cao, Y. Yu, H. Yu, J. Li, J. Ye, H. Chen, Y. Zhou, B. Liu, J. Wang, J. Ye, H. Ji, S. Li, P. Ni, J. Zhang, Y. Zhang, H. Zheng, B. Mao, W. Wang, C. Ye, S. Li, J. Wang, G-K. Wong, and H. Yang; Biology Analysis Group, A draft sequence for the genome of the domesticated silkworm (*Bombyx mori*), *Science* 306(5703), 1937–1940 (2004).
22. For sequence information on the *Bombyx mori* genome available at: http://silkworm.genomics.org.cn/ or http://papilio.ab.a.u-tokyo.ac.jp/genome/index.html
23. Honey Bee Genome Project, The HGSC is currently sequencing the honey bee, *Apis mellifera*. The version 4.0 assembly was released in March 2006. Available at ftp://ftp.hgsc.bcm.tmc.edu/pub/data/Amellifera/fasta/
24. P. Knight and T. A. Grigliatti, Diversity of Lepidopteran G proteins; partial sequences of six novel G protein alpha subunits from insect cell lines, *Arch. Insect Biochem. Physiol.* 57(3), 142–150 (2004).
25. L. Harvey, R. E. Reid, C. Ma, P. J. K. Knight, T. A. Pfeifer, and T. A. Grigliatti, Human genetic variations in the 5HT2A receptor: A single nucleotide polymorphism identified with altered response to clozapine, *Pharmacogenetics*, 13, 107–118 (2003).

18. TAC–TICS: TRANSPOSON-BASED BIOLOGICAL PEST MANAGEMENT SYSTEMS

Thomas A. Grigliatti,* Gerald Meister, and Tom A. Pfeifer
Department of Zoology, Life Sciences Institute, University of British Columbia, 2350 Health Sciences Mall, Vancouver, BC, Canada V6T 1Z3

Abstract. A system in which a specific pest insect population is targeted for management by making that population susceptible to control is described. The four key components of this genetic control system are: (a) transformation of the targeted insect, (b) dissemination of the engineered construct throughout the targeted population, (c) an inducible promoter to activate the expression of, (d) the incapacitating gene or genes. The progress made in transformation of insects other than Drosophila is described and using model organisms, we show that transposon constructs can spread quite rapidly through a targeted population. The multiplicative transposition process, which drives the spread of the engineered transposon construct, is not highly error prone and thus the use of transposon armed cassettes (TAC) in insect populations is feasible. Examples of genes that might be used as relatively insect-specific incapacitating genes if over-expressed, mis-expressed or inactivated in specific insects or insect tissues are discussed, and finally we discuss the possibility of horizontal transfer: that is, the transfer of a TAC-type construct from a genetically engineered insect to a reproductively isolated species by non-reproductive mechanisms.

18.1. Introduction

Two different molecular genetic approaches have been used to control pest insects over the last several decades. One approach has been to manipulate the genes of entomopathogenic bacteria, viruses and fungi to produce more efficient bio-pesticides.[1–3] The second approach has been to clone a gene encoding an insecticidal protein, such as the Bt toxin (*Bacillus thuringiensis*), and to place it into the genome of a plant on which the insect feeds resulting in a transgenic crop plant that produces its own bio-pesticide.[4,5] Advances in biotechnology will undoubtedly lead to improvements and refinements of

*To whom correspondence should be addressed, e-mail: grigliat@zoology.ubc.ca

M. Vurro and J. Gressel (eds.), Novel Biotechnologies for Biocontrol Agent Enhancement and Management, 327–351.

these two bio-control methodologies, and they will likely become the predominant bio-control strategies of the next few decades.

A third emerging approach to pest management is genetic modification of the pest insect to target it for bio-control. Genetic transformation of insects allows direct manipulation of the genetic constitution of the organism and provides an opportunity for bio-control under specific circumstances. Genetic manipulation of the pest genome should have very low environmental impact, since it targets a single species (the pest). In fact, assuming simple geographical constraints, genome manipulation may provide population specificity. Thus genetic manipulation offers some advantages over the traditional spraying or dispersal of chemical pesticides. Genetic manipulation, in the form of release of sterile insects, has been used quite successfully for several decades to control some insect vectors of animal disease. The sterile male technique relies on introducing massive numbers of irradiated, or otherwise sterilized, males into a population of insects. Variations on the sterile male technique use chromosome abnormalities, such as translocations, to cause massive genetic imbalance and consequently cellular dysfunction and lethality early in embryogenesis.[6,7] Releasing insects that have been genetically engineered to facilitate pest management is simply a variant of the sterile male technology.

Using genetic engineering and transgenic technologies, it should be possible to produce insects that are predisposed or highly susceptible to specific management protocols, and thus target specific insect populations (not whole species) for management. Indeed, it is possible to apply the management only in those years when large outbreaks of the pest threaten crops or human health, and in low-density years one can elect not to interfere with the population dynamics. However, genetic manipulation entails risk from the release of bio-engineered organisms, and consequently the potential of releasing bio-engineered constructs, into other species. In addition, it is a long-term approach that requires a dedicated bio-control program.

In this chapter, we focus our discussion on engineering insects as pest insects damage approximately one-third of the agricultural crops in the world, imparting particularly devastating consequences in regions of the world that are less technologically developed. Similarly, about one-third of the human population suffers from diseases carried by insect vectors. Nevertheless, we emphasize that these types of control strategies are not limited to insects. Indeed, similar strategies have been proposed to control fish[8] and parasitic weeds.[9] In fact, we believe control strategies based on genetically engineered expression cassettes may be more easily applied to, manipulated in, and more easily managed and contained by engineering plants to provide their own protection when stimulated to do so. Hence, while this chapter uses insects as examples of what we have named the TAC–TICs bio-control strategy, these pest management strategies have global applications.

18.2. Transposable Elements

Transposable elements are the foundation for both inserting a cloned, gene expression cassette into the genome of a target insect and dispersing the engineered construct through the target population. As their name implies, transposable, or mobile, genetic elements are capable of moving from one location to another within the genome. Transposable elements have been found in all organisms, yet the function of any particular transposable element is generally limited to specific genera. The latter feature is useful as it provides specificity and some level of containment should horizontal transfer occur. There are two general categories of transposable elements: DNA elements, and retro-transposable elements. They differ by their method of replication. We limit our examples to DNA elements, but both types of transposable elements can be used as transformation agents and dispersal agents. Transposable elements are generally quite small, about 2 to 5kb in size and are easily engineered to carry gene expression cassettes. Most DNA elements have inverted repeat sequences located at their termini. These inverted repeats are about 15–50 bps in length, and are necessary for both transformation into, and mobilization within, the genome. Removal or mutation of either repeat disables movement. Most organisms contain a number of different transposable elements, distinguished by DNA sequence and size. There are many copies of each type of transposable element in any individual, usually 20–100, but for some elements the number can be several thousand. Some copies of these transposable elements are defective, that is, they cannot move. The multiple copies of each element are generally dispersed throughout the genome, and different individuals within the same population often have the transposable elements located at different sites within their respective genomes. The last observation is the foundation for the hypothesis that the elements can increase in number and move to occupy different positions within the genome, via a replicative transposition mechanism.

18.3. Targeted Insect Control Strategies

We have previously proposed introducing the genetic characteristic "*susceptibility to management*" into a specific insect population.[10,11] We refer to this method of insect control as Targeted Insect Control Strategies (TICS). There are many different genes that can be engineered and added to the genome of a pest to make it quite susceptible to management strategies. Some examples will be provided later. The genetically engineered construct must be integrated into the genome of the target pest. The most common methods of introducing one or more engineered constructs into an insect target genome is via mobile genetic elements or via a host virus. Mobile or transposable elements are capable of moving and integrating into different locations within the genome. These elements can be cloned and then genetically engineered by recombinant DNA technologies to function as transformation agents, in which case they are called transposons. If transposons carrying genetically engineered, conditionally expressed pest incapacitating genes are used as transformation agents, then the resulting transgenic pest insect and its descendants will be "susceptible" to population size management. We call these genetically engineered recombinant DNA constructs Transposons with Armed Cassettes, or TAC for short. Hence, we refer to the management system as TAC–TICS. The use of mobile genetic elements in some form of pest control is a relatively new concept, but not novel; as the characteristics of mobile elements were defined, their potential application to various management strategies has been discussed by Kidwell and Ribeiro,[12] Miller,[13] Crampton et al.,[14] Salvado et al.,[15] ourselves[10,11] and others. The eventual goal of such strategies, regardless of the acronym used to describe it, is to develop a genetic control method that may be applied, either alone or in conjunction with other pest control strategies, to the long-term management of pests that attack and destroy important crops as well as those that act as disease vectors.

The following is a simplistic summary of TAC–TICS applied to pest insects in an agricultural or forestry setting. A number of transformed insects containing the TAC construct are released when the pest insect population density is very low, i.e., non-outbreak years. The numbers of insects that need to be released is considerably lower than the numbers used for sterile insect release, as the engineered bio-control construct will spread through the population by a combination of its encoded replicative transposition mechanism and normal mating. Control during the release year(s) is not necessary, as the population densities remain below the economically important threshold. The released insects would mate with native insects. In the next generation, the effect would multiply because (1) the TAC construct increases in copy number per individual due to the replicative transposition process and the new copies occupy new sites within the genome of each "contaminated" individual, and (2) normal chromosome segregation during gamete formation assures that at least one, if not more, of the TAC constructs is distributed to each offspring. Consequently, after several generations of random mating, most individuals within the population would contain several copies of the TAC construct, with each copy inserted at a different site in the genome. The multiple copies per genome, each located at a different locus within the genome, enhances the effectiveness of the bio-control system and assures a level of stability. If necessary, the rate of dispersal of the TAC through the targeted population can be enhanced by repeated releases of transformed insects in successive generations. During outbreak years, the incapacitating gene(s) would be activated by application of an inducer, and the genetically engineered, and thus susceptible, members of the population would be incapacitated. Other insects in the food chain would not respond to the otherwise benign inducing agent. Hence, there would be minimal impact on the food chain. The incapacitating gene does not have to immediately kill the insect; it need only prevent it from feeding and thus minimize crop damage. It could do this by paralyzing or immobilizing the insect, arresting its development, altering its gustatory response, and so forth. The incapacitated insect then simply becomes a food source for other organisms. The managed population would collapse to low numbers, but it would not be eradicated, since there would always be pockets where the inducing agent is absent, or present in concentrations too low to be effective. Genetically modified insects in these small niches would survive, and these insects would pass the engineered TAC construct to their descendants and to any TAC-untainted immigrants that "recolonized" the region.

In the case of insects that are vectors of human or animal disease, the TAC construct could carry a gene driven by a non-conditional promoter (instead of an inducible promoter) that disrupts reproduction and targets the pest for more immediate reduction. TAC constructs might express genes that would disrupt gamete formation or function. Such constructs would spread through the population via members of the unaffected sex, which act as hosts. Approximately half of all of the offspring from the first set of matings would be sterile and after a few generations the population density would collapse. An alternative approach would be to use a TAC that contains a gene that disrupts the ability of the disease-causing vector to reproduce in the insect, but does not substantively reduce the fitness of the insect itself, assuming the insect itself is rather benign.

18.4. Is the TAC–TICS System feasible—Are the Components Available and Can They Be Assembled to Function Effectively?

There are four requirements for the TAC–TICS system: (1) *Transformation*: a method of introducing the TAC construct into the insect; (2) *Dissemination*: a method for rapidly spreading the TAC construct through the targeted insect population; (3) *Incapacitation*: a gene, or set of genes, whose products are capable of incapacitating the insect; and (4) *Controllable Switch*: conditional promoter that responds when a very specific external agent is applied or when the insect comes in contact with a plant or animal expressing this compound, and thus allows expression of the incapacitating gene(s) in a given tissue type or sex.

18.4.1. TRANSFORMATION

Transformation of insects is absolutely crucial to the TAC–TICS system. Indeed, many laboratories are focusing their efforts on establishing the appropriate conditions for successful transformation in a variety of insects. Thus, it seems inevitable that a variety of different transformation systems will be forthcoming. In the past decade, at least 20 different insects species have been transformed and hence, while the specifics of the transformation protocols may differ, transformation is certainly well demonstrated. Hence, we will discuss it rather briefly. Researchers have been introducing genes into the model organism *D. melanogaster*, a non-pest fruitfly, for nearly 25 years using the *P* transposable element.[16] This allowed detailed genetic analyses of genes and mutations in vivo, and many thousands of genes have been introduced into, and expressed in, *D. melanogaster*. Indeed, *P*-mediated transformation was so successful and allowed so many experiments to be undertaken that there was virtually no need to develop other methods of transformation in *Drosophila*. Nonetheless, in recent years a number of laboratories have shown that *Drosophila* can also be transformed with the *hobo*, *mariner*, *Hermes*, and *I* transposable elements.[17–20] Thus, the properties of P-mediated transformation are not limited to P elements alone.

P element mediated transformation was also successful in several other *Drosophila* species including the very closely related *D. simulans*,[21] and the more distantly related *D. hawaiiensis*.[22] In contrast, discouraging results were obtained from attempts to utilize *P* elements to transform mammalian cells and, more importantly, non-drosophilid insects including mosquitoes, tephritid fruitflies, grasshoppers and houseflies,[23–30] and R. Lansman, H. Brock, and T. Grigliatti, unpublished results. This inability of *D. melanogaster P* elements to efficiently transform species outside of the Drosophilidae, probably reflects a requirement for specific host encoded functions in the transposition process. This level of specificity likely exists for many transposable elements. While a limited host-range requires that transposable elements be isolated for each of the target insects, this specificity is very advantageous in addressing bio-safety concerns.

Thirteen years after *D. melanogaster* was transformed, the Mediterranean fruit fly *Ceratitis capitata*, a true pest, was transformed using the transposable element *Minos*.[31] Between 1995 and 1999, several other dipteran insects were transformed with *Minos, Hermes*, and *mariner*[32] (see Table I). These newly discovered mobile elements are typical Class 2 mobile elements, consisting of short inverted terminal repeats and one or more open reading frames coding for a transposase protein. Attempts to transform lepidopteran insects with these transposable elements failed, and it appeared that these transposon vectors would only work in a narrow sub-set of diptera (flies). Lepidopteran

TABLE I. Reported transformations of agriculturally or medically important insects

Insect species	Transposon used	Ref.	Common names
Diptera			
Aedes aegypti	Hermes	35	Yellow fever Mosquito
		36	
	Mariner	37	
	PiggyBac	See Meister et al.[33]	
Anopheles stephensii	Minos	38	Mosquito
Annopheles albimanus	piggyBac	See Meister et al.[33]	Mosquito
Culex quiqueasciatus	Hermes	See Meister et al.[33]	Mosquito
Ceratitis capitata	Hermes	See Meister et al.[33]	Med fly
	Minos	32	
	piggyBac	39	
Anastrepha suspensa	piggyBac	40	Caribbean fruit fly
Bactrocera dorsalis	piggyBac	41	Oriental fruit fly
Stomoxys calcitrans	Hermes	43	Stable fly
Musca domestica	Piggyback	100	House fly
Coleoptera			
Tribolium castaneum	Hermes	42	Red flour beetle
	piggyBac	42	
Lepidoptera			
Bombyx mori	piggyBac	44	Silkworm
Pectinophora gossypiella	piggyBac	45	Pink bollworm

insects, moths, were not transformed until the piggyBac transposon, based on a lepidopteran transposable element, was developed in 2000 (Table I). Surprisingly this lepidopteran based transposon also works in some dipteran insects. Thus, piggyBac, along with the *mariner* transposable element, which also functions in a number of different insect genera, may represent a more broad-spectrum insect transformation system. The broad-spectrum capability of these transformation constructs may be considered a liability from a bio-safety perspective. Conversely, these transformation vectors may at least allow testing of engineered constructs under controlled laboratory conditions, while transposable elements with a more limited host range are developed for a specific pest target.

There are a number of challenges that must be overcome in the development of a transformation system for any given insect species. Obviously one must have basic knowledge of the physiology of the insect, such as when and how the germ-line tissue is formed. In addition, one must find and engineer a transposable element that is mobile within the insect species that is targeted for management. Several techniques have been developed that facilitate the identification of new transposable elements,[33] including the isolation of an active element from a closely related species.[34,35] The development of the inter-plasmid transposition assay was a very important achievement in the development of transformation assays.[36] It is an in vivo test for the ability of an engineered element to transpose between two plasmids that have been co-injected into insect embryos. Thus, while insect transformation has not yet become routine, these and other advancements make it likely that a much wider variety of insects can and will be transformed in the near future.

18.4.2. DISPERSAL OF THE TAC CONSTRUCT

The second critical requirement of the TAC–TICS system is a method of spreading the transformed construct through a target insect population. Native, intact *P* elements are capable of spreading very rapidly through experimental populations of *D. melanogaster* established with very low frequencies of *P* element containing individuals.[37–40] It is therefore plausible that transposons may be used as vectors for transformation as well as for dispersal of engineered DNA through a target population. In addition to the potential for rapid rates of dispersal of engineered DNA, transposons often have a fairly narrow host range, that is relatively genus specific, and thus spread is usually restricted to vertical inheritance within a single interbreeding population. However, several fundamental questions must be answered before transposons can be seriously considered as dispersal vectors.

First, does increasing the size of the transposon, e.g., by adding a gene expression cassette, impede its dispersal? Several independent lines of evidence indicated that transposition rates may decrease as element size increases,[41–43] suggesting that transposons that contain "passenger" genes, and are often three to four times the size of native elements, might disperse through insect populations at a significantly lower rate than unmodified elements. Secondly, would the dispersed passenger genes retain their ability to produce a functional protein product, or is the transposition mechanism error prone, thus causing the expression cassette to acquire mutations rendering it non-functional? Indeed, many transposons, including *P* elements, accumulate internal deletions that probably arise by incomplete copying of a template element, or mis-repair of the target DNA site, during the replication/integration process.[44,45] Furthermore, Daniels et al.[21] demonstrated that when deletions occurred in a *P* element transposon that carried a *rosy*$^+$ gene as the "passenger," one or both ends of the deletion occasionally extended into and thus interrupted the *rosy* gene. Obviously higher than average mutation rates, could result in loss of expression of passenger genes during transposition. Therefore it is paramount to know if the transposition process is highly error prone. Are passenger genes replicated with good fidelity and thus retain their ability to encode an active protein after dispersal?

To answer these two questions, *D. melanogaster* laboratory populations were established with female flies that lacked a functional alcohol dehydrogenase gene (genotypically Adh^-/Adh^-) and contained no *P* elements. Most of the females were mated to males of the same strain; however, 1% or 10% of the females were mated to males from a strain that had previously been transformed with both a helper *P* element (to supply transposase activity) and a *P* element transposon containing an *alcohol dehydrogenase*$^+$ allele (P element-Adh^+ construct), which simulates a TAC construct in size and function). The founding populations were comprised of the offspring of this mating scheme and thus each founding population contained either 0.5 or 5% *P* genomes (the P element containing individuals within each population were heterozygous for the P element-Adh^+ and P-helper constructs). The dispersal of *P* elements to new genomes was monitored at each subsequent generation, by randomly selecting females and performing DNA hybridization assays on dissected ovarian tissue. The assay allowed us to detect the helper *P* elements and the Adh^+loaded *P* elements (TAC-like construct) separately, and showed that the TAC-like *P* element constructs dispersed rapidly and were present within virtually all individuals of all populations after 8–10 generations.[46] Sample data for one population are shown with open circles in Figure 1. After the ovaries were removed for the hybridization assays, each carcass was tested for ADH activity using a simple histochemical assay. Again, this ADH assay was performed on each female sampled; therefore, we could correlate presence or absence of P element with ADH activity in each individual sampled. Virtually all individuals had ADH activity after 10 generations or less. These results clearly demonstrated that, despite an approximate threefold increase in size, the *P* element based constructs, containing a functioning gene, were still capable of rapid dispersal through the experimental populations. The rate of this dispersal was equivalent to the rate of dispersal exhibited by unmodified elements.[38] Secondly, many, if not all, of the dispersed *alcohol dehydrogenase* genes still encoded an active product. Thus, the multiplicative transposition process does not appear to be highly error prone. These experiments support the notion that transposons might act as efficient vectors for dispersal and eventual expression of TAC cassettes.

A second series of experiments asked whether rapid dispersal is a property that is peculiar to *P* elements, or if other elements, such as *hobo* elements, can also disperse rapidly. The apparent recent invasion of natural populations of *D. melanogaster* by *hobo* elements provides circumstantial evidence that *hobo* elements are capable of rapid dispersal and accumulation from a few individuals to entire populations.[47–50] However, experiments monitoring the accumulation of *hobo* elements within lines of *hobo*-transformed

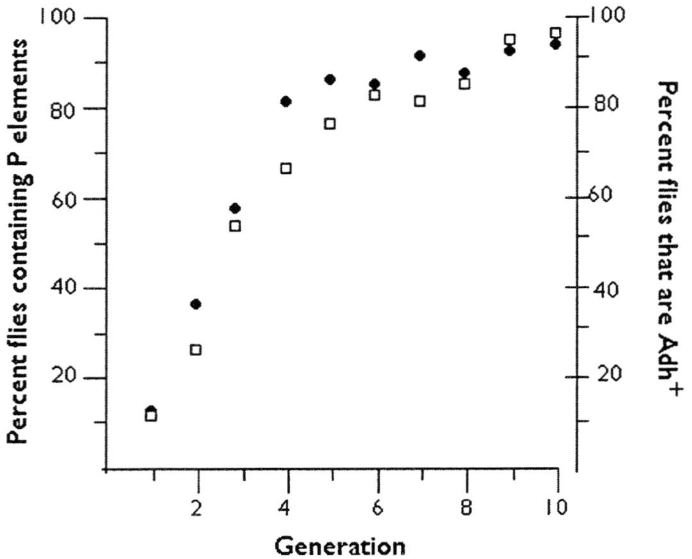

Figure 1. Dispersal of P element and P transposons in experimental populations of *D. melanogaster*. The data for a population in which 0.5% of the genomes of the founders contained a P transposon with an *adh*$^+$ passenger gene and a P "helper" element and the remaining 99.5% contained no P elements of any kind. The filled circles show the percentage of flies containing *P* elements as determined by single fly dot blots, while the open boxes show the percentage of flies that exhibit ADH activity. A minimum of 200 individuals were sampled at each generation. Six separate populations were established and examined at each generation

D. melanogaster cast some doubt on this conclusion.[51–53] These lines originally contained no *hobo* elements and had been transformed by injection of embryos with autonomous (complete) *hobo* elements. After 52 generations, individuals from each of six lines had an average of only four to six *hobo* insertion sites. In contrast, individuals from natural *hobo* containing strains have 50–100 copies of the *hobo* element,[54,55] and previous results from *D. melanogaster* lines transformed with autonomous *P* elements had accumulated about 110 copies per individual within a similar time frame.[43] Since an increase in copy number per genome is expected to contribute to the ability of transposons to spread through populations, these experiments on transformed lines suggest that *hobo* elements may be far less capable of spreading through populations than *P* elements.

Given the questionable ability for *hobo* elements to disperse, *hobo* elements may not seem to be a logical choice for investigation. However, it is important to know whether many different mobile elements can function as dispersal vectors or whether this property is limited to a few, select elements. There is a number of compelling practical reasons to study *hobo* elements. First, inter-plasmid transposition assays have indicated that *hobo* and *hobo*-like elements transpose, and therefore might act as transformation vectors, in a broad range of insect species,[56] including at least one lepidopteran, the corn earworm *Helicoverpa armigera*.[57] Indeed, the *Hermes* element, isolated from the housefly *Musca domestica,* has already been used to transform several agriculturally and medically important insects (see Table I). Secondly, unlike *P* and most other mobile genetic elements, there is evidence that *hobo* can undergo significant transpositional activity in at least some populations that already contain many elements.[58,59] The ability of *hobo*-like elements to move and multiply within element containing individuals might allow *hobo*-like elements to be used as vectors for transformation and dispersal even in insect populations that already contain them. In practical terms, this may mean that *hobo*-like elements could be used to disperse a second engineered construct through a given insect population. Such successive rounds of dispersal, which would provide a "second chance" if an initial dispersal did not provide the desired population control, is simply not possible with most elements since their transposition is severely restricted in element containing flies.

A series of experimental populations were initiated with mated females to investigate the dispersal ability of *hobo* elements. Either 2% or 20% of the females that were used to establish these populations had been previously mated to males containing *hobo* elements, while the remainder had been mated to non-*hobo* males; thus 1 and 10%, respectively, of the founding members of each population contained *hobo* elements. Both dot blots and Southern blots of DNA prepared from single flies showed that *hobo* elements spread rapidly and were present within virtually all individuals of all populations within less than eight generations.[60] Typical results are shown in Figures 2(A) and 2(B). Quantification of the dot blots revealed that, among those flies that contain elements, the mean amount of *hobo* hybridizing DNA per individual decreased in the first few generations. This suggests that the initial dispersal of the element depended primarily on Mendelian segregation. However, the dot blots revealed that the *hobo* DNA per individual increased after the first few generations and that by generation ten most flies had approximately 50% of the amount of *hobo* DNA present within individuals of the original element donating strain. Clearly the number of *hobo* elements had increased within, and dispersed throughout, these populations. Single fly Southern blot analyses were done on individuals sampled from each generation, and these analyses demonstrated that the *hobo* elements were located at different sites within the individual genomes. These date indicate that movement and dispersal occurred by a multiplicative transposition process. If *hobo*-like elements loaded with TAC sized constructs still disperse rapidly, then they might make particularly versatile dispersal vectors.

The movement of mobile genetic elements to new locations and increases in their number within the genome requires chromosome breakage and repair

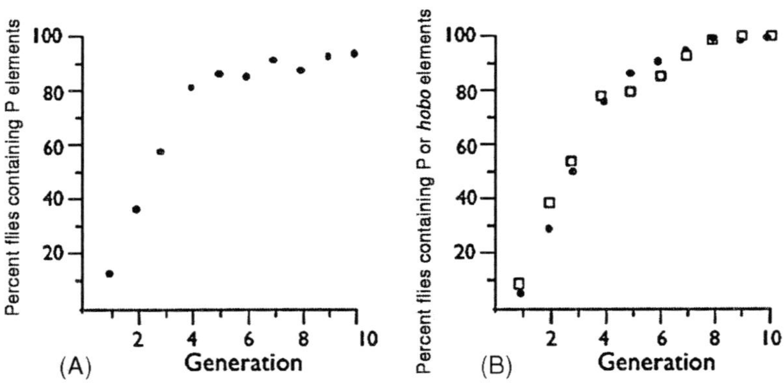

Figure 2. Dispersal of P elements (●) alone and the simultaneous dispersal of P and *hobo* elements (□) in an untainted population of *D. melanogaster*. Graph A shows the dispersal of P elements in a population in which 1% of the founding females had been mated to P element containing males while the remaining 99% of the females were mated to non-P containing males (M strain males). Graph B shows a similar experiment in which both P and hobo elements are spreading simultaneously. The population was founded with females in which 1% had been mated to P-containing males (P strain), 1% had been mated to *hobo*-containing males and the remaining 98% were mated to males that contained neither P nor hobo elements. Six separate populations were established for each set of experiments. A minimum of 200 flies were sampled at each generation from each population, and assessed for the presence or absence of P elements and/or *hobo* elements

that must occur during the excision and integration processes. Thus mobility is thought to create a significant genetic load on a population.[61–65] The rapid dispersal of either *P* or *hobo* elements, as discussed above, suggests that multiplicative transposition of these elements is capable of counteracting this negative selection. But could two elements disperse simultaneously or would they combine to create too much of a genetic load? To address this question, we analyzed a series of populations in which both *P* and *hobo* elements were introduced.[60] Dot blots of DNA from single flies revealed that both elements dispersed concurrently within these populations. Moreover, the rate of dispersal of each element was very comparable to its rate of "spread" when it alone was moving. Hence, the dispersal of one element had little or no affect on the dispersal of the other element. Obviously, significant additional flexibility accrues to the TAC–TICS system if different TAC transposons are able to disperse simultaneously. This allows several different constructs to be added simultaneously or sequentially.

18.4.3. INCAPACITATING GENES THAT MAY BE MORE DISCRIMINATORY

Several different "kill enhancing genes" have been incorporated into baculoviruses in an effort to enhance the potency of a particular virus against its target host. These include two different genes that encode neurotoxins, one from a scorpion[66] and another from a mite,[67] in addition to the crystal protein genes from *Bacillus thuringiensis*.[1] Their broad spectrum of action makes them good choices for use in model systems in laboratory settings. However, they may be poor choices for use in native populations, since they may be toxic or injurious to other insect populations within the ecosystem that consume the dead insect. A number of genes and pathways that control specific physiological and developmental events have been found. The genes encoding these proteins should provide a large number of more rational choices for use in a TAC–TICS type system. Examples of these are: (1) receptors that are responsible for most of the coordination of physiological responses of the organism to its physical environment, as well as the orchestration and control of intercellular communication; (2) various intra-cellular signal transduction pathways; (3) genes associated with hormone or neuropeptide production and function; (4) paralytic peptides; and (5) genes that regulate cell cycle or specific developmental processes. The mis-expression or over-expression of any of these genes could severely hamper or arrest the development, feeding behavior, migration, mating behavior, or reproductive capacity of the targeted "engineered" insect. Alternatively the TAC construct could express an antagonist or an engineered anti-sense construct to inhibit or down regulate gene function or protein expression. The aberrant expression, and thus the biological effects of these gene products, would be limited to the targeted pest insect and perhaps a few closely related species and thus represent more "environment friendly" choices than broad-spectrum bio-toxins. An example of how mis-expression of a standard developmentally important gene, and therefore its protein product, can be used to incapacitate an insect can be found in a very simple, yet elegant experiment in which a construct that was able to produce anti-sense RNA to an essential moth cell cycle control gene was introduced into the target moth. When the anti-sense construct was expressed, the production of the cell cycle protein was dramatically reduced, as expected. More importantly, the caterpillar ceased feeding within a few hours, and its development was arrested when feeding ceased.[68] Death occurred within a couple of days, although that is less important as feeding had ceased.

Studies on the molecular genetics of physiological control processes in insects are gaining momentum, and a number of potential candidates for incapacitating genes have already been identified. These include genes encoding allatotropins, allotostatins, ecdysone,[69] diuretic hormone,[70] juvenile hormone esterase,[71] and genes that control the synthesis and release of molting hormones. Expression of these hormones would have a dramatic effect on the insect, causing either premature molting or a failure to molt, or disruption of growth and feeding. Furthermore, studies on the structure and function of the receptors for these hormones would allow screening for, or synthesis of, agonists or antagonists that might be very effective, highly specific bio-pesticides with minimal negative environmental impact. In addition to hormones that control developmental and physiological events, a number of 20–30 amino acid paralytic peptides have been identified from the hemolymph of a variety of insects. In a modified form, these paralytic-type peptides are capable of specifically paralyzing several insect species including some lepidopterans.[72] Since these genes are small and their product is rather fast acting, they are excellent candidates for use in a TAC–TICS-type system.

With insects that go through population booms, such as some moths and locusts, it may be desirable to disrupt the mating cycle during outbreak years. Pheromone biosynthesis and release by females, and detection and destruction of the pheromone by their cognate receptors on the males are all critical steps in mate detection. If one or more of these pathways is disrupted or modified, the mating process would be disrupted. The gene for pheromone biosynthesis activating neuropeptide[73] is one good candidate. Many of the genes that are responsible for olfaction, which encompasses both mate detection and host or plant target detection, as well as the genes for gustatory response are G-protein coupled receptors that have been identified in several insects including the model organism *Drosophila melanogaster*, the malaria bearing mosquito *Anopholes gambia*, and the silk moth *Bombyx mori*. In addition, these genes are moderately conserved in nature, and hence it is often possible to isolate their orthologs by standard molecular biological techniques (see Chapter 17 for a discussion of G-protein coupled receptors). Consequently they are both interesting targets and accessible genes for use in TAC–TICS-type systems.

Much work still needs to be done to decipher the physiological pathways that control responses to environmental cues and thus govern mating, swarming, host or food recognition, as well as the physiological control and integration of the various tissue functions. The genes that encode both the regulatory and signal transduction of these various physiological and behavioral responses need to be identified, cloned and their function examined, from a wide variety of pest insects. There is a huge opportunity for physiologists and molecular geneticists to examine the evolution of various physiological and behavioral systems. These endeavors would also provide the basis for the development of far more precise and thus environmentally rational means of pest control. The pathology associated with under-expression, over-expression and mis-expression of genes that regulate various developmental events would be particularly useful for developing precise pest management expression cassettes. A number of developmental and sex-related traits are being considered in mosquitos.[74] The genes that encode the receptors and signal pathways that regulate behavioral response are equally important as targets and tools for population control. Disrupting the control pathways for these cellular processes should quickly incapacitate an insect. Our understanding of the intracellular protein kinase cascades, at the moment based principally on studies from mammalian cells and tissues, brings a wealth of knowledge with regard to control points. These phosphorylation cascades are often conserved as enzymatic pathways. Thus it is time to use mammalian systems as models for the identification of important physiological control points in insects that could be disrupted either by over-expression, mis-expression, or down-regulation.

In summary, while there appear to be a plethora of genes that could potentially act as incapacitating genes in TAC cassettes, there is much for insect physiologists, pathologists and molecular geneticists to resolve before we can engineer and test effective, expeditious and specific "insecticidal" gene constructs. No single gene-protein system will be useful for all pest insects. Therefore, it will be useful to test several systems where gene expression cassettes can be developed and their in vivo function examined prior to placing the construct into an insect.

18.4.4. PROMOTERS—THE SWITCH THAT ACTIVATES PEST MANAGEMENT

Promoters that activate the molecular control agent are the final part of the TAC–TICS type of bio-management system. They should be quick acting, allow a self-perpetuation of the bio-engineered organism and thus provide long-term genetic control. The type of promoter depends on what the activating agent is, and how and when it needs to be applied. A promoter that is sex, tissue or developmentally specific would be useful for insect populations that contribute to disease in either humans or agriculturally important animals. For example, using a sex-specific promoter to activate the incapacitating gene would dramatically reduce the numbers of one sex. This would cause the population to become unbalanced and collapse after a few generations. One sex would survive and breed with new immigrants to the population, keeping the population growth in check. These types of promoters are also beneficial if costs of activating an inducible promoter were prohibitive. In the mosquito, a number of developmental and sex-related traits are being examined, and promoters of these genes will be useful. For example, the promoter of the mosquito vitellogenin gene which is activated in females after a blood meal, was used to drive the expression of the Defensin-A gene in transgenic females.[75] The Defensin gene encodes a major insect immune factor, that defends the mosquito against infections. The Defensin gene was strongly activated in fat bodies (the insect equivalent of a liver) of the transgenic females after a blood meal and large amounts of Defensin protein was produced. The question remains whether Defensin will disrupt the life cycle of the disease causing *Plasmodium*. It may be preferable to activate the incapacitating system with an inducible promoter for pests of agricultural or forest crops. In those years when the insect population expands and control is economically warranted, the control gene can be activated for the insects within a specific area. Many other promoters would meet the criteria described above, and the various insect genome projects will provide many potentially useful and adaptable sex or developmental specific promoters for regulating gene expression.

Much research needs to be devoted to identifying and testing inducible promoters. Inducible promoter systems have been constructed and used successfully in bacteria, yeast, and even mammalian and insect cell-lines. The tetracycline inducible system is a very tightly controllable expression system that was designed for use in mammalian cell-line expression[76] and which has potential for use in other systems. The metallothionein promoter allows tight regulation of genes placed under its control in *Drosophila* cells[77] but uses metal as an inducing agent; and thus, it might not be a good choice for environmental reasons. We must now focus on how to create a novel inducible promoter, one that is tightly controlled within a particular pest insect. Clearly the inducing agent must be a benign compound, i.e., one that activates a response in the engineered insects, those carrying the TAC constructs, but has no deleterious effect on other plants or animals. In addition, it must be a compound or agent that is not normally present in the native environment. A variation of an inducible system would be a chimeric promoter, i.e., a hybrid between two promoters, or a system that contained two independent but complementary expression cassettes. These types of genetic constructs should provide very tight control of the incapacitating gene. One half of the system would contain the on/off switch, while the other half may confer tissue or developmental specificity for expression. Therefore gene expression would rely on two conditions occurring simultaneously, correct developmental stage and presence of the exogenous inducer. Conditional chimeric promoters are limited only by imagination and cloned material available for genetic engineering. Our current limitation to engineering such systems is based solely on the paucity of well-defined promoters to use as the basic building blocks. These promoters exist; they simply need to be analyzed, defined, and collected. Major agrochemical companies would benefit from this research, since they could supply the inducer for activation of the incapacitating gene in a chimeric, or dual expression cassette, system. This would be analogous to their current chemical sales, except it would be more "environmentally friendly." Currently, there are no reported examples of these types of inducible systems in insects. Obviously, they must be constructed and thoroughly tested prior to any consideration of their use in population control.

Finally, instead of being externally applied to an infested field or geographic region, the gene encoding the inducer could be cloned into the appropriate crop plants and these engineered crops planted in geographical locales where the engineered insect species is known or predicted to replicate. This modification of the TAC–TICS scheme would simply combine a TAC engineered pest population with an engineered crop. In this case, the engineered crop would express only a benign inducer compound, instead of a bio-toxin or pesticide. Certainly these would be of interest to seed companies.

18.5. Horizontal Transmission—Can Mobile Elements Be Transmitted to Other Species?

Mobile genetic elements (transposable elements) are found in virtually all organisms, plants and animals, from microbes to man. Each species usually houses several, often dozens, of different transposable elements. In addition, there are usually many copies of each type of element in every genome. The number of copies of each element per genome seems to be a property of the particular element; some elements have a dozen copies, others a hundred or more, and some elements have thousands copies per genome. Hence, there appears to be an upper limit to the density of a particular element within the genome, and thus the mobile element-host interaction must somehow restrict the replication and spread of the element within the individuals of the population. Since virtually all organisms contain transposable elements, but genera often differ in the types of elements they contain, one must ask how these disparate mobile elements arose in nature, and how they “arrived” in a particular species. There are only three possibilities: (1) they arose *de novo* by assembling appropriate segments of DNA within the genome; (2) they arose, or arrived, in a particular species at various times during evolution and once present they are passed down from generation to generation in a mating dependent fashion (vertical or orthologous transmission), but diverge rapidly within each genus as the genera evolve; or (3) they pass from one species or genus to another by a mechanism that does not involve mating (horizontal or xenologous transmission). Vertical transmission (the mating dependent mode of transmission) makes two predictions. First, the distribution of an element within a genus should be virtually continuous. All descendants of an ancestral species should contain homologues of the element, unless the element was somehow lost from a particular species, in which case, its descendants should lack the element and a gap, or patch of loss, should be apparent in phylogenetic studies. Second, the level of DNA sequence conservation of homologous elements should be congruent with the phylogeny of the species radiation. Horizontal transmission makes the diametrically opposite set of predictions. First, the distribution of the elements should be discontinuous or patchy. Second, the sequence conservation of the element need not be related

TABLE II. Classification of the subgenus *Sophophora* of the *melanogaster* species, their geographic origins, and the distribution of transposable element homologues

Species group	Sub-group	Species	Geographic origin	P	I	Copia	Gypsy
melanogaster	*ananassae*	*ananassae*	Mexico	0	H	M	M
	suzukii	*lucipennis*	Taiwan	L	N/A	H	L
	montium	*kikkawai*	Colombia	L	H	H	M
	elegans	*elegans*	Philippines	0	H	M	M
	eugracilis	*eugracilis*	New Guinea	0	H	0	M
	takahashii	*takahashii*	Nepal	0	0	H	M
	ficusphila	*ficusphila*	Taiwan	L	H	H	M
obscura[46]	*affinis*	*affinis*	Nebraska	M	0	H	H
	obscura	*pseudoobscura*	Arizona	M	0	M	H
willistoni[55]	*wllistoni*	*willistoni*	Nicaragua	H	0	M	L
		equinoxialis	Honduras	H	0	M	L
		tropicalis	El Salvador	H	0	M	L
		paulistorum	Mesitas	H	0	M	M
		paulistorum-like	Mexico	H	0	M	M
		nebulosa	Colombia	H	0	M	L
		succinea	Colombia	H	0	M	L
		copricorni	Colombia	H	0	M	L
		fumipennis	Colombia	H	0	M	M
saltans[55]	*saltans*	*saltans*	Costa Rica	H	0	M	M
		australosaltans	Brazil	H	N/A	M	L
		prosaltans	Costa Rica	0	0	M	L
	cordata	*neocordata*	Brazil	0	0	M	0
	sturtevantii	*sturtevantii*	Costa Rica	H	0	M	L
	emarginata	*emarginata*	Costa Rica	0	0	M	L

Notes: The categories on the right show the highest stringency wash at which hybridization to *D. melanogaster* transposable element (P, I, copia, or gypsy) could be detected. Distribution of homologues as follows: 0 = No hybridization detected: L = hybridization readily detected at low-stringency wash only; M = hybridization readily detected at both low- and moderate-stringency washes: H = hybridization readily detected at low-, medium-, and high-strigency washes. N/A = data not available. The number within the brackets [ ] is the estimated divergence times of the Drosophila species groups, using the melanogaster species group as the root.

to the phylogeny of the host; in fact, the degree of sequence conservation is more likely to reflect the geographic distribution or niche relatedness of the hosts that contain the homologous elements rather than phylogenetic relationships.

Various transposable elements show different distribution patterns (Table II). Some elements, like the I element, are highly conserved within the *melanogaster* species group, but not found in any of the other species examined (Table II). Other elements, such as the P element have a very patchy distribution within the subgenus *Sophophora*.[78,79] These analyses were extended to a wider variety of the subgenus Drosophila with very similar results.[80,81] Both the very limited pattern observed with the I element and the discontinuous pattern, with a large number of discontinuities, such as seen with the P element, are consistent with horizontal transmission. The conclusion is strongly supported by the level of conservation observed among the various species. The sequences homologous to the *D. melanogaster* P-element probe are more closely related (bind under high stringency conditions = H) to the phylogenetically distant *willistoni* and *saltans* species groups than they are to the closely related *melanogaster* species groups (no P elements = 0, or very low sequence similarity = L). This was verified by cloning and sequencing the P-element homologues from *D. willistoni, D. nebulosa* and *D. saltans*. DNA sequence comparisons with the intact and fully functional P element from *D. melanogaster* demonstrated that the homologous elements differed by <3%,[80] Indeed, an element isolated from *D. willistoni* differed from its *D. melanogaster* homologue by only a dozen or so base pairs even though the two species have been reproductively isolated for about 50 million years. The geographic ranges of *D. melanogaster* and *saltans* and *willistoni* overlap in South America and suggest that the P element invasion of *D. melanogaster* may have occurred in South America. The mechanism of horizontal transfer between *D. willlistoni* and *D. melanogaster* remains unknown. Speculation has focused on vector organisms, such as viruses and mites.[82,83]

Of course these predictions about the continuity of distribution and the congruence of sequence conservation with phylogeny depend on the assumption that sequence divergence, or loss of homologous elements, occurs at the same rate in different species, and little is known about this. However, results from experiments on the spread of mobile elements within naïve populations, under laboratory conditions, suggest that the multiplicative transposition process is not exceptionally error prone. The divergence of non-coding DNA in *Drosophila* has been estimated at 2.5%/Myr.[84] If the P elements were spread strictly by vertical transmission, in order to produce the distribution pattern we observed, we estimate that the divergence rates in some species of *Drosophilidae* must be 8–19 times higher than it is in the *willistoni*, *saltans*, or *nasuta* species. Such wide variances in divergence rates among relatively similar species are highly unlikely. In fact, those rates are so high that they compare with the variance or divergence seen between non-conserved sequences. In contrast to the I and P elements, the distribution and conservation of sequences homologous to the F and copia elements is consistent with proposed *Drosophila* phylogenies[85,86] and support vertical transmission of these elements. Hence, horizontal transmission need not be invoked for all transposable elements.

If we accept that horizontal transmission has occurred for some transposable elements, the question remains: how frequently does it occur? Is it frequent enough to be of concern for containment of engineered constructs to the target species? The numbers of documented examples of possible horizontal transmission are rare. The *mariner* transposable element appears to be the most peripatetic transposable element.[87] However, horizontal transfer events involving mariner, while frequent on the evolutionary time scale, are typically separated by millions of years. Hence it appears that horizontal transfer does not occur that frequently, but perhaps we have not searched meticulously enough for evidence of horizontal transfer in metazoans.

Clearly, horizontal transfer occurs frequently in bacteria; in fact, other than plasmid transfer, transformation and transduction may be the more common methods of DNA transfer in prokaryotes.

18.6. Risk Analysis

Clearly all of the basic components for a TAC–TICS type system are available, and the number of choices of incapacitating genes, transposable elements, and even promoters and regulatory elements are increasing yearly. Therefore a TAC–TICS type control system appears to be technically feasible.

The question now becomes what are the implications and risks of introducing such a system into a natural setting? There are many. The possibility of horizontal transfer is among the most obvious and consequential of these risk factors. Horizontal transfer has been noted for several insect transposons. It is estimated that in the last 3 million years, *P* elements have undergone 11 horizontal transfer events among the 18 species surveyed.[88] The advantages and risks of releasing engineered insects, whether beneficial or pest insects, is a topic that requires serious consideration and certainly demands extensive model and laboratory testing prior to field use. By combining mobile elements with narrowly acting incapacitating genes and species specific and/or inducible promoters, one should be able to dramatically reduce the likelihood that a TAC construct would survive outside its targeted insect, even if horizontal transfer occurred. Nonetheless, the risk can never be eliminated. While we are interested in the mechanisms, biology and genetics of TAC–TICS type systems, our intent in investigating model systems is to provide molecular and empirical data that can be used in the risk assessment studies undertaken by others.

References

1. B. M. Ribeiro and, N. E. Crook, Expression of full length and truncated forms of crystal protein genes from *Bacillus thuringiensis* subsp. *kurstaki* in a Baculovirus and pathogenicity of the recombinant virsuses, *J. Invertebr. Pathol.* 62, 121–130 (1993).

2. L. H. Miller, R. K. Sakai, P. Romans, R. W. Gwadz, P. Kantoff, and H. G. Coon, Stable integration of a bacterial gene in the mosquito *Anopheles gambiae, Science* 237, 779–781 (1987).
3. R. J. St. Leger, D. C. Frank, D. W. Roberts, and R. C. Staples, Molecular cloning and regulatory analysis of the cuticle-degrading-protease structural gene from the entomopathogenic fungus *Metarhizium anisopliae, Eur. J. Biochem.* 204, 991–1001 (1992).
4. D. A. Fischhoff, K. S. Bowdish, F. J. Perlak, P. G. Marrone,, S. M. McCormick, J. G. Niedermeyer, D. A. Dean, , K. Kusano-Kretzmer, E. J. Mayer, D. E. Rochester, S. G. Rogers, and R. T. Fraley, Insect tolerant transgenic tomato plants, *Bio/Technolgy* 5, 807–813 (1987).
5. M. Vaeck, A. Reynaerts, H. Hofte, S. Jansens, M. Beuckeleer, C. Dean, M. Zabeau, M. Montagu, and J. Leemans, Transgenic plants protected from insect attack, *Nature* 238, 33–37. (1987).
6. G. G. Foster, M. J. Whitten, T. Prout, and R. Gill, *Science* 176, 875 (1972).
7. M. Fitz-Earle, D. G. Holm, and D. T. Suzuki, Genetic control of insect populations, I: Cage studies of chromosome replacement by compound autosomes in *Drosophila melanogaster, Genetics* 74, 461–475 (1973).
8. S. A. Davis, E. A. Catchpole, and G. R. Fulford, Periodic triggering of an inducible gene for control of wild population, *Theor. Popul. Biol.* 58, 95–106 (2000).
9. J. Gressel and A. Levy, Giving *Stiga hermonthica* the DTs, in *Breeding for Striga Resistance in Cereals*, edited by B. I. G. Haussmann, D. E. Hess, M. L. Koyama, L. Grivet, H. F. W. Rattunde, and H. H. Geiger (Margraf Verlag, Weikersheim, 2000), pp. 207–224.
10. T. A. Pfeifer and T. A. Grigliatti, Future perspectives on insect pest management: Engineering the pest, *J. Invertbr. Pathol.* 67. 109–119 (1996).
11. T. A. Pfeifer and T. A. Grigliatti, Genetic pest management strategies: A view of targeted pest insect management in the 21st century, *Agro-Food-Ind. Hi-Tech* 8, 29–35 (1997).
12. M. G. Kidwell and, J. M. C. Ribeiro, Can transposable elements be used to drive disease refractoriness genes into vector populations? *Parasitol. Today* 8, 325–329 (1992).
13. L. H. Miller, The challenge of malaria, *Science* 257, 36–37 (1992).
14. J. M. Crampton, A. Warren, G. J. Lycett, M. A. Hughes, I. P. Comley, and P. Eggleston, Genetic manipulation of insect vectors as a strategy for the control of vector-borne diseases, *Ann. Trop. Med. Parasitol.* 88, 3–12 (1993).
15. J. C. Salvado, N. Bensaadi-Merchermek, and C. Mouches, Transposable elements in mosquitoes and other insect species, *Comp. Biochem. Physiol.* B 109, 531–544, (1994).
16. G. M. Rubin and A. C. Spradling, Genetic transformation of Drosophila with transposable element vectors, *Science* 218, 348–353 (1982).
17. R. K. Blackman, M. M. Koehler, R. Grimaila, and W. M. Gelbart, Identification of a fully-functional *hobo* transposable element and its use for germ-line transformation of Drosophila, *EMBO J.* 8, 211–217 (1989).
18. D. Lidholm, A. R. Lohe, and, D. L. Hartl, The transposable element mariner mediates germline transformation in *Drosophila melanogaster Genetics* 134, 859–868 (1993).
19. P. W. Atkinson, W. D. Warren, and D. A. O'Brochta, The *hobo* transposable element of *Drosophila* can be cross-mobilized in houseflies and excises like the *Ac* element of maize, *Proc. Natl. Acad. Sci. USA* 90, 9693–9697 (1993).
20. M. A. Pritchard, J.-M. Dura, A. Pelisson, A. Bucheton, and D. J. Finnegan, A cloned I-factor is fully functional in *Drosophila melanogaster, Mol. Gen. Genet.* 214, 533–540 (1988).
21. S. B. Daniels, M. McCarron, C. Love, and A. Chovnick, Dysgenesis-induced instability of*rosy* locus transformation in *Drosophila melanogaster*: Analysis of excision events and the selective recovery of control element deletions, *Genetics* 109, 95–117 (1985).

22. M. D. Brennan, R. G. Rowan, and W. J. Dickinson, Introduction of a functional *P* element into the germ-line of *Drosophila hawaiiensis, Cell* 38, 147–151 (1984).
23. D. W. Clough, H. M. Lepinske, R. L. Davidson, and R. V. Storti, Drosophila *P* element-enhanced transfection in mammalian cells, *Mol. Cell. Biol.* 5, 898–901 (1985).
24. J. S. Khillan, P. A. Overbeek, and H. Westphal, Drosophila *P* element integration in the mouse, *Dev. Biol.* 109, 247–250 (1985).
25. L. H. Miller, Genetically engineered insect virus pesticides: Present and future, *J. Invertbr. Pathol.* 65, 211–216 (1995).
26. V. McGrane,, J. O. Carlson, B. R. Miller, and B. J. Beaty, Microinjection of DNA into *Aedes triseriatus* ova and detection of integration, *Am. J. Trop. Med. Hyg.* 39, 502–510 (1988).
27. D. A. O'Brochta and A. M. Handler, Mobility of *P* elements in drophilids and non-drosophiids, *Proc. Natl. Acad. Sci.* 85, 6052–6056 (1988).
28. A. C. Morris, P. Eggelston, and J. M. Crampton, Genetic transformation of the mosquito *Aedes aegypti* by micro-injection of DNA, *Med. Vet. Entomol.* 3, 1–7 (1989).
29. Walker, V. K., Gene transfer in insects, *Adv. Cell Culture* 7, 87–124 (1989).
30. A. M. Handler, S. P. Gomez, and D. A. O'Brochta, A functional analysis of the *P*-element gene-transfer vector in insects, *Arch. Insect Biochem. Physiol.* 22, 373–384 (1993).
31. T. G. Loukeris, I. Livadaras, B. Arca, S. Zabalou, and C. Savakis, Gene transfer into the medfly, *Ceratitis capitata*, with a *Drosophila hydei* transposable element, *Science* 270(5244), 2002–2005 (1995).
32. A. M. Handler, A current perspective on insect gene transformation, *Insect Biochem. Mol. Biol.* 31, 111–128 (2001).
33. G. A. Meister, R. A. Lansman, and T. A. Grigliatti, A simple differential DNA hybridization method for the detection of transposable elements, *Biotechniques* 18, 250–255 (1995).
34. J. H. Whalen and T. A. Grigliatti, Molecular characterization of a retrotransposon with target site preference in *Drosophila melanogaster, nomad,* and its relationship to other retrovirus-like mobile elements, *Mol. Gen. Genet.* 260, 401–409 (1998).
35. T. Pfeifer, M. Ring, and T. A. Grigliatti, Molecular analysis of a new transposable element from *Lymantria dispar, Insect Mol. Biol.* 9, 349–356 (2000).
36. D. A. O'Brochta, W. D. Warren, K. J. Saville, and P. W. Atkinson, Interplasmid transposition of Drosophila hobo elements in non-drosophilid insects, *Mol. Gen. Genet.* 244, 9–14 (1994).
37. G. A Meister, Spread of *P* elements in *Drosophila melanogaster,* M.Sc. Thesis (University of British Columbia, Vancouver, 1992).
38. A. G. Good, G. A. Meister, H. W. Brock, T. A. Grigliatti, and D. A. Hickey, Rapid spread of transposable *P* elements in experimental populations of *Drosophila melanogaster, Genetics* 122, 387–396 (1989).
39. M. G. Kidwell, J. B. Novy, and S. M. Feeley, Rapid unidirectional change of hybrid dysgenesis potential in Drosophila, *J. Hered.* 72, 32–38 (1981).
40. P. K. Kiyasu and M. G. Kidwell, Hybrid dysgenesis in *Drosophila melanogaster:* The evolution of mixed P and M populations maintained at high temperature, *Genet. Res.* 44, 251–259 (1984).
41. A. C. Spradling, *P* element-mediated transformation, in *Drosophila: A Practical Approach*, edited by D. B. Roberts (IRL Press, Oxford, 1986), pp. 175–198.
42. H. M. Robertson, C. R. Preston, R. W. Phillis, D. Johnson-Schlitz, W. K. Benz, and W. R. Engels, A stable genomic source of *P* element transposase in *Drosophila melanogaster, Genetics* 118, 461–470 (1988).
43. S. B. Daniels, S. H. Clark, M. G. Kidwell, and A. Chovnick, Genetic transformation of *Drosophila melanogaster* with an autonomous *P* element: Phenotypic and molecular analyses of long-established transformed lines, *Genetics* 115, 711–723 (1987).

44. W. R. Engels, D. M. Johnson-Schlitz, W. B. Eggleston, and J. Sved, High-frequency *P* element loss in *Drosophila* is homology dependent, *Cell* 62, 515–525 (1990).
45. G. B. Gloor, N. A. Nassif, D. M. Johnson-Schlitz, C. R. Preston, and W. R. Engels, Targeted gene replacement in Drosophila via *P* element-induced gap repair, *Science* 253, 1110–1117 (1991).
46. G. A. Meister and T. A. Grigliatti, Rapid spread of a *P* element/*Adh* gene construct through experimental populations of *Drosophila melanogaster, Genome* 36, 1169–1175 (1993).
47. B. R. Calvi, T. J. Hong, S. D. Findley, and W. M. Gelbart, Evidence for a common evolutionary origin of inverted repeat transposons in Drosophila and plants: *hobo, Activator,* and *Tam3, Cell* 66, 465–471 (1991).
48. L. Pascual and G. Periquet, Distribution of *hobo* transposable elements in natural populations of *Drosophila melanogaster, Mol. Biol. Evol.* 8, 282–296 (1991).
49. G. M. Simmons, Horizontal transfer of *hobo* transposable elements within the *Drosophila melanogaster* species complex: Evidence from DNA sequencing, *Mol. Biol. Evol.* 9, 1050–1060 (1992).
50. G. Periquet, F. Lemeunier, Y. Bigot, M. H. Hamelin, and C. Bazin, The evolutionary genetics of the*hobo* transposable element in the *Drosophila melanogaster* complex, *Genetica* 93, 79–90 (1994).
51. V. Ladeveze, M. I. Galindo, L. Pascual, G. Periquet, and F. Lemeunier, Invasion of the *hobo* transposable element studied by in situ hybridization on polytene chromosomes of *Drosophila melanogaster, Genetica* 93, 91–100 (1994).
52. V. Ladeveze, I. Galindo, N. Chaminade, L. Pascual, G. Periquet, and F. Lemeunier, Transmission pattern of*hobo* transposable element in transgenic lines of *Drosophila melanogaster, Genet. Res.* 71, 97–107 (1998).
53. M. I. Galindo, V. Ladeveze, F. Lemeunier, R. Kalmes, G. Periquet, and L.Pascual, Spread of the autonomous transposable element *hobo* in the genome of *Drosophila melanogaster, Mol. Biol. Evol.* 12, 723–734 (1995).
54. R. D. Streck, J. E. MacGaffey, and S. K. Beckendorf, The structure of *hobo* transposable elements and their insertion sites, *EMBO J.* 5, 3615–3623 (1986).
55. G. Yannopoulos, N. Stamatis, M. Monastirioti, P. Hatzopoulos, and C. Louis, *hobo* is responsible for the induction of hybrid dysgenesis by strains of *Drosophila melanogaster* bearing the male recombination factor 23.5MRF, *Cell* 49, 487–495 (1987).
56. A. Sarkar, C. J. Coates, S. Whyard, U. Willhoeft, P. W. Atkinson, and D. A. O'Brochta, The *Hermes* element from *Musca domestica* can transpose in four families of cyclorrhaphan flies *Genetica* 99, 15–29 (1997).
57. A. C. Pinkerton, K. Michel, D. A. O'Brochta, and P. W. Atkinson, Green fluorescent protein as a molecular marker in transgenic *Aedes aegypti, Insect Mol. Biol.* 9, 1–10 (2000).
58. R. K. Blackman and W. M. Gelbart, The transposable element *hobo* of *Drosophila melanogaster,* in *Mobile DNA*, edited by D. E. Berg and M. M. Howe (American Society for Microbiology, Washington, DC, 1989), pp. 523–530.
59. K. Harada, K. Yukuhiro, and T. Mukai, Transposition rates of movable genetic elements in *Drosophila melanogaster, Proc. Natl. Acad. Sci. USA* 87, 3248–3252 (1990).
60. G. A. Meister, Dispersal of transposable elements, Ph.D. thesis (University of British Columbia, Vancouver, 1999).
61. K. Yukuhiro, K. Harada, and T. Mukai, Viability mutations induced by the *P* elements in *Drosophila melanogaster, Jpn. J. Genet.* 60, 531–537 (1985).
62. T. F. Mackay, Transposable elements and fitness in *Drosophila melanogaster, Genome* 31, 284–295 (1989).

63. B. Fitzpatrick and J. A. Sved, High levels of fitness modifiers induced by hybrid dysgenesis in *Drosophila melanogaster*, *Genet. Res.* 48, 89–94 (1986).
64. W. J. Eanes, C. Wesley, J. Hey, D. Houle, and J. W. Ajioka, The fitness consequences of *P* element insertion in *Drosophila melanogaster, Genet. Res.* 52, 17–26 (1988).
65. V. N. Bolshakov, A. P. Galkin, L. Z. Kaidanov, V. A. Gvozdev, and C. Louis, Closely related *Drosophila melanogaster* strains with altered fitness also show changes in their *hobo* element properties, *Genet. Sel. Evol.* 26, 205–216 (1994).
66. L. F. Carbonell, M. R. Hodge, M. D. Tomalski, and L. K. Miller, Synthesis of a gene coding for an insect-specific scorpion neurotoxin and attempts to express it using baculovirus vectors, *Gene* 73, 409–418 (1988).
67. M. D. Tomalski and L. K. Miller, Insect paralysis by baculovirus-mediated expression of a mite neurotoxin gene, *Nature* 352, 82–85 (1991).
68. S. Y. Lee, X. Qu, W. Chen, A. Poloumienko, N. MacAfee, B. Morin, C. Lucarotti, and M. Krause, Insecticidal activity of a recombinant baculovirus containing an antisense c-myc fragment,*J. Gen. Virol.* 78, 273–281 (1997).
69. T. J. Sliter, Imaginal disk-autonomous expression of a defect in sensory bristle patterning caused by the lethal ecdysoneless mutation of *Drosophila melanogaster, Development* 106, 347–354 (1989).
70. S. Maeda, Increased insecticidal effect by a recombinant baculovirus carrying a synthetic diuretic hormone gene, *Biochem. Biophys. Res. Commun.* 165, 1177–1183 (1989).
71. R. Eldridge, D. R. O'Reilly, B. D. Hammock, and L. K. Miller, Insecticidal properties of genetically engineered baculoviruses expressing an insect juvenile hormone esterase gene, *Appl. Environ. Microbiol.* 58, 1583–1591 (1992).
72. W. S. Skinner, P. A. Dennis, J. P. Li, R. M. Summerfelt, R. L. Carney, and G. B. Quistad, Isolation and identification of paralytic peptides from hemolymph of the lepidopteran insects *Manduca sexta, Spodoptera exigua,* and *Heliothis virescens, J. Biol. Chem.* 266, 12873–12877 (1991).
73. M. B. Davis, V. N. Vakharia, J. Henry, T. G. Kempe, and A. K Raina, Molecular cloning of the pheromone biosynthesis-activating neuropeptide in *Helicoverpa zea,Proc. Natl. Acad. Sci.* 89, 142–146 (1992).
74. P. Eggleston, The control of insect-borne disease through recombinant DNA technology, *Heredity* 66, 161–172 (1991).
75. V. Kokoza, A. Ahmed, W. Cho, N. Jasinskiene, A. A. James, and A. Raikhel, Engineering blood meal-activated systemic immunity in the yellow fever mosquito, *Aedes aegypti, Proc. Natl. Acad. Sci. USA* 97, 9144–9149 (2000).
76. K. Forster, V. Helbl, T. Lederer, S. Urlinger, N. Wittenburg, and W. Hillen, Tetracycline-inducible expression systems with reduced basal activity in mammalian cells, *Nucl. Acids Res.* 27, 708–710 (1999).
77. T. A. Bunch, Y. Grinblat, and L. S. B. Goldstein, Characterization and use of the *Drosophila* metallothionein promoter in cultured *Drosophila melanogaster* cells, *Nucl. Acids Res.* 16, 1043–1061 (1988).
78. J. F. Brookfield, E. Montgomery, and C. H. Langley, Apparent absence of transposable elements related to P elements in other species of *Drosophila, Nature* 310, 330–332 (1984).
79. R. A. Lansman, S. N. Stacey, T. A. Grigliatti, and H. W. Brock, Sequences homologous to the P mobile element of *Drosophila melanogaster* are widely distributed in the subgenus Sophophora, *Nature* 318(6146), 561–562 (1985).
80. S. N. Stacey, R. A. Lansman, H. W. Brock, and T. A. Grigliatti, Distribution and conservation of mobile elements in the genus *Drosophila, Mol. Biol. Evol.* 3(6), 522–534 (1986).

81. R. A. Lansman, R. O. Shade, T. A. Grigliatti, and H. W. Brock, Evolution of *P* transposable elements: Sequences of *Drosophila nebulosa P* elements, *Proc. Natl. Acad. Sci. USA* 84, 6491–6495 (1987).
82. W. R. Engels, The origin of P elements in *Drosophila melanogaster, BioEssays* 14, 681–686 (1992).
83. M. A. Houck, J. B. Clark, K. R. Peterson, and M. G. Kidwell, Possible horizontal transfer of Drosophila genes by the mite *Proctolaelaps regalis, Science* 253, 1125–1128 (1991).
84. C. H. Langley, E. Montgomery, and W. F. Quattlebaum, Restriction map variation in the Adh region of Drosophila, *Proc. Natl. Acad. Sci. USA* 79, 5631–5635 (1982).
85. L. H. Throckmorton, The phylogeny, ecology, and geography of *Drosophila,* in *Handbook of Genetics*, edited by R. C. King vol. 3 (Plenum Press, New York, 1975).
86. S. M. Beverly and A. C. Wilson, Molecular evolution in *Drosophila* and the higher Diptera, II: A time scale for fly evolution, *J. Mol. Evol.* 21, 1–13 (1984).
87. H. M. Robertson, The mariner transposable element is widespread in insects, *Nature* 362, 241–245 (1993).
88. J. C. Silva and M. G. Kidwell, Horizontal transfer and selection in the evolution of P elements, *Mol. Biol. Evol.* 17, 1542–1557 (2000).

19. FAILSAFE MECHANISMS FOR PREVENTING GENE FLOW AND ORGANISM DISPERSAL OF ENHANCED MICROBIAL BIOCONTROL AGENTS

Jonathan Gressel*
Plant Sciences, Weizmann Institute of Science, Rehovot 76100, Israel

Abstract. A science-based regulatory system is expected to have three key elements in requirements for enhanced biocontrol agents: no off-site dispersal, poor long-term environmental persistence, and limited possibility of recombination with other pathogens. These can be achieved by using appropriate combinations of some of the following elements: synergists that are present for a single generation; organisms that are permanently asporogenic; and by inserting genes in tandem with virulence-enhancing genes that would render recombined offspring to be unfit to compete in the environment.

Keywords: asporogenic, dispersal, failsafes, fitness, persistence, recombination, synergists, tandem constructs, virulence

19.1. The Need for Failsafe Mechanisms

Some regulatory authorities have become extremely stringent in the regulation of biocontrol agents, especially "classical" agents, out of fear of belatedly discovering new hosts for an imported organism. The native species that might be used in inundative biocontrol might be easier to register, if they are enhanced. Science-based regulation of the biosafety aspects of a biocontrol agent would be expected to deal with three basic risks:

(a) There should be no off target effects. As with pesticides, application should be to the target site only. The effects of drifts of biocontrol agents should usually be less than with pesticides, due to narrow host ranges.
(b) The agent should not persist in the environment. Many view this requirement as a cynical way to help commercial entities to continue to obtain repeated sales. Still, we do not wish chemical pesticides to remain in the environment, and there are more reasons not to wish a biological to persist. The longer it persists, the greater the likelihood of change (see next criterion). Such changes are often to lowered virulence, and if attenuated

*To whom correspondence should be addressed, e-mail: jonathan.gressel@weizmann.ac.il

M. Vurro and J. Gressel (eds.), Novel Biotechnologies for Biocontrol Agent Enhancement and Management, 353–362.

organisms persist, they can compete with the next application of the biocontrol agent. Persistence of chemical pesticides can limit crop rotations, and the same holds for biologicals, especially if the biocontrol agents replicate in the environment. Chemicals cannot replicate.

(c) The biocontrol agent should not change host range or increase virulence in the environment. Loss of virulence may result in loss of persistence if the pathogen is incapable of reproducing outside the host. Pathogens that are dependent on the host for dispersal or do not have a resting stage may lose virulence, so the host lives longer. However, genetic changes are regarded as undesirable because their effects are unpredictable. Such problematic genetic changes may result from mutation or from recombination with other organisms, and both must be prevented.

19.1.1. LIKELIHOOD OF MUTATIONS THAT CHANGE HOST RANGE

Multitudes of mutational differences related to virulence or host specificities are found when the DNA of virulent pathogens is compared with their wild apathogenic relatives. This is as true of "killer" *E. coli* O157:H7 that split off from the common ancestor of common enteric types a few million years ago,[1] and for different forma speciales of *Fusarium oxysporum*[2] diverging from each other at least 50,000 years ago (L. Hornok, personal communication). The number of mutations required to attain virulence or change host range renders it nigh impossible to achieve in human times, and such changes are likely only on an evolutionary time scale.

19.1.2. THE LIKELIHOOD OF RECOMBINATION BETWEEN ORGANISMS

There is always a possibility that the hypervirulence genes (mutations or transgenes) will move from a biocontrol agent into a pathogen that attacks crops or beneficial insects. This possibility is finite but exceedingly remote, if the transfer is "horizontal," i.e., between fungi in different families or kingdoms. Horizontal transfer of genes is well documented among prokaryotic organisms, but that should not be extrapolated to eukaryotes. Still, there is indirect evidence that points to the possibility that the horizontal transfer of clustered genes has taken place in evolutionary times[3,4] The risk that this can happen in human time is exceedingly low, many orders of magnitude lower than risks regulators normally take with any item to be regulated, and thus should not be considered further with fungi.

Vertical gene transfer, i.e., between sexually or even asexually compatible organisms is of course an issue that must be dealt with.

This author has coined the term "diagonal" gene transfer[5] to denote the sporadic ability to transfer genes among ostensibly incompatible members of the same family, genus, or even formae speciales, or strains within a species. Diagonal recombination often occurs under severe selection pressure, e.g., when two different incompatible auxotrophics are placed together on a minimal medium (see review[6] and recent laboratory studies on *Cryphonectria*,[7] on *Fusarium*,[8] and *Colletotrichum*[9]). The likelihood of diagonal gene transfer also increases when two organisms are attacking the same target; e.g., when two *Beauveria* strains are in the same insect (e.g., Castrillo et al.[10]). Such recombinations can be due to forced heterokaryon formation; with later nuclear recombinations, parasexual recombination,[11] etc.

There is also indirect but convincing evidence that this can indeed happen in the field. Sometime before 1941 the wheat pathogen *Pyrenophora tritici-repentis* (but not other *Pyrenophora* spp.) acquired the *tox*A virulence gene, most probably from another wheat pathogen *Staganospora nodorum*.[12] Both are members of the same family, but distantly related (based on DNA evidence, as well as morphology). The gene has no diversity in *Pyrenophora* but is highly diverse in the putative source, *Staganospora*. It is quite probable that conidial anastomosis tubes, known to form between these two species[6] occurred in wheat co-infected by both pathogens.[12] Even though this transfer occurred recently, it appears to be a rare event, as this is the only example in the literature of changes in virulence of a crop pathogen due to diagonal gene movement. Regulators are still bound to be mindful of this recent event. While there is no evidence that the wheat pathogen changed host range, regulators may assume that this too could happen. Thus, there may be a demand that failsafe mechanisms be instated that will preclude such unlikely recombination events.

Two types of failsafe mechanisms will be discussed, based on the two forms of enhancing biocontrol agents; chemical enhancement with synergists, and transgenic enhancement. Mutational enhancement (Chapter 14) will not be discussed, as the organisms that are enhanced by mutation and selection do not possess new factors and are based on pre-existing genes.

19.2. Auto Failsafes of Chemically Synergized Biocontrol Agents

Chemicals, especially anti-metabolites have been used to synergistically enhance the virulence of biocontrol agents (see Chapter 15). This can include those that were specifically chosen to overcome host defenses, e.g., the addition of the shikimate pathway inhibitor glyphosate to preclude synthesis of phytoalexins derived from that pathway;[13] chelators of calcium to suppress calcium mediated host defenses such as callose synthase,[14,15] the addition of pectinase, proteases, cellulases or chitinases to facilitates rapid penetration into weed, pathogens or insect pests, etc. (see Chapters 6, 9, and 11).

These synergists enhance virulence of the biocontrol agents, but only the first generation, i.e., only when applied. Thus, propagules arising from the initial synergized infection are not hypervirulent, being no different from the wild type. Thus, there is an automatic failsafe that the chemically-rendered hypervirulent material has a next generation that is wild type.

The hydrolyzing enzymes described above are often added to biocontrol agents to ascertain whether they confer hypervirulence as a prelude to engineering the genes encoding the hydrolases into the biocontrol agent. The rationale is: if an exogenous application of the hydrolase does not synergize, why put in the gene? Still, there is a failsafe advantage of using added hydrolases, over transgenically produced hydrolases, and one should investigate the practicalities of using the hydrolases.

19.3. Failsafes for Transgenic Biocontrol Agents

A scheme to obviate the spread of native or host-range mutated agents, and an additional scheme to mitigate introgression are described (Figure 1). The

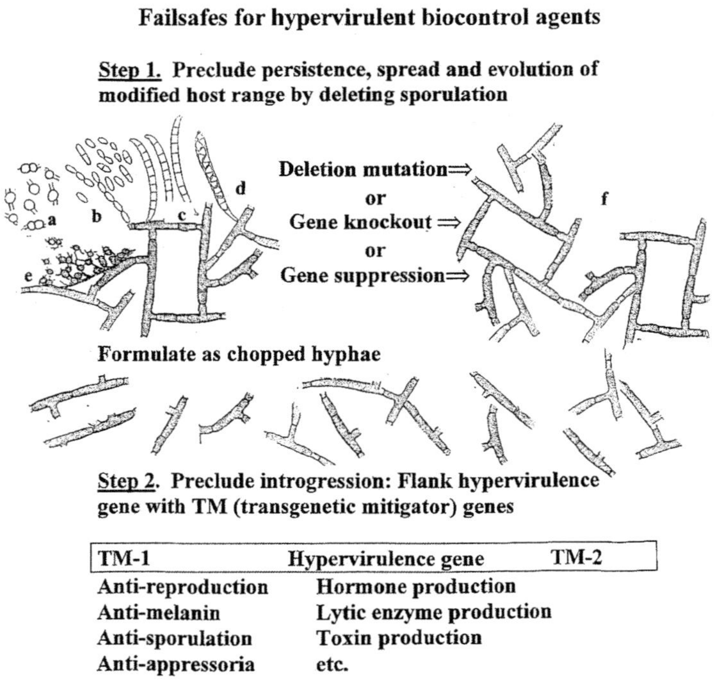

Figure 1. Dual failsafes to prevent (Step 1) spread of biocontrol agents, and (Step 2) their introgression into other organisms. a: chlamydospores; b: microconidia; c: macroconidia; d: ascus with ascospores; e: sclerotia; f: asporogenic mycelia. Source: From Gressel[16] by permission (copyright 2001 Elsevier Science)

general concepts described may be broadly applicable to many agents including the parasitic insects used to control insect pests. The specific examples presented are more limited to fungal pathogens of weeds (mycoherbicides).

19.3.1. PREVENTION OF PERSISTENCE AND SPREAD OF TRANSGENIC BIOCONTROL AGENTS

Fungi and bacteria typically spend the dormant part of their life cycles as spores or analogous structures that are resistant to heat, cold, desiccation, etc. These dormant propagules are a major form of dispersal. The suppression of resting structure formation in hypervirulent biocontrol agents can prevent both persistence and spread (Figure 1a). Non-sporulating mutants are not hard to isolate by deletion-causing mutations in organisms with deeply pigmented spores, but are hard to isolate when the spores are hyaline. Point mutations can revert, whereas deletions cannot. A similar concept was proposed to prevent the persistence of a wide host range (non-transgenic) pathogen, *Sclerotinia*, that was proposed for general weed control.[17] That proposal additionally suggested using auxotrophic mutants of the biocontrol agent that could exist only on the culture medium and on the pest host, but would have trouble existing away from them.[17] This concept was never really brought to fruition because it was assumed that efficacious inoculation with any biocontrol agent had to be with dormant spores, but it is now documented that chopped mycelia could be dried, stored for over a year and rehydrated.[18] The rehydrated mycelia were more virulent than spores of the same species, because the mycelia establish more quickly in the pest.[18] It is usually far more efficient to produce mycelia in liquid culture than spores in liquid or on solid media.

The spread of microbial biocontrol agents can also be prevented by rendering them transgenically asporogenic (Figure 1a). This could be performed by antisense/RNAi type strategies or preferably by gene targeting[19] and knockout. Many pathogenic species seem to require melanized spores for pathogenicity.[20] The germinating spore develops a melanized appressorium that attaches tightly to the host, forming an infection peg that penetrates the host. Occasionally, the same species can attack both via melanized appressoria, as well as by mycelial penetration through stomates in the leaves.[21]

Mycelia themselves are not always pathogenic. Where only spores are pathogenic, the spread of a transgenic hypervirulent biocontrol agent could be prevented by a more complex strategy akin to the "terminator" strategy.[22] Transgenes that could potentially suppress sporulation could be engineered into the biocontrol organism under the control of a chemically-inducible promoter. Sporulation genes in antisense configuration or in high overexpression so as to cause co-suppression would suppress sporulation. Spores or mycelia to be used as inoculum could then be treated with the chemical inducer, before application to the target pest. The chemical inducer could be in the micropellet used for application,[18] or the chemical inducer could be an endogenous, specific compound in the pest host. Thus, the biocontrol agent could be contained to the single, purposely infested pest population. Transgenic suppression of sporulation might best be performed by suppression of more than one sporulation gene because of the possibility of transgene silencing, allowing the organism to revert back to wild type.

19.3.2. OBVIATING RECOMBINATION

The above strategies can be used to prevent persistence and spread, but would not preclude diagonal gene flow of hypervirulence genes to other organisms. Thus, means are needed to mitigate the possibility that recombined, introgressed hypervirulent organisms become "superbugs" attacking non-targeted species. The hypervirulence gene could be flanked with transgenetic mitigator (TM) genes that are positive or neutral to the biocontrol agent, but would be detrimental to any recombinant (Figure 1b). At its simplest, the hypervirulence gene could be flanked by one or two of the genes affecting reproduction, appressorium formation, spore stalk formation, viable spore formation, spore germination, melanin biosynthesis, mineral nutrition, etc. (see Table 2 in Gressel[23]). These genes in the RNAi, antisense, or co-suppressive form would affect one of the processes leading to the ability to recombine, to form viable spores, or to make efficient infection structures. There are already many candidate genes that could be tested as mitigators, and there are probably many additional yet to be discovered genes that would be appropriate additions to this list.

An antisense gene suppressing sporulation should prevent sporulation in a heterokaryon or other recombinant organisms, serving a dual purpose of also preventing dispersal. The genes that control melanin biosynthesis and/or conidiation might only be applicable for biocontrol agents that do not need spores or melanin for pathogenicity. Spores without melanin are less viable, particularly in the light and in harsh environments. Thus, any spores that do form, would have reduced vigor. The concept of using analogous TM constructs has been extensively tested as a failsafe to mitigate introgression of transgenes from crops to weeds, where it has been quite successful.[24–26]

19.4. Risk Analysis of These Failsafes

Risk analysis must be separately performed for each transgenic biocontrol agent and must consider two types of issues: (a) the limitations on the failsafes that can be used; and (b) the biology of the pathogen and its relatedness to other pathogens.

19.4.1. LIMITATIONS OF VARIOUS FAILSAFE MEASURES

A safety aspect that must be clearly ascertained is that all types of sporulation are suppressed. For example, only light-induced conidiation is precluded in some asporogenic fungal mutants, but not starvation-induced sporulation.[27] Some organisms make more than one type of spore; many *Fusarium* species can produce micro and macro conidia as well as chlamydospores. Each is produced under different environmental conditions. It would be interesting to ascertain whether each of the genes that control spore stalk development can (when antisensed) suppress all types of spore forms. The *stuA* transcription factor does control both sexual and asexual reproduction in *Aspergillus nidulans*.[28] It will be easier to load more or simpler failsafes into organisms that do not require appressoria for penetration. This includes mycoherbicides that attack through stomates or other inter- or intra-cellular penetration or bacterial pathogens that attack through cut surfaces. It includes most of the biocontrol agents used against insects that attack through the alveoli.

More complex methods such as the modified "terminator" technology will have to be considered for organisms that use melanized appressoria, or where hyphae are not typically pathogenic. Not all organisms (e.g., *Alternaria alternata*) utilize melanization of appressoria as part of infection.[29] One can consider using a spore-specific promoter for anti-melanin genes where hyphae form appressoria requiring melanin. Thus, failsafes will perforce be more complex with the melanized appressoria-utilizing *Colletotrichum* species.[20] Still, some *Colletotrichum* spp. can attack plants by stomatal penetration.[21] Many regulatory genes that are activated during sporulation are known.[29] Such genes could be used to activate melanin suppressing or other anti-sporogenesis genes to render them more spore-specific. Another advantage of asporogenic mutants is worker safety. Spores are often allergenic and spores can initiate opportunistic infections of humans.

19.4.1.1. *Pathogen Biology*

Background knowledge about the possibility of a pathogen mutating its host specificity and its ability or inability to sexually or asexually conjugate "diagonally" with related organisms will govern the required number and level of failsafe mechanisms that must be instated. Thus, one must consider the possibility of mutation of pathogenicity to a broader host spectrum of a host-specific biocontrol agent, e.g., a specific pathovar of *Fusarium oxysporum*. While there are no documented cases of a member of this species mutating its host range, the species is sub-divided into hundreds of known *forma speciales*, each with its own host specificity. There must have been evolution to different hosts, even if not documented cases. Still, the possible existence of alternate crop hosts must be considered.

Some species easily conjugate with close relatives forming heterokaryotic mycelia with mixed nuclei (e.g., *Trichoderma*), and the mycelia have mixed properties. Problems might ensue where spores are multinucleate or where there is recombination among nuclear chromosomes. Further generations will carry the heterokaryotic complemented properties. In a multigeneration experiment with hundreds of millions of (uninucleate) spores, there was no recombination among complementing nuclei that allowed a heterokaryon of two different *Trichoderma* auxotrophs to live on minimal media. However, uni-nucleate spores forming on these heterokaryons were not viable on this minimal medium, suggesting no DNA exchange (E. Galun, unpublished results). This demonstrates that even closely related or con-specific organisms have impenetrable barriers that prevent recombination with "alien" genomes in nature, even when the traits could be beneficial or even vital for existence. Imperfect (asexual) fungi have less capacity to transfer traits than perfect (sexual) fungi. Still, it is impossible to "prove" that an imperfect fungus does not have a rare sexual form that appears only in highly special conditions.

Transgenic biocontrol agents have much to offer agriculture and human health and welfare. There are probably many cases where a minimum of anti-introgressional failsafe mechanisms introduced into asporogenic deletion mutants would lower the risk to an infinitesimal level. The resulting products would combine fitness with a level of safety and containment greater than that of the spore forming but inefficient non-transgenic biocontrol agents.

References

1. S. F. Elena, T. S. Whittam, C. L. Winkworth, M. A. Riley, and R. E. Lenski, Genomic divergence of *Escherichia coli* strains: Evidence for horizontal transfer and variation in mutation rates, *Int. Microbiol.* 8, 271–278 (2005).
2. Z. Amsellem, Y. Kleifeld, Z. Kerenyi, L. Hornok, Y. Goldwasser, and J. Gressel, Isolation, identification, and activity of mycoherbicidal pathogens from juvenile broomrape plants, *Biol. Control* 21, 274–284 (2001).
3. U. L. Rosewich and H. C. Kistler, Role of horizontal gene transfer in the evolution of fungi, *Annu. Rev. Phytopath.* 38, 325–363 (2000).
4. J. D. Walton, Horizontal gene transfer and the evolution of secondary metabolite gene clusters in fungi: An hypothesis, *Fungal. Genet. Biol.* 30, 167–171 (2000).
5. J. Gressel, *Molecular Biology of Weed Control* (Taylor & Francis, London, 2002).
6. M. G. Roca, N. D. Read, and A. E. Wheals, Conidial anastomosis tubes in filamentous fungi, *FEMS Microbiol. Lett.* 249, 191–198 (2005).

7. J. C. McGuire, J. E. Davis, M. L. Double, W. L. MacDonald, J. T. Rauscher, S. McCawley, and M. G. Milgroom, Heterokaryon formation and parasexual recombination between vegetatively incompatible lineages in a population of the chestnut blight fungus, *Cryphonectria parasitica*, *Mol. Ecol.* 14, 3657–3669 (2005).
8. H. A. S. Teunissen, J. Verkooijen, B. J. C. Cornelissen, and M. A. Haring, Genetic exchange of avirulence determinants and extensive karyotype rearrangements in parasexual recombinants of *Fusarium oxysporum*, *Mol. Genet. Genom.* 268, 298–310 (2002).
9. E. A. Souza-Paccola, L. C. L. Favaro, C. R. Casela, and L. D. Paccola-Meirelles, Genetic recombination in *Colletotrichum sublineolum*, *J. Phytopath.* 151, 329–334 (2003).
10. L. A. Castrillo, M. H. Griggs, and J. D. Vandenberg, Vegetative compatibility groups in indigenous and mass-released strains of the entomopathogenic fungus *Beauveria bassiana*: likelihood of recombination in the field, *J. Invertebr. Pathol.* 86, 26–37 (2004).
11. A. Molnar, L. Sulyok, and L. Hornok, Parasexual recombination between vegetatively incompatible strains in *Fusarium oxysporum*, *Mycol. Res* 94, 393–398 (1990).
12. T. L. Friesen, E. H. Stukenbrock, Z. H. Liu, S. W. Meinhardt, H. Ling, J. D. Faris, J. B. Rasmussen, P. S. Solomon, B. A. McDonald, and R. P. Oliver, Emergence of a new disease as a result of interspecific virulence gene transfer, *Nat. Genet.* 38, 953–956 (2006).
13. A. Sharon, Z. Amsellem, and J. Gressel, Glyphosate suppression of an elicited defense response, *Plant Physiol.* 98, 654–659 (1992).
14. J. Gressel, D. Michaeli, V. Kampel, Z. Amsellem, and A. Warshawsky, Ultralow calcium requirements of fungi facilitate use of calcium regulating agents to suppress host calcium-dependent defenses, synergizing infection by a mycoherbicide, *J. Agric.Food Chem.* 50, 6353–6360 (2002).
15. B. Ahn, T. Paulitz, S. Jabaji-Hare, and A. Watson, Enhancement of *Colletotrichum coccodes* virulence by inhibitors of plant defense mechanisms, *Biocontrol Sci. Technol.* 15, 299–308 (2005).
16. J. Gressel, Potential failsafe mechanisms against the spread and introgression of transgenic hypervirulent biocontrol fungi, *Trends Biotechnol.* 19, 149–154 (2001).
17. D. C. Sands and R. V. Miller, Altering the host range of mycoherbicides by genetic manipulation, in *Pest Control with Enhanced Environmental Safety*, edited by S. O. Duke, J. J. Menn and J. R. Plimmer (American Chemical Society, Washington, DC, 1993), pp. 101–109.
18. Z. Amsellem, N. K. Zidack, P. C. Quimby, Jr., and J. Gressel, Long term dry preservation of active mycelia of two mycoherbicidal organisms, *Crop Protect.* 18, 643–649 (1999).
19. H. Shiotani and T. Tsuge, Efficient gene targeting in the filamentous fungus *Alternaria alternata*, *Mol. Gen. Genet.* 248, 142–150 (1995).
20. S. E. Perfect, H. B. Hughes, R. J. O'Connell, and J. R. Green, *Colletotrichum*—A model genus for studies on pathology and fungal-plant interactions, *Fungal Genet. Biol.* 27, 186–198 (1999).
21. A. O. Latunde-Dada, R. J. O'Connell, C. Nash and J. A. Lucas, Stomatal penetration of cowpea (*Vigna unguiculata*) leaves by a *Colletotrichum* species causing latent anthracnose, *Plant Pathol.* 48, 777–784 (1999).
22. M. L. Crouch. How the terminator terminates: An explanation for the non-scientist of a remarkable patent for killing second generation seeds of crop plants (The Edmonds Institute, Edmond WA, USA, 1998), available at; http://www.bio.indiana.edu/people/terminator/html.
23. J. Gressel, Transgenic mycoherbicides; needs and safety considerations, in *Handbook of Fungal Biotechnology,* edited by D. K. Arora (Dekker, New York, 2004), pp. 549–564.

24. H. Al-Ahmad, J. Dwyer, M. M. Moloney, and J. Gressel, Mitigation of establishment of *Brassica napus* transgenes in volunteers using a tandem construct containing a selectively unfit gene, *Plant Biotech. J.* 4, 7–21 (2006).
25. H. Al-Ahmad, S. Galili, and J. Gressel, Poor competitive fitness of transgenically mitigated tobacco in competition with the wild type in a replacement series, *Planta* 272, 372–385 (2005).
26. H. I. Al-Ahmad, S. Galili, and J. Gressel, Tandem constructs mitigate risks of transgene flow from crops: Tobacco as a model, *Mol. Ecol.* 13, 687–710 (2004).
27. B. A. Horwitz, J. Gressel, S. Malkin, and B. L. Epel, Modified cryptochrome in vivo absorption in *Trichoderma dim* photosporulation mutants in *Trichoderma*, *Proc. Natl. Acad. Sci. USA* 82, 2736–2740 (1985).
28. J. G. Wu and B. L. Miller, *Aspergillus* asexual reproduction and sexual reproduction are differentially affected by transcriptional and translation mechanisms regulating stunted gene expression, *Mol. Cell Biol.* 17, 6191–6201 (1997).
29. T. H. Adams, J. K. Weiser, and J. H. Yu, Asexual sporulation in *Aspergillus nidulans*, *Microbiol. Mol. Biol. Rev.* 62, 35–54 (1998).

EPILOGUE: GETTING FROM HERE TO ETERNITY

David Sands
Plant Sciences and Plant Pathology, Montana State University, Bozeman, MT, 59717 USA

We start a little bit naïve
and use those techniques
handed down by our teachers.

And we learn to tilt the playing field
so that our statistics don't need to lie
but truth is not of the moment.

Only time to write the next grant
and this time the pea is under
a different shell.

Yet if we are diligent
we will have a product
and we find an impasse.

It can stand alone in a synthetic world
but the true added value in the market place
is when it is used synthetic free.

So those who depend on fear
have a fear-free food.

But if only they knew
that they too are loaded with transposons
each looking for a new market niche.

So regulations and industries being what they are
will force independent people
unto co-dependency
in this way

So in 20 years
the safest most nutritious pest and toxin free
and most delicious foods
will be organically grown
genetically modified

M. Vurro and J. Gressel (eds.), Novel Biotechnologies for Biocontrol Agent Enhancement and Management, 363–365.

INDEX